Two week loan

Please return on or before the last
date stamped below.
Charges are made for late return.

Nuclear Organization, Chromatin Structure, and Gene Expression

Edited by
Roel van Driel
and
Arie P. Otte

E.C. Slater Instituut
University of Amsterdam
Amsterdam
The Netherlands

Oxford   New York   Tokyo
OXFORD UNIVERSITY PRESS
1997

Oxford University Press, Great Clarendon Street, Oxford OX2 6DP

Oxford   New York

Athens   Auckland   Bangkok   Bogota   Bombay   Buenos Aires
Calcutta   Cape Town   Dar es Salaam   Delhi   Florence   Hong Kong
Istanbul   Karachi   Kuala Lumpur   Madras   Madrid   Melbourne
Mexico City   Nairobi   Paris   Singapore   Taipei   Tokyo   Toronto   Warsaw
and associated companies in
Berlin Ibadan

Oxford is a trade mark of Oxford University Press

Published in the United States
by Oxford University Press Inc., New York

© Oxford University Press, 1997

A catalogue record for this book is available from the British Library

Library of Congress Cataloging in Publication Data

Nuclear organization, chromatin structure, and gene expression/
edited by Roel van Driel and Arie P. Otte.
Includes bibliographical references and index.
1. Chromatin.   2. Gene expression. 3. Genetic regulation.
4. Cell nuclei.   I. Driel, Roeland van.   II. Ote, Arie P.
QH599.N83 1997   572.8'7—dc21   97–15623 CIP

ISBN 0 19 854923 7 (Hbk)
ISBN 0 19 854922 9 (Pbk)

Typeset by Interactive Sciences Ltd., Gloucester

Printed in Great Britain by Bookcraft Ltd., Midsomer
Norton, Avon

Contents

Contributors

Marisa S. Bartolomei
Howard Hughes Medical Institute, Department of Cell and Developmental Biology, University of Pennsylvania School of Medicine, Philadelphia, PA 19104, USA

Miguel Beato
Institut für Molekularbiologie und Tumorforschung, I.M.T., University of Marburg, E.-Mannkopff-Str. 2, D-35037 Marburg, Germany

Constanze Bonifer
Institut für Biologie III der Universtät Freiburg, Schänzlestr.1, D–79104 Freiburg i.Br., Germany

E. Morton Bradbury
Department of Biological Chemistry, School of Medicine, University of California at Davis, Davis, California 95616, U.S.A. and Life Sciences Division MS M888, Los Alamos National Laboratory, Los Alamos, New Mexico 87545, USA

Miriam Braunstein
Department of Molecular Biology, Princeton University, Princeton, NJ 08544, USA

Achim Breiling
Center for Molecular Biology, University of Heidelberg, Im Neuenheimer Feld 282, D69120 Heidelberg, Germany

James R. Broach
Department of Molecular Biology, Princeton University, Princeton, NJ 08544, USA

Reyes Candau
Present address: The Wistar Institute, 3601 Spruce Street, Philadelphia, PA 19104–4268, USA

Sebastián Chávez
Departamento de Genética, Facultad de Biología, Universidad de Sevilla, Apdo, 1085, 41080 Sevilla, Spain

Victor G. Corces
Department of Biology, The Johns Hopkins University, 3400 N. Charles St., Baltimore, MD 21218, USA

Amy K. Csink
Fred Hutchinson Cancer Research Center, Howard Hughes Medical Institute, 1124 Columbia Street, A1–162, Seattle, WA 98104–2015, USA

Nicole Faust
Institut für Biologie III der Universität Freiburg, Schänzlestr. 1, D–79104 Freiburg i.Br., Germany

Peter Fraser
Department of Cell Biology and Genetics, Erasmus University Rotterdam, PO Box 1738, 3000 DR Rotterdam, The Netherlands

Tatiana I. Gerasimova
Department of Biology, The Johns Hopkins University, 3400 N. Charles St., Baltimore, MD 21218, USA

Frank Grosveld
Department of Cell Biology and Genetics, Erasmus University, Rotterdam, PO Box 1738, 3000 DR Rotterdam, The Netherlands

Steven Henikoff
Fred Hutchinson Cancer Research Center, Howard Hughes Medical Institute, 1124 Columbia Street, A1–162, Seattle, WA 98104–2015, USA

Maty Hershkovitz
InSight, Kiryat Weizmann Science Park, Rehovot 76326, Israel

Scott G. Holmes
Department of Molecular Biology, Princeton University, Princeton, NJ 08544, USA

Matthias C. Huber
Institut für Biologie III der Universität Freiburg, Schänzlestr.1, D–79104 Freiburg i.Br., Germany

M. E. Eva Ludérus
E.C. Slater Instituut, University of Amsterdam, Plantage Muidergracht 12, 1018TV Amsterdam, The Netherlands

Laura Manuelidis
Yale Medical School, 310 Cedar Street, New Haven, Ct. 06510, USA

Antonius J. M. Matzke
Austrian Academy of Sciences, Institute of Molecular Biology, Billrothstrasse 11, A–5020 Salzburg, Austria

Marjori A. Matzke
Austrian Academy of Sciences, Institute of Molecular Biology, Billrothstrasse 11, A–5020 Salzburg, Austria

Geert Meersseman
Laboratorium voor Moleculaire Biologie, Katholieke Universiteir Leuven, Campus Gasthuisberg, B–3000 Leuven, Belgium

Christian Möws
Institut für Molekularbiologie und Tumorforschung, I.M.T., University of Marburg, E.-Mannkopff-Str. 2, D-5037 Marburg, Germany

Timothy E. O'Neill
Department of Food Service and Technology, University of California at Davis, Davis, California 95616, USA

Arie P. Otte
E.C. Slater Instituut, University of Amsterdam, Plantage Muidergracht 12, 1018TV Amsterdam, The Netherlands

Renato Paro
Center for Molecular Biology, University of Heidelberg, Im Neuenheimer Feld 282, D69120 Heidelberg, Germany

Sari Pennings
Department of Biochemistry, University of Edinburgh, George Square, Edinburgh, EH8 9XD, UK

Craig L. Peterson
Program in Molecular Medicine and Department of Biochemistry and Molecular Biology, University of Massachusetts Medical Centre, Worcester MA 01605, USA

Janet Quinn
Departments of Biochemistry & Genetics and Medicine, University of Newcastle upon Tyne, Newcastle, NE2 4HH, UK

Arthur D. Riggs
Biology Department, Beckman Research Institute of the City of the Hope, Duarte CA 91010, USA

Georgette L. Sass
Fred Hutchinson Cancer Research Center, Howard Hughes Medical Institute, 1124 Columbia Street, A1–162, Seattle, WA 98104–2015, USA

Harald Saueressig
Institut für Biologie III der Universität Freiburg, Schänzlestr.1, D–79104 Freiburg i.Br., Germany

Albrecht E. Sippel
Insitut Für Biologie III dev Universität Freiburg, Schänzlestr, 1, D–79104 Freiburg i.Br., Germany

David L. Spector
Cold Spring Harbor Laboratory, PO Box 100, Cold Spring Harbor, New York, NY 11724, USA

Mathias Truss
Institut für Molekularbiologie und Tumorforschung, I.M.T., University of Marburg, E.-Mannkopff-Str.2, D–35037 Marburg, Germany

Roel van Driel
E.C. Slater Instituut, University of Amsterdam, Plantage Muidergracht 12, 1018TV Amsterdam, The Netherlands

Alan P. Wolffe
Laboratory of Molecular Embryology, Nationall Institute of Child Health and Human Development, NIH, Bldg 18T, Rm. 106, Bethesda, MD 20892–2710, USA

Introduction: general perspective and key questions

Our understanding of the molecular mechanisms that underlie the regulation of the expression of individual genes has increased dramatically in the last decades. Many aspects of regulation, such as promoters, enhancers, and components of the transcription machinery, have been uncovered. However, the picture that emerges is still incomplete and leaves many questions unanswered. Here are two well-known examples that illustrate our lack of knowledge concerning gene control in higher eukaryotes. First, problems that have been encountered with the expression of transgenes. The expression levels of transgenes have been found to vary dramatically between individual transformants. These transgenes are under the control of the same promoters and enhancers. Therefore, the variability in their transcription levels can only be explained by differences in regulation that go beyond the level of promoters and enhancers. In support of this idea there is accumulating evidence that differences in the chromatin environment in which transgenes integrate are responsible for their unpredictable and variable expression levels. The second example involves the epigenetic regulation of X-chromosome inactivation. In X-chromosome inactivation one of the two X-chromosomes in the somatic cells of females becomes inactivated during early embryonic development. A prevailing model is that the genes on the inactivated X-chromosome become inaccessible for the transcription machinery due to compaction of chromatin, and are therefore inactive. Strikingly, in the same nuclei, corresponding genes that are located on the other, active X-chromosome are accessible for the transcription machinery and are therefore active. Also in this example an important regulatory role for chromatin structure in regulation of gene activity is envisioned. This level of regulation goes beyond regulation of gene activity at the level of individual genes, involving promoters and enhancers.

The incentive to compile this book was our desire to explore current insights into various aspects of regulation of gene activity at the level of chromatin structure. The molecular mechanisms underlying these levels of regulation are operational in all eukaryotic cells, ranging from yeast to flies and mammals. Understanding such molecular mechanisms will significantly increase our knowledge about this important, fundamental level of regulation of gene activity. It may further help to solve problems of unpredictable and variable transgene expression. Solving this biotechnological problem will have a substantial economical impact.

The book is divided in three closely related parts. Part I focuses on the first level of the organization of chromatin beyond the naked double helix structure of

DNA. This includes the organization of DNA into nucleosomes. Structural aspects of nucleosome organization, as well as the role of nucleosomes in the regulation of gene activity, are discussed in Chapters 1 and 2. Post-transcriptional modifications of histones are of importance for the involvement of these proteins in the regulation of gene activity (Chapter 3). The role of a general activator complex, the SNF/SWI complex, in transcription regulation and its function in improving the accessibility of nucleosomal DNA for the transcription machinery is discussed in Chapter 4.

Part II focuses on chromatin organization beyond the level of nucleosomes. We have only limited knowledge about these higher order levels of chromatin structure. There is, however, increasing evidence that chromatin is divided in functional domains (Chapters 5–8). These domains form functional units in which genes or gene clusters are coordinatedly regulated, independent of neighbouring domains. The domains are shielded from the surrounding chromatin, which may imply the existence of genomic sequences that serve as boundaries of the domains. The molecular nature and the involvement of these boundary elements in the regulation of gene activity are discussed in Chapters 5 and 6. Another type of genetic element, the locus control region (LCR), defines an independent, functional chromatin domain for the ß-globin gene cluster. The LCR is a crucial factor in the transcriptional regulation of the globin gene cluster (Chapter 8). Much less is known about higher order chromatin structure beyond these functionally defined domains in the chromatin fibre. A view on the organization of the chromatin fibre in the interphase nucleus is described in Chapter 9. The distribution of active genes and their transcripts in the interphase nucleus is discussed in Chapter 10.

Part III focuses on examples that highlight the involvement of chromatin structure in the stable transmission to progeny cells of the on/off state of genes. There is, for instance, ample evidence that methylation of DNA is involved in the inheritance of the silenced state of genes from one generation to another (Chapter 11). Methylation of DNA negatively influences the ability of genes to become transcribed. Possible molecular mechanisms underlying the inheritably stable inactivation of one female X-chromosome are discussed in Chapter 12. At the light microscopical level the inactive X-chromosome appears to be highly compacted chromatin. This cytologically recognizable chromatin has been defined as heterochromatin. Genes located in heterochromatin are transcriptionally silent. Hetero-chromatinization of genes forms an important mechanism that underlies the transmission of the silenced state of genes from one generation to another. Molecular mechanisms that involve heterochromatinization of genes are evolutionary highly conserved. Insights from studies in *Drosophila* and yeast are described in Chapters 13, 14, and 15. Related phenomena in plants are described in Chapter 16.

The role of chromatin structure and epigenetic aspects in the regulation of gene activity have been encountered in diverse areas that study the regulation of eukaryotic genes. This not only underlines their importance, but also makes it difficult to obtain a good overview that covers all different aspects of these processes. This problem is enhanced by the broad variety of model systems that are employed. By combining contributions that cover different aspects of epigenetic aspects of gene control in various model systems we expect that this book provides a comprehensive overview that summarizes the current state of our knowledge.

We hope that this book provides a stimulating overview for investigators who already work on specific epigenetic aspects of regulation of gene activity and who are interested in related fields. We also hope that this book attracts the attention of molecular geneticists who are involved in transgenic research. Many of the examples described in this book are probably related to problems that they encounter. In particular, however, we expect that the book finds its way to investigators and students that may become interested in epigenetic aspects of regulation of gene activity. It would be highly rewarding if this book provokes their curiosity and encourages them to enter this exciting field of research. The contributions are, therefore, aimed to be within reach of students at advanced phases in their studies.

Finally, we wish to thank the authors for their contributions and their patience during the editing process and the production of this book.

Amsterdam   R.v.D. and A.P.O.
January 1997

Part I

Nucleosome dynamics and
transcription

1

Nucleosomes: dynamic repressors of transcription

Sari Pennings, Timothy E. O'Neill, Geert Meersseman, and E. Morton Bradbury

Introduction

The structure of the nucleosome, which was elucidated from early work on enzyme digestion patterns, neutron scatter, X-ray diffraction data, and histone–DNA crosslinking (reviewed by Van Holde 1988), is essentially a static picture of an average structure. This is because these experimental approaches average out the heterogeneities across the largely native bulk nucleosome and chromatin samples, as well as any fluctuations in the structure over time. Most notably, the model of the nucleosome rarely includes the long N-terminal and short C-terminal 'tails' of the histones, which are highly dynamic and contain the sites of reversible chemical modifications associated with chromatin functions (reviewed by Bradbury 1992; also see Chapter 3). Studies of higher order structures of chromatin also hold this caveat and here the paucity of data allows for several models of the 30 nm fibril model (reviewed by Van Holde 1988). However useful these structural models are in understanding the packaging of chromatin, they fail to convey how chromatin structure can be flexible enough not only to accommodate the wide range of DNA sequences in the genome, but also to allow dynamic processes such as transcription and replication to take place. An insight into the dynamic nature of chromatin is essential in order to understand how chromatin can be a functional component of the active eukaryotic nucleus.

A corollary of packaging the genome into chromatin structure is the reduced accessibility of the DNA. A large part of the genome in eukaryotes is silenced by the formation of inaccessible densely packed chromatin, which may also provide structural components within the nucleus. This chromatin structure repression of unwanted gene expression plays an important role in differentiation. It can involve linker histone variants such as H5, or differentiation-specific accessory proteins such as Polycomb. Although the generally repressive effect of chromatin structure has long been appreciated, it has become increasingly clear in recent years that chromatin structure also participates actively in gene regulation at the level of nucleosomes and possibly at higher orders of chromatin structure.

Although active regions of the genome are in a more open chromatin conformation, the nucleosomal organization can mask DNA binding sites to *trans*-acting factors (reviewed by Wolffe 1995). Gene regulation includes a fine control of access to these DNA recognition sequences in promoter regions. In this control mechanism the initial positioning of particular nucleosomes on the DNA sequence is crucial in determining how *trans*-acting factor recognition sites are presented (Venter *et al.* 1994; Li and Wrange 1995; Fragoso *et al.* 1995). Thus, far from being a passive obstruction these nucleosomes are intimately involved in gene activation through their interaction with certain transcription factors (reviewed by Wolffe 1995). In addition, transcriptional activator complexes have been identified recently that interact with histones rather than the promoter sequence itself (reviewed by Peterson and Tamkun 1995; Tsukiyama and Wu 1995; also see Chapter 4), illustrating the direct involvement of histones in gene regulation. These interactions alter these nucleosomes to make them more dynamic. Substantial nucleosome rearrangements have been observed in promoter regions in response to gene activation (Tsukiyama *et al.* 1994). This eventually leads to clearance of nucleosomes from the promoter region for the subsequent binding of transcription factors (Almer *et al.* 1986) by a nucleosome displacement mechanism that is not yet understood.

Although transcription initiation has been a major focus, the dynamic behaviour of nucleosomes need not be limited to promoters. It has been established that genes retain their histones when being transcribed (De Bernardin *et al.* 1986; Nacheva *et al.* 1989; Ericsson *et al.* 1990). At least some dynamic transitions of the nucleosomes along the gene are required during the process of transcription elongation by RNA polymerase, which is at odds with the canonical nucleosome structure. Various modes of nucleosome disruption, unfolding and transfer have been proposed (reviewed by Thoma 1991; Van Holde *et al.* 1992; Adams and Workman 1993). It has been long known that actively transcribing chromatin contains hyperacetylated histones and is DNAase I sensitive. Recently it has been reported that the histones in the 33 kb chromatin domain containing the chicken globin genes become acetylated when these genes are activated. This domain of acetylated histones corresponds with the region of active globin gene DNAase I hypersensitivity (Hebbes *et al.* 1994). Previously it had been shown that hyperacetylated chromatin was selectively digested by DNAase I (Vidali *et al.* 1978). Histone acetylation is thought to perturb chromatin structure to a more open form that makes it accessible to DNAase I or to *in vivo* transacting factors. This perturbation, however, extends over long chromatin domains and will affect a large number of contiguous nucleosomes. How these chromatin perturbations affect nucleosome stability and dynamics is not understood.

The possibility that nucleosomes can have a dynamic behaviour has been suggested by *in vivo* studies of DNA repair and gene activation. Following DNA damage, labelled DNA repair patches were found to be incorporated into nucleosomes. This behaviour was attributed to the migration of nucleosomes along the DNA in the newly repaired DNA regions of intact, non-dividing cells.

It was proposed that a natural process of random fluctuations in the nucleosome positions along the genome was responsible for these movements extending over 40 bp or more (Nissen *et al.* 1986). The nucleosomal spacing on a number of genes, as assessed by an endonuclease cleavage of DNA *in vivo*, is shorter in active chromatin compared to inactive chromatin (Villeponteau *et al.* 1992). The close spacing was found to extend far beyond the regions disrupted by transcription and therefore seems a domain property of active chromatin. The results were taken to suggest that the nucleosomes in active chromatin are mobile *in vivo* and, when not being constrained by linker histones, freely move closer together. These reorganizations of nucleosomes on promoters and genes are closely related to the issue of nucleosome positioning.

Nucleosome positioning

Nucleosomes adopt non-random locations along the DNA sequence. The precise position of a nucleosome on the DNA is likely to be determined by several factors. An important determinant is the anisotropy of flexibility of the DNA sequence that is required to accommodate tight bending around the nucleosome (Satchwell *et al.* 1986). Almost any DNA can be assembled into nucleosomes, but with various affinities. Because the binding affinity is the cumulative result of many interactions that induce small bends, nucleosome DNA positions can be rotationally unique but translationally degenerate (Shrader and Crothers 1990). This translates into the observations of multiple overlapping positions spaced by 10 bp (the DNA helical repeat) or multiples thereof (Lowman and Bina 1990; Meersseman *et al.* 1991; Davey *et al.* 1995). In addition, there are translational determinants of nucleosome positioning (reviewed by Simpson 1991). Certain DNA sequences constitute a boundary in that they are less favoured for assembly into nucleosomes. An extreme example of this is long DNA homopolymer tracts (Prunell 1982). Similarly, certain features of core particle structure may promote the incorporation of particular DNA sequence patterns. These features could include specific histone–DNA sequence contacts, or deviations from smooth coiling in the DNA topology. DNA 'kinks' near the dyad axis have been suggested to direct nucleosome positioning (Satchwell *et al.* 1986). Similar to DNA bendability, translational positioning most probably results from structural features of DNA, rather than DNA sequence specificity. It is significant that DNA sequences can be designed that bind histone octamers stronger than genomic DNA (Shrader and Crothers 1989). It would appear that the less than optimal binding strengths of histones for DNA sequences are required for histones to package the enormous variety of DNA sequences found in eukaryotic genomes. Weaker binding could also confer the potential for nucleosome movement or removal.

DNA sequence-directed nucleosome positioning will often allow for several alternative overlapping sites (Davey *et al.* 1995). The actual positions of nucleosomes can further be influenced by the binding of linker histones

(Meersseman *et al*. 1991; Lauderdale and Stein 1993), the presence of neighbouring nucleosomes (Fedor *et al*. 1988), the formation of higher order chromatin structures, or by the binding of DNA sequence-specific proteins. High resolution mapping of the nucleosome positions over the entire chicken beta-globin gene indicates that the nucleosomal spacing on the inactive gene may be encoded in the strong positioning sites in the DNA sequences flanking the genes and in the introns (Davey *et al*. 1995). Following the initiation of transcription the nucleosomal repeat length becomes shorter (Villeponteau *et al*. 1992), which suggests an alternative use of available positioning sites. Changes in nucleosome positioning could be most simply interpreted in terms of nucleosome movement.

Nucleosome mobility

Dynamic behaviour of positioned nucleosomes was first identified on sea urchin 5S rDNA (Pennings *et al*. 1991). The nucleosome positioning on this DNA is well documented. The sequence contains a strong nucleosome positioning site (Simpson and Stafford 1983) on which the histone octamers assemble in a dominant position flanked by minor positions 10 bp apart (Meersseman *et al*. 1991). It was found that histone octamers redistribute between the possible 5S rDNA nucleosome positions, which share the same rotational setting of the DNA around the octamer. This rearrangement, which is in dynamic equilibrium, was termed nucleosome mobility (Pennings *et al*. 1991). It is a temperature-dependent process that occurs at relatively low ionic strength with all histone–DNA interactions intact (Meersseman *et al*. 1992).

The assay employed in these studies makes use of the finding that the polyacrylamide gel electrophoretic mobility of positioned nucleosomes is a function of the proximity of the octamer to the ends of the DNA fragment (Meersseman *et al*. 1992; Pennings *et al*. 1992). For monomers and head-to-tail dimers of the 5S rDNA sequence each of the possible nucleosome positions has been assigned to a band in the nucleoprotein gel pattern. This direct visualization of the native nucleosome positions on a DNA fragment has provided a sensitive tool to follow nucleosome mobility using two-dimensional gel electrophoresis (Meersseman *et al*. 1992). The running conditions for both dimensions are the same, so that the two-dimensional nucleoprotein gel displays the first dimension pattern on a diagonal line, unless between the runs there occurs a redistribution of the nucleosomes relative to the DNA fragment. Such a redistribution can occur when the gel strip is incubated at an elevated temperature of 37 °C between the runs. These mobile nucleosomes have an altered electrophoretic mobility when electrophoresed in the second dimension and therefore are observed as off-diagonal dots in the two-dimensional gel pattern (Fig. 1.1a). Nucleosome mobility is a general phenomenon observed also for bulk nucleosomes (Meersseman *et al*. 1992). The generality of short range nucleosome mobility provides a

A

B

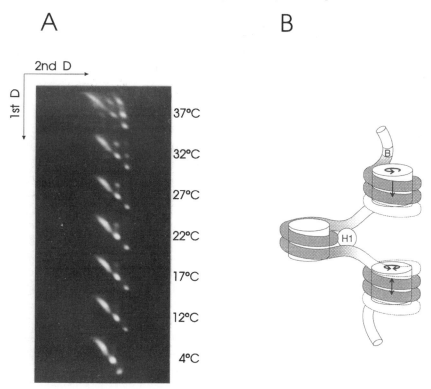

Fig. 1.1 (a) Visualization of nucleosome mobility by two-dimensional gel electrophoresis. Different possible nucleosome positions on a DNA fragment migrate as separate bands on a polyacrylamide nucleoprotein gel (shown are the three bands corresponding to nucleosome positions on a 5S rDNA fragment). Separation in the second dimension (identical conditions, 4 °C) brings the bands on a diagonal line, unless redistribution between nucleosome positions has occurred between both runs. This alters the electrophoretic mobilities of these nucleosomes, which then migrate as off-diagonal dots. The gel shows the development of off-diagonal elements as a function of temperature incubation between the first and second dimension gel electrophoresis (see Meersseman *et al.* (1992) for technical details). (b) Model for nucleosome mobility. Nucleosomes in motion follow a corkscrew movement within the superhelical path of the DNA. A nucleosome will be allowed to move if the coiling path of the DNA can be continued beyond its immediate location (dashed coils). Note that DNA not associated with the histone octamer will be extended and accessible. B, boundary of incompatible DNA structure or bound protein; H1, a bound linker histone that prevents nucleosome mobility.

compelling argument that the position of a nucleosome should be regarded as a probability rather than a static factor type of DNA sequence binding. It means that although all possible positions of nucleosomes can be precisely defined, the actual locations of nucleosomes are in dynamic equilibrium over a range of sites. This supports a more dynamic view of the nucleosomal organization.

Nucleosomes in motion may be visualized as following a corkscrew movement within the superhelical path of the DNA coiled around the histone octamer (Fig. 1.1b). Whereas small translational variations in position within the same rotational setting of the DNA around the octamer differ little in binding affinity, the accommodation of a DNA sequence that is more resistant to nucleosome formation holds a larger energy penalty (Shrader and Crothers 1990). The mobility of a nucleosome is likely to depend on the bendability of sequences flanking its position. A nucleosome would be allowed to move as long as the coiling path of the DNA could be continued beyond its immediate location (Fig. 1b). If there are two overlapping nucleosome positions with different rotational settings of the DNA around the histone octamer, both may not be able to satisfy this condition and only one nucleosome position may be mobile (Meersseman *et al.* 1992).

Mobility may be limited by the same elements that act as boundaries to nucleosome positioning (reviewed by Simpson 1991). We think that here could lie the difference between the short-range mobilities of nucleosomes at low salt discussed above and nucleosome 'sliding' at higher non-physiological ionic strength. Early reports of nucleosome movement usually refer to the *in vitro* observation of long-range sliding along the DNA that occurs at higher ionic strengths (Beard 1978). The non-physiological ionic conditions required for this seemingly related process would be expected to reduce interactions between the histones and DNA such that any barriers to nucleosome movement would no longer be effective.

In vivo, local ionic conditions differ and a number of other factors may assert their effects, such as linker histones, histone modifications, DNA binding proteins, or neighbour nucleosome interactions. Nucleosomes may be immobilized by any of these factors. This hypothesis was tested for the linker histones H1 and H5. H1, the linker histone that closes two turns of DNA around the nucleosome, plays an essential role in the formation and stabilization of the 30 nm higher order chromatin fibre (Thoma *et al.* 1979; Allan *et al.* 1980). It has been identified as a general repressor of transcription (Wolffe 1989; Laybourn and Kadonaga 1991). H1 and H5 can be faithfully reconstituted into 5S rDNA chromatin (Meersseman *et al.* 1991). The mobility of histone octamers positioned on constructs of sea urchin 5S rDNA was found to be suppressed by binding of histone H1 or H5 to the nucleosome (Pennings *et al.* 1994). This implies that if nucleosome mobility is required for access to the DNA, then H1 could function as a gene repressor, in addition to its structural role in the 30 nm fibre, through its immobilization of nucleosome cores. Histone H5 is the more potent inhibitor of nucleosome mobility, in accordance with its higher affinity for chromatin (Pennings *et al.* 1994). The dynamic nature of nucleosomes seems therefore to depend on their affinity for linker histones. These results have been reproduced using the *Xenopus* 5S rDNA sequence repeat. Inclusion of linker histones in the chromatin inhibited nucleosome mobility and repressed transcription, demonstrating that stable states of gene repression can be established even at the nucleosomal level (Ura *et al.* 1995). All of the above studies raise the possibility that during development the redistribution of very lysine-rich histone subtypes provides another mechanism for the silencing of genes not required at a particular state of development.

The mechanisms governing nucleosome (chromatin) dynamics *in vivo* are apt to be much more complex, involving a range of cofactors. One convincing demonstration of nucleosome rearrangement was reported in a study using chromatin assembled in a cell-free *Drosophila* embryo extract. Introduction of transcription factors resulted in disruption of the nucleosomal array, specifically at the hsp70 and hsp26 gene promoters (Tsukiyama *et al.* 1994; Wall *et al.* 1995; see also Pazin *et al.* 1994). Rearrangement was dependent on ATP hydrolysis. A nucleosome remodelling cofactor (NURF) was subsequently identified in the extracts (Tsukiyama and Wu 1995). The 500 kDa protein complex is proposed to act in concert with transcription factors to perturb nucleosome structure in an energy-requiring process. It is distinct from the earlier discovered SWI/SNF transcription activator complex (reviewed by Peterson and Tamkun 1995). This 2 MDa multiprotein complex, first brought to light by genetic evidence in yeast, also synergizes transcription by alleviating chromatin repression but it seems to affect nucleosome structure by a different mechanism (Cote *et al.* 1994; also see Chapter 13).

Apart from the spontaneous mobility of nucleosomes, at least two ATP-dependent systems seem to be involved in remodelling chromatin for transcription. It would be interesting to know if their effects on nucleosome structure result in more mobile nucleosomes, bringing about the observed nucleosome rearrangements. The *Drosophila* embryonic extract produces arrays of nucleosomes that can be mobile even when H1 is present. This ATP-dependent dynamic state of chromatin globally increases the accessibility of nucleosomal DNA to incoming proteins (Varga-Weisz *et al.* 1995). It remains possible that the early embryonic extract contains stage-specific accessories to alleviate any chromatin repression in the transcriptionally active nuclei of a developing embryo. Nevertheless, the demonstration of nucleosome mobility in a variety of conditions relevant to *in vivo* situations supports its biological relevance.

Transcription through nucleosomes

Although chromatin structural changes can be correlated with gene activation or repression in *in vivo* studies (reviewed in Gross and Garrard 1987; van Holde 1988; Wolffe 1995), the specific mechanisms that initiate and control the process of gene activation remain largely unknown. The use of defined systems in *in vitro* studies of transcription of chromatin has provided important insights into this process. Early studies with *Escherichia coli* RNA polymerase (Mathis *et al.* 1978; Williamson and Felsenfeld 1978; Meneguzzi *et al.* 1979; Waslyk *et al.* 1979; Waslyk and Chambon 1979) and eukaryotic RNA polymerases I and II (Mathis *et al.* 1978; Waslyk and Chambon 1979) demonstrated that rates of nucleotide incorporation into RNA transcripts synthesized from nucleosomal templates were greatly decreased resulting in shorter transcripts, compared to those obtained from free DNA templates. Further studies demonstrated the ability of bacteriophage SP6 RNA polymerase and eukaryotic RNA polymerases II and III to transcribe through one nucleosome (Lorch *et al.* 1987; Losa and Brown 1987) or short stretches of nucleosomes (Morse 1989; Felts *et al.* 1990). However, the templates used were limited as models for *in vivo* chromatin transcription, because they were either incompletely defined as regards protein composition, nucleosome spacing and positioning, or they contained only one or a few nucleosomes.

The studies of O'Neill *et al.* (1992, 1993, 1995) used an improved *in vitro* transcription system based on a tandemly repeated array of nucleosome positioning sequence constructed by Simpson *et al.* (1985) from the 5S rRNA gene of *Lytechinus variegatus*. A DNA fragment containing 18 repeats of a 207 bp nucleosome positioning sequence was inserted into a plasmid adjacent to promoters for bacteriophage RNA polymerases T7 and SP6. Nucleosome cores were reconstituted onto this supercoiled, closed circular DNA construct by salt dialysis or dilution, using purified histone octamers. Through the use of double labels in transcription experiments the effects of nucleosome cores on the initiation and elongation of transcripts by T7 RNA polymerase were determined simultaneously (O'Neill *et al.* 1992). Both transcript initiation and elongation were markedly inhibited by the presence of nucleosome cores. The extent of this inhibition was directly proportional to the number of nucleosome cores reconstituted onto the pT207-18 DNA templates. To demonstrate that transcription elongation proceeds through nucleosome cores, reconstituted templates were first digested with a restriction enzyme, which recognized a unique site within the nucleosome locating sequence. This site was masked in nucleosome cores but accessible in the free repeat. Thus, digestion with this restriction enzyme provided templates consisting of continuous linear arrays of nucleosomes. *In vitro* transcription of these templates with T7 RNA polymerase produced an RNA ladder with a 207 nucleotide spacing, indicating that transcription had occurred through continuous arrays of positioned nucleosome cores. It was estimated that the efficiency of passage of the transcribing polymerase through each nucleosome core had an upper limit of approximately 85 per cent.

Additional studies have confirmed these results with regard to the effects of nucleosome cores on the efficiency of transcription elongation by prokaryotic RNA polymerases (Pfaffle *et al.* 1990; Kirov *et al.* 1992) and eukaryotic RNA polymerase II (Izban and Luse 1991) and RNA polymerase III (Hansen and Wolffe 1992). However, controversy remains regarding the fate of nucleosomes during transcription. Lorch *et al.* (1987) observed the displacement of histone octamers upon transcription elongation through short (mononucleosome length) reconstituted linear plasmid DNA templates. In contrast, Losa and Brown (1987) observed no net displacement of histone octamers assembled onto a 5S rDNA fragment. This apparent controversy was resolved by Lorch *et al.* (1988), who concluded that net displacement was dependent on the nature of the short DNA sequence being transcribed. The consensus of published reports indicates that the displacement of histone octamers during transcription *in vitro* may be dependent largely upon other experimental conditions. For example, Kirov *et al.* (1992), using electron microscopy, observed no net displacement of histone octamers from a linear 1400 bp DNA template containing either four or seven reconstituted nucleosome cores upon transcription with T7 RNA polymerase. In addition, no apparent net displacement of histone octamers was observed from closed circular DNA templates as a result of transcription with T7 RNA polymerase in *in vitro* experiments in the presence of low levels of topoisomerase I activity (Pfaffle *et al.* 1990; O'Neill and Bradbury, unpublished results). However, significant loss of DNA supercoiling was observed when reconstituted circular DNA templates were transcribed in the presence of high levels of topoisomerase I activity (Pfaffle *et al.* 1990). Paradoxically, Ten Heggeler-Bordier *et al.* (1995) observed disruption of nucleosomes on linear or closed circular DNA templates when transcribed by T7 RNA polymerase, but not when transcribed by RNA polymerase II in rat liver nuclear extracts.

These experiments indicate clearly that RNA polymerases can transcribe through nucleosome cores, although to some extent the nucleosome core presents an obstacle to this process. As the net displacement of nucleosomes does not appear to be absolutely necessary to allow passage of the polymerase, it is unclear what mechanisms allow the separation of DNA strands and their access to RNA polymerase that is necessary for the progress of transcription elongation. The nucleosome model has inspired a number of proposals in which nucleosomes open along the dyad axis or along the interface of the H3/H4 tetramer and H2A/H2B dimers, or unfold in other ways during transcription. There is experimental support for this in reports of half-nucleosomes (Lee and Garrard 1991), increased accessibility to groups inside the histone octamer (Chen and Allfrey 1987), or selective loss of H2A/H2B in transcribed nucleosomes (Baer and Rhodes 1983). However, transcription studies on crosslinked chromatin have demonstrated that there is no need for histone octamers to unfold *per se* during elongation, as transcription was not affected by irreversible crosslinking of histones within histone octamers (Gould *et al.* 1980, O'Neill *et al.* 1993). Recently, transcriptional activators that interact with nucleosomes rather than with DNA at the promoter of genes have been suggested to induce such transitional states in nucleosomes, in which for instance H2A/H2B dimers could be removed (Cote *et al.* 1994; see also Chen *et al.* 1994).

Although the techniques used cannot exclude the occurrence of rapid transitions, most studies, including those using electron microscopy (Ericsson *et al.* 1990), suggest the presence of nucleosomal structures on transcribed genes (with the exception of the heavily transcribed RNA genes). It has therefore also been proposed that during elongation histone octamers are transferred away from the progressing RNA polymerase to the DNA behind the polymerase. One experimentally tested hypothesis was based on the idea that positive DNA supercoiling is generated ahead of the progressing RNA polymerase and negative supercoiling in its wake (Liu and Wang 1987), which would result in a decreased or increased affinity, respectively, for nucleosomes and hence promote their transfer (Clark and Felsenfeld 1991, 1992; O'Donohue *et al.* 1994). More recently, it has been demonstrated that nucleosomes are transferred during transcription elongation on short DNA fragments (Studitsky *et al.* 1994). In this case, it was proposed that the nucleosome was translocated by a spooling mechanism, with the RNA polymerase looping out the DNA from the nucleosomal surface and the DNA in its wake catching on to the histone octamer of the unwrapping nucleosome. It is not certain whether the model can be extrapolated to RNA elongation on arrays of nucleosomes by eukaryotic RNA polymerase. Evidence suggests that in this case transfer relies on an accessory factor (Ten Heggeler-Bordier *et al.* 1995). However, it is difficult to reconcile mechanisms which involve the transfer of an intact histone octamer with the observation that histone H2A/H2B dimers are exchanged between histone octamers in active chromatin *in vivo* (Jackson 1990). The many genes that undergo multiple rounds of transcription, often with several RNA polymerases engaged simultaneously (Cavalli and Thoma 1993), pose an additional conceptual problem, which is not addressed in current models. The question remains whether genes undergoing transcription are organized differently to when they are in an inactive or poised state. There have been several examples where nucleosomal arrays seemed to be perturbed upon activation of the gene (Cavalli and Thoma 1993; Tsukiyama *et al.* 1994).

Histone H1 has a profound effect on the stability of nucleosomes and is necessary for the formation of higher order chromatin structure (reviewed in Van Holde 1988). Histone H1 is strongly implicated in the formation of transcriptionally inert chromatin structures and has been demonstrated to be a general repressor of transcription initiation *in vitro* (Wolffe 1989; Laybourn and Kadonaga 1991). When histone H1 is incorporated into linear nucleosome arrays *in vitro*, transcription initiation and elongation by T7 RNA polymerase are virtually extinguished (O'Neill *et al.* 1995). A similar effect is seen on transcription initiation of the *Xenopus* somatic 5S gene *in vitro*. One straightforward interpretation of these results is that linker histones may exert their effects at least in part by stabilizing nucleosomal structures. This would limit mobility of nucleosomes located on promoters, thereby reducing the accessibility of promoter regions to transcription factors and RNA polymerase. Additionally, the stabilization of nucleosome cores may provide further resistance to the progress of RNA polymerase during transcription elongation. Hence, histone H1 provides a further mechanism for regulation of chromatin activity via direct regulation of nucleosome dynamics.

Future directions

Protein–protein interactions are emerging as crucial in gene regulation, in addition to classical protein–DNA interactions. Regulatory DNA elements typically consist of a collection of common as well as tissue-specific factor binding sites, suggestive of a regulatory network. This now seems to be paralleled by a system of protein–protein interactions involving the histones. This concept is changing our view on chromatin structure, which now appears to be a more active participant in gene expression. Clues are surfacing that link the nuclear matrix and chromatin organization in synergizing transcription (Jackson *et al.* 1993). Some chromatin repressor proteins appear to act by compartmentalizing chromatin regions (reviewed by Pirotta and Rastelli 1994). The unravelling of eukaryotic gene expression has so far presented a challenging complex system and many components most probably remain to be discovered.

Currently, large protein complexes that have been discovered to activate transcription (SWI/SNF, NURF) are the focus of major research activities. The mechanism of activation by these complexes will provide important clues for understanding chromatin transcription. This will require knowing how the complexes interact with nucleosomes to promote their accessibility to DNA-binding factors. It should shed some light on the demonstrated nucleosome rearrangements at promoters and their microinvolvement in the mechanism of gene activation. The relevance of nucleosome mobility to nuclear function will need to be further characterized. Apart from ATP-dependent assisted mobility, inhibitors and other modulators of nucleosome mobility could prove important in regulating DNA access. The link between nucleosome positioning and its dynamics may provide additional clues to eukaryotic gene architecture. Further studies of transcription elongation will likely move towards more realistic conditions of arrays of nucleosomes on gene templates in trying to distinguish between different models. Factors expected to influence both the dynamics and interaction of nucleosomes are the histone subtypes and their states of reversible chemical modifications. Although histone hyperacetylation, among other histone modifications, has been linked to the transcriptionally active states of chromatin, we are still struggling with the structure–function relationships of this important reversible modification (reviewed by Bradbury 1992; also see Chapter 1).

Summary

The complexity of gene expression in eukaryotes results from the specific mechanisms required to overcome the influence of the histone proteins that compact the genome through a series of levels of folding. Although the effect of this chromatin structure is generally repressive, evidence has shown that the histone proteins are often specifically involved in controlling the processes that lead to gene expression. Histone octamers, precisely positioned on the DNA, modulate access of transcription factors to the substrate. This invokes nucleosome rearrangement or displacement as a major regulatory step in gene activation. Molecular mechanisms that facilitate transcription elongation within chromatin may involve the transfer of histone octamers during the passing of a transcribing polymerase. The nucleosomal organization itself is more dynamic than previously thought. There exists a general short-range mobility of nucleosomes between possible positions on DNA. Nucleosome mobility appears to be an important aspect of transcription competence. Other binding components, both activators and repressors of gene activity, can mediate their effects on a chromatin substrate by remodelling this dynamic basal nucleosome structure. DNA processing is becoming dauntingly complex.

Acknowledgements

Research support is acknowledged from the Department of Energy (DE-FG03-88ER60673 to E.M.B.), the National Institutes of Health (PHS GM26901 to E.M.B.) and the Wellcome Trust (to S.P.).

References

Adams, C.C. and Workman, J.L. (1993). Nucleosome displacement in transcription. *Cell*, **72**, 305–8.

Allan, J., Hartman, P.G., Crane-Robinson, C., and Aviles, F.X. (1980). The structure of histone H1 and its location in chromatin. *Nature*, **288**, 675–9.

Almer, A., Rudolph, H., Hinnen, A., and Hörz, W. (1986). Removal of positioned nucleosomes from the yeast PHO5 promoter upon PHO5 induction releases additional upstream activating DNA elements. *EMBO Journal*, **5**, 2689–96.

Baer, B.W. and Rhodes, D. (1983). Eukaryotic RNA polymerase II binds to nucleosome cores from transcribed genes. *Nature*, **301**, 482–8.

Beard, P. (1978). Mobility of histones on the chromosome of simian virus 40. *Cell*, **15**, 955–67.

Bradbury, E.M. (1992). Reversible histone modifications and the chromosome cell cycle. *BioEssays*, **14**, 9–16.

Cavalli, G. and Thoma, F. (1993). Chromatin transitions during activation and repression of galactose-regulated genes in yeast. *EMBO Journal*, **12**, 4603–13.

Chen, T.A. and Allfrey, V.G. (1987). Rapid and reversible changes in nucleosome structure accompany the activation, repression and superinduction of murine fibroblast protooncogenes C-Fos and C-Myc. *Proceedings of the National Academy of Sciences of the USA*, **84**, 5252–6.

Chen, H., Li, B., and Workman, J.L. (1994). A histone-binding protein, nucleoplasmin, stimulates transcription factor binding to nucleosomes and factor-induced nucleosome disassembly. *EMBO Journal*, **13**, 380–90.

Clark, D.J. and Felsenfeld, G. (1991). Formation of nucleosomes on positively supercoiled DNA. *EMBO Journal*, **10**, 387–95.

Clark, D.J. and Felsenfeld, G. (1992). A nucleosome core is transferred out of the path of a transcribing polymerase. *Cell*, **71**, 11–22.

Cote, J., Quinn, J., Workman, J.L., and Peterson, C.L. (1994). Stimulation of GAL4 derivative binding to nucleosomal DNA by the yeast SWI/SNF complex. *Science*, **265**, 53–60.

Davey, C., Pennings, S., Meersseman, G., Wess, T.J., and Allan, J. (1995). Periodicity of strong nucleosome positioning sites around the chicken adult beta-globin gene may encode regularly spaced chromatin. *Proceedings of the National Academy of Sciences of the USA*, **92**, 11210–14.

De Bernardin, W., Koller, T., and Sogo, J.M. (1986). Structure of *in vivo* transcribing chromatin as studied in Simian Virus 40 minichromosomes. *Journal of Molecular Biology*, **191**, 469–82.

Ericsson, C., Grossbach, U., Björkroth, B., and Daneholt, B. (1990). Presence of histone H1 on an active Balbiani ring gene. *Cell*, **60**, 73–83.

Fedor, M.J., Lue, N.F., and Kornberg, R.D. (1988). Statistical positioning of nucleosomes by specific protein-binding to an upstream activating sequence in yeast. *Journal of Molecular Biology*, **204**, 109–27.

Felts, S.J., Weil, P.A., and Chalkley, R. (1990). Transcription factor requirements for *in vitro* formation of transcriptionally competent 5S rRNA gene chromatin. *Molecular and Cellular Biology*, **10**, 2390–401.

Fragoso, G., John, S., Roberts, M.S., and Hager, G. L. (1995). Nucleosome positioning on the MMTV LTR results from the frequency-biased occupancy of multiple frames. *Genes and Development*, **9**, 1933–47.

Gould, H.J., Cowling, G.J., Harborne, N.R., and Allan, J. (1980). An examination of models for chromatin transcription. *Nucleic Acids Research*, **8**, 5255–66.

Gross, D.S. and Garrard, W.T. (1987). Poising chromatin for transcription. *Trends in Biochemical Sciences*, **12**, 293–7.

Hansen, J.C. and Wolffe, A.P. (1992). Influence of chromatin folding on transcription initiation and elongation by RNA polymerase III. *Biochemistry*, **31**, 7977–88.

Hebbes, T.R., Clayton, A.L., Thorne, A.W., and Crane-Robinson, C. (1994). Core histone hyperacetylation co-maps with generalized DNase I sensitivity in the chicken β-globin chromosomal domain. *EMBO Journal*, **13**, 1823–30.

Izban, M.G. and Luse, D.S. (1991). Transcription on nucleosomal templates by RNA polymerase II *in vitro*: inhibition of elongation with enhancement of sequence-specific pausing. *Genes and Development*, **5**, 683–96.

Jackson, D.A., Hassan, A.B., Errington, R.J., and Cook, P.R. (1993). Visualization of focal sites of transcription within human nuclei. *EMBO Journal*, **12**, 1059–65.

Kirov, N., Tsaneva, I., Einbinder, E., and Tsanev, R. (1992). *In vitro* transcription through nucleosomes by T7 RNA polymerase. *EMBO Journal*, **11**, 1941–7.

Lauderdale, J. D. and Stein, A. (1993). Effects of plasmid length and positioned nucleosomes on chromatin assembly *in vitro*. *Biochemistry*, **32**, 489–99.

Laybourn, P.J. and Kadonaga, J.T. (1991). Role of nucleosomal cores and histone H1 in the regulation of transcription by RNA polymerase II. *Science*, **254**, 238–45.

Lee, M.S. and Garrard, W.T. (1991). Transcription-induced nucleosome splitting—an underlying structure for DNAse-I sensitive chromatin. *EMBO Journal*, **10**, 607–15.

Li, Q. and Wrange, O. (1995). Accessibility of a glucocorticoid response element in a nucleosome depends on its rotational positioning. *Molecular and Cellular Biology*, **15**, 4375–84.

Liu, L.F. and Wang, J.C. (1987). Supercoiling of the DNA template during transcription. *Cell*, **67**, 833–6.

Lorch, Y., LaPointe, J.W., and Kornberg, R.D. (1987). Nucleosomes inhibit the initiation of transcription but allow chain elongation with the displacement of histones. *Cell*, **49**, 203–10.

Lorch, Y., LaPointe, J.W., and Kornberg, R.D. (1988). On the displacement of histones from DNA by transcription. *Cell*, **55**, 743–4.

Losa, R. and Brown, D.D. (1987). A bacteriophage RNA polymerase transcribes *in vitro* through a nucleosome core without displacing it. *Cell*, **50**, 801–8.

Lowman, H. and Bina, M. (1990). Correlation between dinucleotide periodicities and nucleosome positioning on mouse satellite DNA. *Biopolymers*, **30**, 861–76.

Mathis, D.J., Oudet, P., Waslyk, B., and Chambon, P. (1978). Effect of histone acetylation on structure and *in vitro* transcription of chromatin. *Nucleic Acids Research*, **5**, 3523–47.

Meersseman, G., Pennings, S., and Bradbury, E.M. (1991). Chromatosome positioning on assembled long chromatin: linker histones affect nucleosome placement on 5S rDNA. *Journal of Molecular Biology*, **220**, 89–100.

Meersseman, G., Pennings, S., and Bradbury, E.M. (1992). Mobile nucleosomes—a general behavior. *EMBO Journal*, **11**, 2951–9.

Meneguzzi, G., Chenciner, N., and Milanesi, G. (1979). Transcription of nucleosomal DNA in SV40 minichromosomes by eukaryotic and prokaryotic RNA polymerases. *Nucleic Acids Research*, **6**, 2947–60.

Morse, R.H. (1989). Nucleosomes inhibit both transcriptional initiation and elongation by RNA polymerase III *in vitro*. *EMBO Journal*, **8**, 2343–51.

Nacheva, G.A., Guschin, D.Y., Preobrazhenskaya, O.V., Karpov, V.L., Ebralidse, K.K., and Mirzabek, A (1989). Change in the pattern of histone binding to DNA upon transcriptional activation. *Cell*, **58**, 27–36.

Nissen, K.A., Lan, S.Y., and Smerdon, M.J. (1986). Stability of nucleosome placement in newly repaired regions of DNA. *Journal of Biological Chemistry*, **261**, 8585–8.

O'Donohue, M.-F., Duband-Goulet, I., Hamiche, A., and Prunell, A. (1994). Octamer displacement and redistribution in transcription of single nucleosomes. *Nucleic Acids Research*, **22**, 937–45.

O'Neill, T.E., Roberge, M., and Bradbury, E.M. (1992). Nucleosome arrays inhibit both initiation and elongation of transcripts by T7 RNA polymerase. *Journal of Molecular Biology*, **223**, 67–78.

O'Neill, T.E., Smith, J.G., and Bradbury, E.M. (1993). Histone octamer dissociation is not required for transcript elongation through arrays of nucleosome cores by phage T7 RNA polymerase *in vitro*. *Proceedings of the National Academy of Sciences of the USA*, **90**, 6203–7.

O'Neill, T.E., Pennings, S., Meersseman, G., and Bradbury, E.M. (1995). Deposition of histone H1 onto reconstituted nucleosome arrays inhibits both initiation and elongation of transcripts by T7 RNA polymerase. *Nucleic Acids Research*, **23**, 1075–82.

Pazin, M.J., Kamakaka, R.T., and Kadonaga, J.T. (1994). ATP-dependent nucleosome reconfiguration and transcriptional activation from preassembled chromatin templates. *Science*, **266**, 2007–11.

Pennings, S., Meersseman, G., and Bradbury, E.M. (1991). Mobility of positioned nucleosomes on 5S rDNA. *Journal of Molecular Biology*, **220**, 101–10.

Pennings, S., Meersseman, G., and Bradbury, E.M. (1992). Effect of glycerol on the separation of nucleosomes and bent DNA in low ionic strength polyacrylamide gel electrophoresis. *Nucleic Acids Research*, **20**, 6667–72.

Pennings, S., Meersseman, G., and Bradbury, E.M. (1994). Linker histones H1 and H5 prevent the mobility of positioned nucleosomes. *Proceedings of the National Academy of Sciences of the USA*, **91**, 10275–9.

Peterson, C.L. and Tamkun, J.W. (1995). The SWI–SNF complex: a chromatin remodeling machine. *Trends in Biochemical Sciences*, **20**, 143–6.

Pfaffle, P., Gerlach, V., Bunzel, L., and Jackson, V. (1990). *In vitro* evidence that transcription-induced stress causes nucleosome dissolution and regeneration. *Journal of Biological Chemistry*, **265**, 16830–40.

Pirotta, V. and Rastelli, L. (1994). *white* gene expression, repressive chromatin domains and homeotic gene regulation in *Drosophila*. *BioEssays*, **16**, 549–56.

Prunell, A. (1982). Nucleosome reconstitution on plasmid-inserted poly(dA).poly(dT). *EMBO Journal*, **1**, 173–9.

Satchwell, S.C., Drew, H.R., and Travers, A.A. (1986). Sequence periodicities in chicken nucleosome core DNA. *Journal of Molecular Biology*, **191**, 659–75.

Shrader, T.E. and Crothers, D.M. (1989). Artificial nucleosome positioning sequences. *Proceedings of the National Academy of Sciences of the USA*, **86**, 7418–22.

Shrader, T.E. and Crothers, D.M. (1990). Effects of DNA sequence and histone–histone interactions on nucleosome placement. *Journal of Molecular Biology*, **216**, 69–84.

Simpson, R.T. (1991). Nucleosome positioning: occurrence, mechanisms, and functional consequences. *Progress in Nucleic Acid Research and Molecular Biology*, **40**, 143–84.

Simpson, R.T. and Stafford, D.W. (1983). Structural features of a phased nucleosome particle. *Proceedings of the National Academy of Sciences of the USA*, **50**, 51–5.

Simpson, R.T., Thoma, F., and Brubaker, J.M. (1985). Chromatin reconstituted from tandemly repeated cloned DNA fragments and core histones; a model system for study of higher order structure. *Cell*, **42**, 799–808.

Studitsky, V.M., Clark, D.J., and Felsenfeld, G. (1994). A histone octamer can step around a transcribing polymerase without leaving the template. *Cell*, **76**, 371–82.

Ten Heggeler-Bordier, B., Schild-Pouter, C., Chapel, S., and Whali, W. (1995). Fate of linear and supercoiled multinucleosomic templates during transcription. *EMBO Journal*, **14**, 2561–9.

Thoma, F. (1991). Structural changes in nucleosomes during transcription: strip, split or flip? *Trends in Genetics*, **7**, 175–7.

Thoma, F., Koller, T., and Klug, A. (1979). Involvement of histone H1 in the organization of the nucleosome and of the salt-dependent superstructures of chromatin. *Journal of Cell Biology*, **83**, 403–27.

Tsukiyama, T. and Wu, C. (1995). Purification and properties of an ATP-dependent nucleosome remodeling factor. *Cell*, **83**, 1011–20.

Tsukiyama, T., Becker, P.B., and Wu, C. (1994). ATP-dependent nucleosome disruption at a heat-shock promoter mediated by binding of GAGA transcription factor. *Nature*, **367**, 525–32.

Ura, K., Hayes, J.J., and Wolffe, A.P. (1995). A positive role for nucleosome mobility in the transcriptional activity of chromatin templates: restriction by linker histones. *EMBO Journal*, **14**, 3752–65.

Van Holde, K.E. (1988). *Chromatin*. Springer-Verlag, New York.

Van Holde, K.E., Lohr, D.E., and Robert, C. (1992). What happens to nucleosomes during transcription? *Journal of Biological Chemistry*, **267**, 2837–40.

Varga-Weisz, P.D., Blank, T.A., and Becker, P.B. (1995). Energy-dependent chromatin accessibility and nucleosome mobility in a cell-free system. *EMBO Journal*, **14**, 2209–16.

Venter, U., Svaren, J., Schmitz, J., Schmid, A., and Hörz, W. (1994). A nucleosome precludes binding of the transcription factor Pho4 *in vivo* to a critical target site in the PHO5 promoter. *EMBO Journal*, **13**, 4848–55.

Vidali, G., Boffa, L.C., Bradbury, E.M., and Allfrey, V.G. (1978). Supression of histone deacetylation leads to accumulation of multiacetylated forms of histones H3 and H4 and increased DNase 1 sensitivity of associated DNA sequences. *Proceedings of the National Academy of Sciences of the USA*, **75**, 2239–44.

Villeponteau, B., Brawley, J., and Martinson, H.G. (1992). Nucleosome spacing is compressed in active chromatin domains of chick erythroid cells. *Biochemistry*, **31**, 1554–63.

Sari Pennings et al.

Wall, G., Varga-Weisz, P.D., Sandaltzopoulos, R., and Becker, P.B. (1995). Chromatin remodeling by GAGA factor and heat shock factor at the hypersensitive *Drosophila* hsp26 promoter *in vitro*. *EMBO Journal*, **14**, 1727–36.

Waslyk, B. and Chambon, P. (1979). Transcription by eukaryotic RNA polymerases A and B of chromatin assembled *in vitro*. *European Journal of Biochemistry*, **98**, 317–27.

Waslyk, B., Thevenin, G., Oudet, P., and Chambon, P. (1979). Transcription of *in vitro* assembled chromatin by *Escherichia coli* RNA polymerase. *Journal of Molecular Biology*, **128**, 411–40.

Williamson, P. and Felsenfeld, G. (1978). Transcription of histone-covered T7 DNA by *Escherichia coli* RNA polymerase. *Biochemistry*, **17**, 5695–705.

Wolffe, A.P. (1989). Dominant and specific repression of *Xenopus* oocyte 5S RNA genes and satellite I DNA by histone H1. *EMBO Journal*, **8**, 527–37.

Wolffe, A.P. (1995). *Chromatin: structure and function*. Academic Press, London.

2

Role of a positioned nucleosome in constitutive repression and hormone induction of the MMTV promoter

Miguel Beato, Reyes Candau, Sebastián Chávez, Christian Möws, and Mathias Truss

Introduction

It seems obvious that the packaging and condensation of eukaryotic DNA in chromatin within the cell nucleus should represent an obstacle for the protein machinery involved in replication, recombination, and transcription of the genetic message. This argument probably applies to all levels of chromatin organization, from the basic repeating unit, the nucleosome, over the solenoid superhelix or other condensed arrays of nucleosomes, to chromatin domains and loops. In the past this problem has been circumvented or eluded by invoking *ad hoc* hypothesis. For some years after the discovery of the nucleosomal organization of chromatin, the curious idea was favoured that the histones have evolved to packaged DNA into the cell nucleus in a 'transparent' way, which virtually would not interfere with the processes of replication and transcription. The prevailing opinion was that one could treat the genetic material of eukaryotic cells in very much the same way as that of bacteriophage lambda, ignoring the fact that it is tightly condensed into a regular structure. To make this assumption acceptable, it was claimed that the nucleosomes are randomly located along the DNA, and for this reason do not pose a problem for protein recognizing specific sequences or for enzymes involved in DNA metabolism. In recent years, however, this view has been challenged by the accumulation of experimental evidence demonstrating a precise nucleosomal organization in relevant DNA regions of many genes. Almost 20 different yeast genes, and a similar number of animal genes, have been shown to exhibit positioned nucleosomes in their promoter or enhancer regions, and it is becoming progressively clear that the location of a DNA sequence relative to the histone octamer does influence its affinity for DNA binding proteins (Vettese-Dadey *et al.* 1994). At the same time, and given that virtually all eukaryotic DNA is organized in chromatin, cells must have evolved, along with the packaging tools, a machinery able to handle chromatin-organized DNA. The genes and the molecular mechanism involved in chromatin dynamics are only starting to be identified, but sufficient evidence has been generated to support the claim that the chromatin organization, in particular nucleosomal positioning, plays a general role in the regulated expression of the genetic information (Wolffe 1994; see also Chapters 1, 3, and 4).

In this chapter we will review our knowledge on the role of nucleosomal organization in gene regulation, based on the MMTV (Mouse Mammary Tumor Virus) promoter. We will briefly summarize what is presently known about nucleosome positioning and chromatin remodelling and will sketch potential lines for future research.

The MMTV promoter: *cis*-acting elements and *trans*-acting factors

Regulated transcription from the MMTV promoter is controlled by a region comprising some 200 bp upstream of the transcription start point, which includes the core promoter elements and a complex hormone-responsive unit (HRU). The later encompasses a set of hormone-responsive elements (HREs), a binding site for members of the nuclear factor 1 (NF1) family, and two degenerated octamer motifs (Truss and Beato 1993).

Hormone-responsive elements and hormone receptors

Induction of MMTV transcription by glucocorticoids has been a classical system to study the mechanism of action of steroid hormones (Ringold *et al.* 1975). In cells equipped with the corresponding receptors, hormone administration enhances the rate of transcription of MMTV DNA (Ringold *et al.* 1977; Young *et al.* 1977; Ucker and Yamamoto 1984). The nucleotide sequences relevant for mediating transcriptional activation were localized to the long terminal repeat region (LTR) of the proviral genome, using gene transfer methods (Buetti and Diggelmann 1981; Huang *et al.* 1981; Hynes *et al.* 1981; Lee *et al.* 1981). Fine mapping identified the region between –200 and –50 upstream of the transcription start point as sufficient for hormonal regulation (Buetti and Diggelmann 1983; Hynes *et al.* 1983; Majors and Varmus 1983). Partially purified glucocorticoid receptor (GR) binds preferentially to a region overlapping these functionally relevant sequences (Payvar *et al.* 1981, 1983; Geisse *et al.* 1982; Chandler *et al.* 1983; Scheidereit *et al.* 1983). Using DNAase I footprinting techniques, four main sites sharing the hexanucleotide motif TGTTCT were identified in the hormone-responsive region of the MMTV LTR in the GR mice strain as being protected by the receptor against nuclease digestion (Scheidereit *et al.* 1983). In addition, a cryptic degenerated receptor binding site was identified around position –150 (Scheidereit *et al.* 1983). Mutation of any of the hexanucleotide motifs has a dramatic effect on hormonal induction (Chalepakis *et al.* 1988). This was the first demonstration of the existence of HREs as defined nucleotide sequences in the vicinity of regulated promoters. In addition, these experiments demonstrated the existence of a hormone-responsive region (HRR) on the MMTV LTR, composed of a complex array of HREs, which later was shown to confer hormone responsiveness also to heterologous promoters.

Although MMTV induction was originally used as a classical example of glucocorticoid regulation, later it was shown that the MMTV promoter also responds to progestins (Cato *et al.* 1986), as well as to mineralocorticoids (Arriza *et al.* 1987; Cato and Weinmann 1988) and to androgens (Darbre *et al.* 1986; Cato *et al.* 1987). Binding experiments with purified progesterone receptor (PR) from rabbit uterus demonstrated that the region protected against DNAase I is very similar to that covered by GR (Ahe *et al.* 1985; Chalepakis *et al.* 1988). A similar relevance of the individual TGTTCT motifs for response to androgens has been reported (Ham *et al.* 1988). The fact that the different steroid hormone receptors bind to the same regulatory elements raises the question of how hormone specificity is achieved *in vivo*. Most likely the mechanism involves receptor-specific interactions with other transcription factors required for the hormonal response.

Most HREs recognized by GR and PR are imperfect palindromes spaced by three base pairs (Beato 1989). The steroid hormone receptors bind to these HREs in the form of homodimers in a highly cooperative fashion (Chalepakis *et al.* 1990). The elucidation of the three-dimensional structure of the DNA binding domain of GR bound to various HREs suggested a strong interaction of one monomer with the best conserved half of the palindrome and a weak interaction of the other monomer with the more degenerated half (Luisi *et al.* 1991). Receptor dimers can also cooperate when binding to adjacent HREs (Jantzen *et al.* 1987; Martinez and Wahli 1989; Schmid *et al.* 1989). Though the exact stoichiometry of receptor binding to the MMTV HRR is not completely clear (Truss *et al.* 1992), we and other have detected a strong cooperativity between the individual sites (Chalepakis *et al.* 1988; Perlmann *et al.* 1990). Mutation of any of the TGTTCT motifs has a strong inhibitory effect on receptor binding. In addition, there is a functional synergism between the receptor binding sites, as demonstrated by deletion mutants and by the influence of changing their distance by inserting oligonucleotides of different length (Chalepakis *et al.* 1988). It remains to be established how exactly the particular array of receptor binding sites influences the interaction with the hormone receptor when the MMTV LTR is organized into chromatin (see below).

NF1 binding site

In addition to the hormone receptors, other factors have been found to be involved in transcribing the MMTV promoter. The first to be identified was the transcription factor NF1 (Nowock *et al.* 1985; Miksicek *et al.* 1987; Buetti *et al.* 1989; Gowland and Buetti 1989; Brüggemeier *et al.* 1990). There are several genes coding for proteins that recognize the NF1 sites and differential splicing generates a large family of homologous proteins (Rupp *et al.* 1990). Since these proteins can form homo- and heterodimers that are able to bind the palindromic

NF1 sites, the number of possible combinations is very large and could have implications in the regulation of specific genes (Kruse and Sippel 1994). The transactivation functions of the various NF1 proteins have been recently delineated. An essential core domain between amino acids 463 and 508 contains the sequence motif SPTSPSYSP, which resembles the heptapeptide repeat YSPTSPS in the CTD region of RNA polymerase II (Wendler *et al.* 1994).

The binding site for NF1 in the MMTV promoter is located immediately downstream of the HRR, and mutations of this site that inhibit NF1 binding *in vitro* have been reported to strongly reduce hormone-induced transcription (Miksicek *et al.* 1987; Buetti *et al.* 1989). We have confirmed these findings in various cell lines containing either PR or GR (Brüggemeier *et al.* 1990). Our experiments clearly identify NF1 as a transcription factor in the MMTV promoter, which is needed to obtain optimal hormonal induction.

It has been shown previously that NF1 and steroid hormone receptors can cooperate in transactivation of artificial promoters, carrying binding sites for the two proteins in the correct distance (Schüle *et al.* 1988a; Strähle *et al.* 1988). It was, therefore, predicted that the NF1 site in the MMTV promoter would participate in similar interactions. However, we found that purified hormone receptors do not cooperate, but rather compete with NF1 for binding to the MMTV promoter, and *vice versa* (Brüggemeier *et al.* 1990). Under no experimental conditions have we detected a synergism between the two proteins in terms of interaction with the MMTV promoter. In fact, the observed competition is not surprising as the area protected against DNAase I by the hormone receptors overlaps by several bases with the footprint generated by NF1. Given the observed requirement of sequences flanking the HRE for efficient binding of the receptor (Chalepakis *et al.* 1990), a steric hindrance in the interaction of both proteins with the MMTV promoter would be expected (Brüggemeier *et al.* 1990). Thus, we are faced with the paradox that although NF1 does act as an essential transcription factor for the MMTV promoter and is required for optimal induction, there is no direct cooperation between steroid hormone receptors and NF1 in terms of DNA binding. Either the functional synergism between the two proteins is mediated by an adaptor molecule that is missing in our partially purified fractions, or mechanisms other than DNA binding synergism may be involved in NF1-mediated transcription in response to steroid hormones.

The octamer motifs

MMTV promoters lacking the NF1 binding site are still able to respond to hormone administration, albeit with much lower efficiency than the wild-type promoter (Brüggemeier *et al.* 1990). Thus, in addition to NF1 there must be other factors that can also mediate induction of the MMTV by steroid hormones. A search for other possible factors involved in transcriptional activation of the MMTV promoter led to the identification of two octamer motifs between the NF1 binding sites and the TATA box (Brüggemeier *et al.* 1991). Mutations at these sites resulted in a significant reduction of the hormonal induction of the MMTV promoter in gene transfer experiments (Toohey *et al.* 1990; Brüggemeier *et al.* 1991). A third, more degenerated octamer motif, which overlaps the two main sites, appears to be functionally irrelevant (Huang *et al.* 1993). The two main sites are not functionally equivalent. The promoter distal site exhibits a single mismatch and binds OTF1 with an affinity similar to that of the consensus octamer motif. Mutations in this site have a significant effect on MMTV transcription in hormone-treated cells. This effect depends on the cell type and is three- to four-fold in HeLa cells and five- to six-fold in T47D cells. The promoter proximal site exhibits two mismatches, and mutations at this site are virtually silent in HeLa cells and show only a 50 per cent reduction in activity in T47D cells (Brüggemeier *et al.* 1991). This site is nevertheless binding OTF1 in the wild-type MMTV promoter, as is demonstrated by the presence of two molecules of OTF1 bound to the corresponding oligonucleotide that contains both octamer motifs (Brüggemeier *et al.* 1991). In fibroblasts that are stably transfected with MMTV constructs, the octamer motifs appear to be important also for the basal expression of the promoter (Buetti 1994).

Contrary to the results with NF1, OTF1 binds weakly to the MMTV promoter in the absence of receptor. However, when either PR or GR is preincubated with the MMTV DNA, binding of OTF1 is strongly enhanced, as demonstrated in DNAase I footprinting experiments (Brüggemeier *et al.* 1991). Since these experiments were performed with highly purified preparations of receptor and OTF1, it is unlikely that the cooperativity of DNA binding is mediated by additional factors. Therefore, with respect to OTF1, the steroid hormone receptors behave as expected in terms of their cooperative binding to their cognate sites on the MMTV promoter. As we will see below, this enables the reproduction of the functional synergism under cell-free conditions.

Core promoter elements

In addition to the hormone receptors, NF1 and OTF1, general transcription factors are involved in the regulated transcription of the MMTV promoter. In *in vivo* experiments it has been shown that mutations of the TATA box region diminish the activity of the MMTV promoter in response to glucocorticoids (Toohey *et al.* 1990). After hormone administration exonuclease digestion of intact nuclei detects a stop at position +1 of the MMTV promoter, suggesting the presence of proteins bound to the region of the TATA box and possibly to the initiator region (Cordingley *et al.* 1987). In other systems it has been shown that a general transcription factor TFIID, composed of the TATA box binding protein (TBP) and transcription accessory factors (TAFs), is responsible for both binding to the TATA box region and for its functional utilization in cell-free transcription (Horikoshi *et al.* 1989). Minimal promoters containing only a TATA box and binding sites for the hormone receptors immediately upstream are able to respond to hormone treatment in gene transfer experiments (Strähle *et al.* 1987; Schatt *et al.* 1990). These findings suggest that under certain conditions the hormone receptors are able to interact directly or indirectly with TFIID or other general transcription factors. Direct evidence has been reported for an interaction of the oestrogen receptor with TBP (Sadovsky *et al.* 1995) and with TFIIB (Ing *et al.* 1992). Whether this interaction plays a role in induction of the wild-type MMTV promoter, or whether the effect in this case is exclusively mediated by sequence-specific transcription factors (NF1 and/or OTF1), remains to be studied.

Negative and other elements

The existence of negative regulatory sites within the MMTV LTR has been repeatedly reported. In transfection experiments it has been shown that the MMTV enhancer can block the action of the HaMuSV (Harvey murine sarcoma virus) enhancer upon a reported gene (Ostrowski *et al.* 1984). Most reports localize the negative elements to regions upstream from the HRR, and these elements will not be discussed here. However, there are also reports suggesting the existence of negative regulatory elements in the region between the two blocks of HREs (Langer and Ostrowski 1988; Tanaka *et al.* 1991; Härtig *et al.* 1993). In genomic footprinting experiments we have not obtained evidence for the existence of such repressor binding sites in the cells lines used for our studies (see below and Truss *et al.* 1995).

Sequences immediately upstream of the HREs, from −290 to −184, have also been reported to be required for optimal hormonal induction in transient transfection studies (Gouillet *et al.* 1991; Cavin and Buetti 1995). Two regions recognized respectively by a ubiquitous and a tissue-specific factor have been identified, which in combination with the distal GRE (Glucocorticoid Responsive Element) are responsible for the tissue selectivity of MMTV expression in transgenic mice (Ross and Sloter 1985; Mok *et al.* 1992; Rollini *et al.* 1992; Cavin and Buetti 1995). The exact nature of the factors involved in these interactions is unknown, but the nucleotide sequence between −223 and −201 contains a CACC box, which is compatible with binding of members of the Sp1 family of transcription factors (Hagen *et al.* 1992). It has been shown that these elements synergize with GREs in induction of the tyrosine aminotransferase and the tryptophan oxygenase genes (Schüle *et al.* 1988b; Strähle *et al.* 1988).

Recently a binding site for the transcription factor Ear3/COUP has been identified between +104 and +120 in the MMTV LTR (Kadowaki *et al.* 1995). Intriguingly, this site, which is a direct repeat with a spacing of 5 bp, does not respond to retinoic acid. Its role in expression of the provirus remains to be established.

Behaviour of the naked promoter in cell-free transcription

In cell-free extracts from HeLa cell nuclei GR and PR are able to activate transcription from the MMTV promoter in an HRE-dependent fashion (Kalff *et al.* 1990; Möws *et al.* 1994). Mutation of the NF1 binding sites reduces basal transcription, but has no influence on the extent of induction by PR (Kalff *et al.* 1990; Brüggemeier *et al.* 1991). Mutation of the octamer motifs on the other hand has relatively little effect on basal transcription, but completely inhibits the effect of added PR (Brüggemeier *et al.* 1991). Thus, we can reproduce on naked DNA templates the synergism between receptor and OTF1, as observed in intact cells, but are unable to demonstrate a functional synergism between receptors and NF1. In addition, OTF1 and NF1 cannot occupy simultaneously their adjacent binding sites on the MMTV promoter, suggesting the existence of a steric hindrance between the two factors (Möws *et al.* 1994). These results confirm the DNA binding data and demonstrate that a full loading with sequence-specific transcription factors cannot be achieved on naked DNA templates.

GRIP170

Evidence has accumulated during the last two years that the transcriptional effects of nuclear receptors may be mediated by transcription intermediary factors (TIFs) or co-activators (Haynes *et al.* 1992; Cavailles *et al.* 1995; Guichon-Mantel *et al.* 1995; LeDouarin *et al.* 1995; Vom Baur *et al.* 1995). The lack of functional synergism between hormone receptors and NF1 *in vitro* could be attributed to the lack of such co-activators in the cell-free system. Recently, a protein has been identified that interacts with the DNA binding domain of GR bound to an HRE and activates transcription by GR in a cell-free system. However, this protein, called GRIP170 (glucocorticoid receptor interacting protein 170 kDa), does not change the lack of synergism between GR and NF1 in cell-free transcription and seems to act by enhancing the interaction of GR with the general transcription factors. We conclude that the use of cell-free transcription assays based on naked DNA template does not reproduce the behaviour of the MMTV promoter observed in the intact cell. One possible explanation could reside in the organization of the MMTV promoter in chromatin.

Nucleosome organization *in vivo* and *in vitro*

The chromatin structure over the MMTV LTR is highly ordered, with individual nucleosomes being translationally positioned, and one of them covering the HRR (Richard-Foy and Hager 1987). Although a variety of positions have been recently reported for this nucleosome (Fragoso *et al.* 1995), we observed a dominant nucleosome frame over the promoter in multicopy episomal vectors and in cells carrying a single copy integrated into the host chromosome. In both cases the positioned nucleosome covers the HRU almost completely from position -190 to position -43, encompassing the HREs, the NF1 site and the distal octamer motif (Truss *et al.* 1995). On the surface of this nucleosome the

DNA double helix is rotationally phased in such a way that only the two external HREs have their major grooves exposed for receptor binding, while the two central HREs, as well as the NF1 binding site and the distal octamer motif, are positioned with the major groove pointing towards the interior of the nucleosome, inaccessible for protein binding (Truss *et al.* 1995). Exactly this rotational positioning of the nucleosome was found in chromatin reconstitution experiments with histone octamers and linear or circular MMTV promoter fragments of various lengths (Piña *et al.* 1990a,b,c). In reconstituted MMTV nucleosomes GR and PR bind only to the two external HREs, whereas the central HREs are much less accessible. No binding to the NF1 site is detected even at very high concentration of protein (Perlmann and Wrange 1988; Piña *et al.* 1990b; Perlmann 1992). These results support the concept that the nucleotide sequence of MMTV promoter is the main determinant of translational and rotational nucleosome positioning (Piña *et al.* 1990c).

The nucleosomal organization of the MMTV promoter is responsible for basal repression

In cells carrying a single chromosomally integrated copy of the MMTV promoter no binding of sequence-specific factors can be detected prior to hormonal stimulation, thus eliminating the possibility that lack of MMTV transcription is due to the action of a specific repressor that binds to the HRU (Truss *et al.* 1994). Therefore, the constitutive repression of the MMTV promoter could be due to its precise organization into a nucleosome, which hinders binding of essential transcription factors (Piña *et al.* 1991). In agreement with this hypothesis, inhibition of chromatin assembly in *Xenopus* oocytes seems to correlate with derepression of microinjected MMTV promoter (Perlmann and Wrange 1991).

The general validity of the concept that positioned nucleosomes repress promoter activity is suggested by the behaviour of other genes in nucleosome-depleted yeast strains. Although these manipulated cells are still viable, they show alterations in several regulatory pathways, such as the mating-type control and in the expression of some regulated promoters, including GAL4 and PHO5 (Han and Grunstein 1988). The chromatin structure of the PHO5 regulatory region is distorted and this is accompanied by expression of the PHO5 gene under conditions of repression at high phosphate (Han and Grunstein 1988). For this and other regulated genes transcriptional stimulation by nucleosome depletion is observed even in the absence of UAS (upstream activating sequence), suggesting that one of the functions of the UAS complex is to remove repression due to chromatin structure.

The use of NF1–VP16 to determine accessibility of the NF1 site

To test this hypothesis we have used co-transfection of MMTV reporters with expression vectors for a chimera of NF1 and the strong viral transactivator VP16.

Under conditions that lead to organization of the transfected reporters in chromatin the wild-type MMTV promoter is poorly transactivated by NF1–VP16, whereas deletion of the HRR leads to a much better access of NF1–VP16 to the promoter (Candau *et al.* 1996). A similar result was obtained in *Saccharomyces cerevisiae*, in which the MMTV promoter is also precisely organized in nucleosomes and transactivated by a synergistic interaction between GR and NF1 (Chávez *et al.* 1995). In this system depletion of nucleosomes does increase the effect of NF1 or NF1–VP16 on transcription from the MMTV promoter in the absence of GR (S. Chávez and M. Beato, unpublished). Thus, it seems very likely that in the systems we have studied the low activity of the MMTV promoter may be accounted for by its organization in nucleosomes. However, in other cell types occupancy of the NF1 binding sites has been reported in the absence of hormones (Härtig and Cato 1994; Mymryk *et al.* 1995), suggesting that additional mechanism may exist to enable constitutive access of NF1 to the promoter.

Optimal hormonal induction requires maintenance of a rearranged nucleosome

In cells carrying episomal copies of the MMTV LTR, hormone treatment induces NF1 binding to the promoter (Cordingley *et al.* 1987) without altering the amount of NF1 in nuclear extracts from induced and control cells (Cordingley and Hager 1988). In cells containing a single copy of the MMTV promoter, integrated in their genome, hormone induction leads to binding of a full complement of transcription factors to the MMTV HRU. All five HREs, the NF1 binding site and the octamer motifs are occupied in the majority of MMTV promoters (Truss *et al.* 1995). These findings are in apparent contradiction with the results obtained on free DNA (steric hindrance between receptor and NF1 and between NF1 and Oct1/OTF1) and with the results obtained on reconstituted nucleosomes (receptor binding only to the two external HREs, no binding of NF1). Most unexpectedly, the nucleosome covering the MMTV HRU is not displaced or removed after induction, but appears to remain in place, as determined by low and high resolution micrococcal nuclease digestion data (Truss *et al.* 1995). A similar binding of transcription factors on the surface of a positioned nucleosome has been reported for the enhancer of the albumin gene in rat liver (McPherson *et al.* 1993). We therefore postulate that the assembly of a full complement of transcription factors is facilitated by their binding to the surface of a positioned nucleosome. This statement presupposes that the organization of the DNA on the surface of the nucleosome must be altered after induction to permit factor binding to the major groove of sites that are originally masked.

This idea is compatible with preliminary results obtained in yeast strains carrying the MMTV reporter system (see above) and a histone H4 gene driven by the GAL promoter and UAS (Han and Grunstein 1988). In the presence of

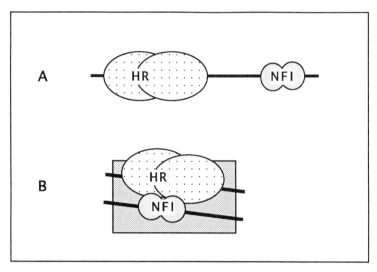

Fig. 2.1   The proximity of the two superhelical turns in nucleosomes may facilitate interactions between proteins bound to DNA sites separated by 70–80 bp. (a) A homodimer of hormone receptors (HR) bound to a hormone-responsive element (HRE) on linear DNA 70–80 bp upstream of a binding site for a homodimer of nuclear factor 1 (NF1). The two proteins are too far apart to interact efficiently. (b)When the DNA is wrapped around a histone octamer the two binding sites are on the same face of the nucleosome separated only by 2 nm and the bound proteins can interact more efficiently.

galactose these yeast strains respond to hormone administration with an NF1-dependent induction of transcription from the MMTV promoter. However, when the cells are grown in the presence of glucose and therefore are partly depleted of nucleosomes, the response to hormone is not better than in wild-type strains, as would be expected if nucleosomes fulfilled an exclusively negative function. On the contrary, nucleosome depletion leads to an impaired hormone response, in agreement with a positive role of nucleosomes in induction (S. Chávez and M. Beato, unpublished). These results reinforce the notion that the positioned nucleosome may be needed for the implementation of the functional synergism between receptors and NF1. One possible explanation may be that the proximity of the two superhelical turns on the histone octamer favours interactions between receptors and NF1 which do not take place on free DNA (Fig. 2.1).

Mechanism of nucleosome rearrangement

Hormone induction of MMTV transcription is accompanied by structural changes of chromatin, as indicated by the appearance of a DNAase I hypersensitive region over the HREs (Zaret and Yamamoto 1984), which is also hypersensitive against chemical nucleases (Richard-Foy and Hager 1987). These findings have been interpreted as indicative of removal or disruption of the positioned nucleosome covering the HRU, an interpretation that contradicts our genomic footprinting results. According to these data, hormone induction does not disrupt the positioned nucleosome but rather generates a very narrow zone of nuclease hypersensitivity coinciding with the nucleosome pseudo-dyad axis (Truss *et al.* 1995). This novel DNAase I-hypersensitive site can only be detected when the cells are treated with nuclease under very gentle conditions and is lost when cell nuclei are purified or manipulated prior to nuclease digestion. Under these conventional conditions a broader DNAase I-hypersensitive region is detected as previously reported (Zaret and Yamamoto 1984). Therefore, the hormone-induced conformational change of chromatin generates a nucleosome that does not resist manipulation of cell nuclei.

Modification of the core histones

The nature of this conformational change remains obscure, but experiments with inhibitors of histone deacetylases suggest a role for modification of the histone tails. Among the changes in nucleosomes that have been associated with transcriptionally active chromatin is hyperacetylation of lysine residues in the N-terminal tails of all four core histones (Hebbes *et al.* 1988). Recently it has been shown that restriction of GAL4 binding to reconstituted nucleosomes containing GAL4 binding sites can be alleviated by proteolytic digestion of the histone tails, suggesting a general repressive role for these highly charged domains of the core histones (Juan *et al.* 1994; Vettese-Dadey *et al.* 1994). Similarly, binding of the transcription factor TFIIIA to a reconstituted nucleosome carrying a 5S RNA gene is enhanced by acetylation of the histone tails (Lee *et al.* 1993). However, in the case of MMTV inhibition of histone deacetylase with 5–10 mM sodium butyrate seems to inhibit hormone induction and nucleosome remodelling (Bresnick *et al.* 1990). We have confirmed these results, but find that lower concentrations of butyrate do indeed activate hormone-independent transcription from a single copy integrated MMTV reporter. A similar response is observed with a more selective inhibitor of histone deacetylase activity, trichostatin A (TSA), which acts at nanomolar concentrations (Yoshida *et al.* 1990). Moreover, inducing concentrations of butyrate or TSA generate the same type of DNAase I hypersensitivity over the pseudo-dyad axis of the regulatory nucleosome that we observed following hormone induction (M. Truss, J. Bartsch, J. Bode, and M. Beato, unpublished). Though we do not know whether hormone treatment alters the acetylation of core histones, these results suggest that a nucleosome remodelling similar to that induced by receptor binding can be generated by changes in the core histone tails. However, alternative changes in nucleosome structure, such as removal of histone H2A/H2B dimers (Hayes and Wolffe 1992), have to be considered as additional possibilities for facilitating factor binding to nucleosomally organized DNA sequences.

It is intriguing that the hormone-induced structural alteration in chromatin structure takes place at the nucleosome pseudo-dyad axis, as this region has been shown to be crucial for control of chromatin-mediated gene expression involving the SWI/SNF complex in yeast (Peterson and Herskowitz 1992; Winston and Carlson 1992). A participation of members of the SWI/SNF complex in glucocorticoid gene induction has been reported in yeast and in animal cells, suggesting a direct protein–protein interaction with the hormone receptor (Yoshinaga *et al.* 1992; Muchardt and Yaniv 1993). Recently, a soluble complex of all SWI/SNF gene products has been identified, which may play an important role in facilitating chromatin transcription (Cairns *et al.* 1994; Peterson *et al.* 1994). This complex enhances binding of GAL4 derivatives to nucleosomally organized GAL4 binding sites in an ATP-dependent manner (Côte *et al.* 1994). Mutations in histones H3 and H4, located near the nucleosome dyad 27 axis, are able to suppress the phenotype of the SWI/SNF mutations (Winston and Carlson 1992), suggesting that the architecture of this region of the nucleosome is an important determinant of transcriptional activity.

Role of the linker histones

As the linker histones are important for condensation of the 10 nm chromatin fibre (Van Holde 1993), they could also influence the accessibility of DNA sequences in chromatin. Moreover, important *cis* elements of the MMTV promoter are located at the edge of the positioned nucleosome or in the linker DNA, which are regions in contact with histone H1 (Hayes and Wolffe 1993). In fact, it has been claimed that upon hormone induction there is a depletion of histone H1 from the MMTV promoter, which could influence the accessibility of promoter sequences (Bresnick *et al.* 1992). Although this remains an attractive possibility, it cannot be an essential prerequisite for hormonal induction of MMTV transcription, since the MMTV promoter is perfectly regulated in budding yeast, which lacks linker histones (Chávez *et al.* 1995). It is of course possible that in animal cells linker histones play a role in modulating promoter accessibility.

Future directions

The positioning of the MMTV DNA on the surface of the histone octamer is an intrinsic property of the nucleotide sequence, since it is maintained under a variety of different conditions *in vivo* and *in vitro*. Even in yeast, where the spacing of nucleosomes is usually 160 bp, the MMTV promoter is organized as in animal cells, suggesting that the nucleotide sequence is dominant over the other parameters determining nucleosome positioning (Chávez *et al.* 1995). As this positioning has profound consequences for the accessibility and function of the promoter, we conclude that, like proteins, DNA has a tertiary structure that determines part of its function. This implies that in addition to coding and regulatory information there is conformational or topological information in DNA that is implemented in chromatin, modulates the accessibility of regulatory information and is, therefore, critical for the realization of the genetic pro-gramme. Understanding the chemical basis of this topological message is one of the important future tasks.

The nucleosome structure described above suggests the existence of two types of DNA binding regulatory proteins: those which are able to interact with nucleosomally organized DNA, and those which are not. Provided the DNA remains in the B form, if a protein needs to contact more than five consecutive base pairs through the major groove it will be unable to see its target sequences organized in a nucleosome, independently of their precise rotational orientation. No matter the phase of the double helix, part of the recognized major groove will be masked. The same will apply if the protein contacts two sets of short sequences that are not on the same face of the double helix. For this kind of protein the essential determinant for DNA binding would be nucleosome translational phasing, namely, whether their binding sites are or are not within the boundaries of a nucleosome. In this case the rotational phasing of the DNA double helix may be relatively irrelevant, but linker histones may influence access to the DNA sites in between nucleosome cores. On the contrary, a protein that contacts short stretches of less than 5 bp, that are located on one side of the double helix, will be able to recognize its cognate sequence on nucleosomes, provided the major grooves are properly oriented. For these proteins the rotational phasing of the double helix on the surface of histone octamer would be the determinant feature for binding site recognition (Li and Wrange 1995), although the location of the site relative to nucleosome pseudo-dyad axis may also modulate binding efficiency (Li and Wrange 1993). In the next years additional examples of this kind of behaviour should be analysed in order to evaluate the general validity of these concepts.

Other mechanisms may also modulate the binding of protein factors to nucleosomally organized DNA. One possibility is a direct interaction with core histones, as has been recently reported for SIR3 and SIR4 and the N-terminal regions of histones H3 and H4 in the context of telomere silencing in yeast (Hecht *et al.* 1995). An involvement of the histone tails in gene regulation has been already postulated in studies with gene strains carrying various mutations in these histone domains (Fisher-Adams and Grunstein 1995; Wan *et al.* 1995). Since the MMTV is regulated in budding yeast in very much the same way as in animal cells (Chávez *et al.* 1995), one could take advantage of the available mutations in the yeast core histones to define the potential role of individual residues in MMTV regulation.

A comprehensive description of the induction process will require knowledge of the chromatin organization of the MMTV promoter at the nucleotide level during the different phases of hormonal induction *in vivo*. To obtain this information we need new techniques for studying the fine structure of regulatory DNA regions with associated proteins in native chromatin. In addition to chemical agents, physical methods, such as UV laser and other kinds of radiation, offer the possibility to follow rapid changes in structure and the formation of multiprotein complexes in unperturbed cells. Ultimately, however, a precise understanding of the transactivation mechanism will only be possible when correctly reconstituted chromatin templates are successfully transcribed *in vitro* using purified receptors, NF1, and OTF1/Oct1, along with all general transcription factors and accessory proteins. These *in vitro* systems will be invaluable to answer the key questions about transcription factor recruitment and assembly, sequential recruitment versus recruitment of holoenzyme complexes, as well as to define the enzymology of chromatin dynamics.

Summary

The MMTV promoter adopts a preferential position on the nucleosome surface, which is relatively independent of flanking sequences and chromosomal context. This structure precludes access of essential transcription factors and is responsible for the low constitutive activity of the MMTV promoter, as demonstrated in yeast by manipulation of the nucleosome density. Hormone induction does not disrupt the nucleosomal organization of the promoter, but leads to a change in nucleosome conformation, which enables simultaneous binding of hormone receptors, NF1 and OTF1. Since a full loading with transcription factors is not observed on free DNA, and yeast strains that are depleted of nucleosomes exhibit suboptimal induction, we postulate that the rearranged nucleosome is required for optimal hormonal induction. Experimental evidence on the nature of the hormone-induced nucleosome changes is discussed in the context of recent developments pointing to the existence of a specialized cellular machinery for chromatin remodelling.

References

Ahe, D., Janich, S., Scheidereit, C., Renkawitz, R., Schütz, G., and Beato, M. (1985). Glucocorticoid and progesterone receptors bind to the same sites in two hormonally regulated promoters. Nature, 313, 706–9.
Arriza, J. L., Weinberger, C., Cerelli, G., Glaser, T.M., Handelin, B.L., Housmann, D.E., *et al.* (1987). Cloning of human mineralocorticoid receptor cDNA: structural and functional kindship with the glucocorticoid receptor. Science, 237, 268–75.
Beato, M. (1989). Gene regulation by steroid hormones. *Cell*, 56, 335–44.
Bresnick, E.H., John, S., Berard, D.S., LeFebvre, P., and Hager, G.L. (1990). Glucocorticoid receptor-dependent disruption of a specific nucleosome on the mouse mammary tumor virus promoter is prevented by sodium butyrate. *Proceedings of the National Academy of Sciences of the USA*, 87, 3977–81.
Bresnick, E.H., Bustin, M., Marsaud, V., Richard-Foy, H., and Hager, G.L. (1992). The

transcriptionally-active MMTV promoter is depeleted of histone H1. *Nucleic Acids Research*, **20**, 273–8.

Brüggemeier, U., Rogge, L., Winnacker, E.L., and Beato, M. (1990). Nuclear factor I acts as a transcription factor on the MMTV promoter but competes with steroid hormone receptors for DNA binding. *EMBO Journal*, **9**, 2233–9.

Brüggemeier, U., Kalff, M., Franke, S., Scheidereit, C., and Beato, M. (1991). Ubiquitous transcription factor OTF-1 mediates induction of the mouse mammary tumour virus promoter through synergistic interaction with hormone receptors. *Cell*, **64**, 565–72.

Buetti, E. (1994). Stably integrated Mouse Mammary Tumor Virus Long Terminal Repeat DNA requires the octamer motifs for basal promoter activity. *Molecular and Cellular Biology*, **14**, 1191–203.

Buetti, E. and Diggelmann, H. (1981). Cloned MMTV virus DNA is biologically active in transfected mouse cells and its expression is stimulated by glucocorticoid hormones. *Cell*, **23**, 335–45.

Buetti, E. and Diggelmann, H. (1983). Glucocortiocd regulation of MMTV: identification of a short essential region. *EMBO Journal*, **2**, 1423–9.

Buetti, E., Kühnel, B., and Diggelmann, H. (1989). Dual function of a nuclear factor I binding site in MMTV transcription regulation. *Nucleic Acids Research*, **17**, 3065–78.

Cairns, L.A., Crotta, S., Minuzzo, M., Moroni, E., Granucci, F., Nicolis, S., *et al.* (1994). Immortalization of multipotent growth-factor dependent hemopoietic progenitors from mice transgenic for GATA-1 driven SV40 tsA58 gene. *EMBO Journal*, **13**, 4577–86.

Candau, R., Chávez, S., and Beato, M. (1996). The hormone responsive region of Mouse Mammary Tumor Virus positions a nucleosome and precludes access of nuclear factor I to the promoter. *Journal of Steroid Biochemistry and Molecular Biology*, **57**, 19–31.

Cato, A.C.B. and Weinmann, J. (1988). Mineralocorticoid regulation of transfected mouse mammary tumour virus DNA in cultured kidney cells. *Journal of Cell Biology*, **106**, 2119–25.

Cato, A.C.B., Miksicek, R., Schütz, G., Arnemann, J., and Beato, M. (1986). The hormone regulatory element of mouse mammary tumour virus mediates progesterone induction. *EMBO Journal*, **5**, 2237–40.

Cato, A.C.B., Henderson, D., and Ponta, H. (1987). The hormone response element of the mouse mammary tumour virus DNA mediates the progestin and androgen induction of transcription in the proviral long terminal repeat region. *EMBO Journal*, **6**, 363–8.

Cavailles, V., Dauvois, S., L'Horset, F., Lopez, G., Hoare, S., Kushner, P.J., *et al.* (1995). Nuclear factor RIP140 modulates transcriptional activation by the estrogen receptor. *EMBO Journal*, **14**, 3741–51.

Cavin, C. and Buetti, E. (1995). Tissue-specific and ubiquitous factors binding next to the glucocorticoid receptor modulate transcription from the mouse mammary tumor virus promoter. *Journal of Virology*, **69**, 3759–70.

Chalepakis, G., Arnemann, J., Slater, E.P., Brüller, H., Gross, B., and Beato, M. (1988). Differential gene activation by glucocorticoids and progestins through the hormone regulatory element of mouse mammary tumor virus. *Cell*, **53**, 371–82.

Chalepakis, G., Schauer, M., Cao, X., and Beato, M. (1990). Efficient binding of glucocorticoid receptor to its responsive element requires a dimer and DNA flanking sequences. *DNA and Cell Biology*, **9**, 355–68.

Chandler, V.L., Maler, B.A., and Yamamoto, K.R. (1983). DNA sequences bound specifically by glucocorticoid receptor *in vitro* render a heterologous promoter hormone responsive *in vivo*. *Cell*, **33**, 489–99.

Chávez, S., Candau, R., Truss, M., and Beato, M. (1995). Constitutive repression and NFI-dependent hormone activation of the MMTV promoter in yeast. *Molecular and Cellular Biology*, **15**, 6987–98.

Cordingley, M.G. and Hager, G.L. (1988). Binding of multiple factors to the MMTV promoter in crude and fractionated nuclear extracts. *Nucleic Acid Research*, **16**, 609–30.

Cordingley, M.G., Riegel, A.T., and Hager, G.L. (1987). Steroid-dependent interaction of transcription factors with the inducible promoter of mouse mammary tumor virus *in vivo*. *Cell*, **48**, 261–70.

Côte, J., Quinn, J., Workmann, J.L., and Peterson, C.L. (1994). Stimulation of GAL4 derivative binding to nucleosomal DNA by the yeast SWI/SNF complex. *Science*, **265**, 53–9.

Darbre, P., Page, M., and King, R.J.B. (1986). Androgen regulation by the long terminal repeat of mouse mammary tumor virus. *Molecular and Cellular Biology*, **6**, 2847–54.

Fisher-Adams, G. and Grunstein, M. (1995). Yeast histone H4 and H3 N-termini have different effects on the chromatin structure of the GAL1 promoter. *EMBO Journal*, **14**, 1468–77.

Fragoso, G., John, S., Roberts, M.S., and Hager, G.L. (1995). Nucleosome positioning on the MMTV LTR results from the frequency-biased occupancy of multiple frames. *Genes and Development*, **9**, 1933–47.

Geisse, S., Scheidereit, C., Westphal, H.M., Hynes, N.E., Groner, B., and Beato, M. (1982). Glucocorticoid receptors recognize DNA sequences in and around murine mammary tumour virus DNA. *EMBO Journal*, **1**, 1613–19.

Gouillet, F., Sola, B., Couette, B., and Richard-Foy, H. (1991). Cooperation between structural elements in hormono-regulated transcription from the mouse mammary tumor virus promoter. *Nucleic Acids Research*, **19**, 1563–9.

Gowland, P.L. and Buetti, E. (1989). Mutation in the hormone regulatory element of mouse mammary tumor virus differently affect the response to progestins, androgens, and glucocorticoids. *Molecular and Cellular Biology*, **9**, 3999–4008.

Guichon-Mantel, A., Savouret, J.F., Qignon, F., Delabre, K., Milgrom, E., and De The, H. (1995). Effect of PML and PML-RAR on the transcription properties and subcellular distribution of steroid hormone receptors. *Molecular Endocrinology*, **9**, 1791–803.

Hagen, G., Muller, S., Beato, M., and Suske, G. (1992). Cloning by recognition site screening of 2 novel GT box binding proteins: a family of Sp1 related genes. *Nucleic Acids Research*, **20**, 5519–25.

Ham, J., Thomson, A., Neddham, M., Webb, P., and Parker, M. (1988). Characterization of response elements for androgens, glucocorticoids and progestins in mouse mammary tumour virus. *Nucleic Acids Research*, **16**, 5263–77.

Han, M. and Grunstein, M. (1988). Nucleosome loss activates yeast downstream promoters *in vivo*. *Cell*, **55**, 1137–45.

Härtig, E. and Cato, A.C.B. (1994). *In vivo* binding of proteins to stably integrated MMTV DNA in murine cell lines: occupancy of NF1 and OTF1 binding sites in the absence and presence of glucocorticoids. *Cellular and Molecular Biology Research*, **40**, 643–52.

Härtig, E., Nierlich, B., Mink, S., Nebl, G., and Cato, A.C.B. (1993). Regulation of expression of mouse mammary tumor virus through sequences located in the hormone response element. Involvement of cell–cell contact and a negative regulatory factor. *Journal of Virology*, **67**, 813–21.

Hayes, J.J. and Wolffe, A.P. (1992). Histones H2A/H2B inhibit the interaction of transcription factor-IIIA with the *Xenopus borealis* somatic 5S RNA gene in a nucleosome. *Proceedings of the National Academy of Sciences of the USA*, **89**, 1229–33.

Hayes, J.J. and Wolffe, A.P. (1993). Preferential and asymmetric interaction of linker histones with 5S DNA in the nucleosome. *Proceedings of the National Academy of Sciences of the USA*, **90**, 6415–19.

Haynes, S.R., Dollard, C., Winston, F., Beck, S., Trowsdale, J., and David, I.B. (1992). The bromodomain: a conserved sequence found in human, *Drosophila* and yeast proteins. *Nucleic Acids Research*, **20**, 2603.

Hebbes, T.R., Thorne, A.W., and Crane-Robinson, C. (1988). A direct link between core histone acetylation and transcriptionally active chromatin. *EMBO Journal*, **7**, 1395–402.

Hecht, A., Laroche, T., Strahl-Bolsinger, S., Gasser, S.M., and Grunstein, M. (1995). Histone H3 and H4 N-termini interact with SIR3 and SIR4 proteins: a molecular model for the formation of heterochromatin in yeast. *Cell*, **80**, 583–92.

Horikoshi, M., Wang, C.K., Fujii, H., Cromlish, J.A., Weil, P.A., and Roeder, R.G. (1989). Cloning and structure of a yeast gene encoding a general transcription initiation factor TFIID that binds to the TATA box. *Nature*, **341**, 299–303.

Huang, A.L., Ostrowski, M.C., Berard, D., and Hager, G.L. (1981). Glucocorticoid regulation of the Ha-MuSV p21 gene conferred by sequences from mouse mammary tumor virus. *Cell*, **27**, 245–55.

Huang, M., Lee, J.W., and Peterson, D.O. (1993). Functional redundancy of octamer elements in the mouse mammary tumor virus promoter. *Nucleic Acids Research*, **21**, 5235–41.

Hynes, N.E., Kennedy, N., Rahmsdorf, U., and Groner, B. (1981). Hormone responsive expression of an endogenous proviral gene of MMTV after molecular cloning and gene transfer into cultured cells. *Proceedings of the National Academy of Sciences of the USA*, **78**, 2038–42.

Hynes, N.E., van Ooyen, A., Kennedy, N., Herrlich, P., Ponta, H., and Groner, B. (1983). Subfragments of the large terminal repeat cause glucocorticoid responsive expression of mouse mammary tumor virus and of an adjacent gene. *Proceedings of the National Academy of Sciences of the USA*, **80**, 3637–41.

Ing, N.H., Beekman, J.M., Tsai, S.Y., Tsai, M.J., and O'Malley, B.W. (1992). Members of the steroid hormone receptor superfamily interact with TFIIB (S300-II). *Journal of Biological Chemistry*, **267**, 17617–23.

Jantzen, H.M., Strähle, U., Gloss, B., Stewart, F., Schmid, W., Boshart, M., *et al.* (1987). Cooperativity of glucocorticoid response elements located far upstream of the tyrosine aminotransferase gene. *Cell*, **49**, 29–38.

Juan, L.-J., Utley, R.T., Adams, C.C., Vettese-Dadey, M., and Workman, J.L. (1994). Differential repression of transcription factor binding by histone H1 is regulated by the core histone amino termini. *EMBO Journal*, **13**, 6031–40.

Kadowaki, Y., Toyoshima, K., and Yamamoto, T. (1995). Dual transcriptional control by Ear3/COUP: negative regulation through the DR1 direct repeat and positive regulation through a sequence downstream of the transcriptional start site of the mouse mammary tumor virus promoter. *Proceedings of the National Academy of Sciences of the USA*, **92**, 4432–36.

Kalff, M., Gross, B., and Beato, M. (1990). Progesterone receptor stimulates transcription of mouse mammary tumour virus in a cell-free system. *Nature*, **344**, 360–2.

Kruse, U. and Sippel, A.E. (1994). Transcription factor nuclear factor I proteins form stable homo- and heterodimers. *FEBS Letters*, **348**, 46–50.

Langer, S.J. and Ostrowski, M.C. (1988). Negative regulation of transcription *in vitro* by a glucocorticoid response element is mediated by a trans-acting factor. *Molecular and Cellular Biology*, **8**, 3872–81.

LeDouarin, B., Zechel, C., Garnier, J.M., Lutz, Y., Tora, L., Pierrat, B., *et al.* (1995). The N-terminal part of TIF1, a putative mediator of the ligand-dependent activation function (AF-2) of nuclear receptors, is fused to B-raf in the oncogenic protein T18. *EMBO Journal*, **14**, 2020–33.

Lee, F., Mulligan, R., Berg, P., and Ringold, G. (1981). Glucocorticoid regulates expression of dihydrofolate reductase cDNA in mouse mammary tumor virus chimeric plasmids. *Nature*, **294**, 228–32.

Lee, D.Y., Hayes, J.J., Pruss, D., and Wolffe, A.P. (1993). A positive role for histone acetylation in transcription factor access to nucleosomal DNA. *Cell*, **72**, 73–84.

Li, Q. and Wrange, O. (1993). Translational positioning of a nucleosomal glucocorticoid response element modulates glucocorticoid receptor affinity. *Genes and Development*, **7**, 2471–82.

Li, Q. and Wrange, O. (1995). Accessibility of a glucocorticoid response element in a nucleosome depends on its rotational positioning. *Molecular and Cellular Biology*, **15**, 4375–84.

Luisi, B.F., Xu, W.X., Otwinowski, Z., Freedman, L.P., Yamamoto, K.R., and Siegler, P.B. (1991). Crystallographic analysis of the interaction of the glucocorticoid receptor with DNA. *Nature*, **352**, 497–505.

Majors, J. and Varmus, H.E. (1983). A small region of mouse mammary tumor virus long terminal repeat confers glucocorticoid hormone regulation on a linked heterologous gene. *Proceedings of the National Academy of Sciences of the USA*, **80**, 5866–70.

Martinez, E. and Wahli, W. (1989). Cooperative binding of estrogen receptor to imperfect estrogen-responsive DNA elements correlates with their synergistic hormone-dependent enhancer activity. *EMBO Journal*, **8**, 3781–91.

McPherson, C.E., Shim, E.Y., Friedman, D.S., and Zaret, K.S. (1993). An active tissue-specific enhancer and bound transcription factors existing in a precisely positioned nucleosomal array. *Cell*, **75**, 387–98.

Miksicek, R., Borgmeyer, U., and Nowock, J. (1987). Interaction of the TGGCA-binding protein with upstream sequences is required for efficient transcription of mouse mammary tumor virus. *EMBO Journal*, **6**, 1355–60.

Mok, E., Golovkina, T.V., and Ross, S.R. (1992). A mouse mammary tumor virus mammary gland enhancer confers tissue-specific but not lactation-dependent expression in transgenic mice. *Journal of Virology*, **66**, 7529–32.

Möws, C., Preiss, T., Slater, E.P., Cao, X., Verrijzer, C.P., Van der Vliet, P., *et al.* (1994). Two independent pathways for transcription from the MMTV promoter. *Journal of Steroid Biochemistry and Molecular Biology*, **51**, 21–32.

Muchardt, C. and Yaniv, M. (1993). A human homologue of *Saccharomyces cerevisiae* SNF2/SWI2 and *Drosophila*-brm genes potentiates transcriptional activation by the glucocorticoid receptor. *EMBO Journal*, **12**, 4279–90.

Mymryk, J.S., Berard, D., Hager, G.L., and Archer, T.K. (1995). Mouse mammary tumor virus chromatin in human breast cancer cells is constitutively hypersensitive and exhibits steroid hormone-independent loading of transcription factors *in vivo*. *Molecular and Cellular Biology*, **15**, 26–34.

Nowock, J., Borgmeyer, U., Påschel, A., Rupp, A.W., and Sippel, A.E. (1985). The TGGCA protein binds to the MMTV-LTR, the adenovirus origin of replication, and BK virus. *Nucleic Acids Research*, **13**, 2045–62.

Ostrowski, M.C., Huang, A.L., Kessel, M., Wolford, R.G., and Hager, G.L. (1984). Modulation of enhancer activity by the hormone respondive regulatory element from mouse mammary tumor virus. *EMBO Journal*, **3**, 1891–9.

Payvar, F., Wrange, Ö., Carlstedt-Duke, J., Okret, S., Gustafsson, J.A., and Yamamoto, K.R. (1981). Purified glucocortiocid receptors bind selectively *in vitro* to a cloned DNA fragment whose transcription is regulated by glucocorticoids *in vivo*. *Proceedings of the National Academy of Sciences of the USA*, **78**, 6628–33.

Payvar, F., DeFranco, D., Firestone, G.L., Edgar, B., Wrange, Okret, S., Gustafsson, J.A., *et al.* (1983). Sequence-specific binding of glucocorticoid receptor to MTV DNA at sites within and upstream of the transcribed region. *Cell*, **35**, 381–92.

Perlmann, T. (1992). Glucocorticoid receptor DNA-binding specificity is increased by the organization of DNA in nucleosomes. *Proceedings of the National Academy of Sciences of the USA*, **89**, 3884–8.

Perlmann, T. and Wrange, Ö. (1988). Specific glucocorticoid receptor binding to DNA reconstituted in a nucleosome. *EMBO Journal*, **7**, 3073–9.

Perlmann, T. and Wrange, Ö. (1991). Inhibition of chromatin assembly in *Xenopus* oocytes correlates with derepression of the Mouse Mammary Tumor Virus promoter. *Molecular and Cellular Biology*, **11**, 5259–65.

Perlmann, T., Erikson, P., and Wrange, Ö. (1990). Quantitative analysis of the glucocorticoid receptor–DNA interaction at the mouse mammary tumor virus glucocorticoid response element. *Journal of Biological Chemistry*, **265**, 17222–9.

Peterson, C.L. and Herskowitz, I. (1992). Characterization of the yeast SWI1, SWI2, and SWI3 genes, which encode a global activator of transcription. *Cell*, **68**, 573–83.

Peterson, C.L., Dingwall, A., and Scott, M.P. (1994). Five SWI/SNF gene products are components of a large multisubunit complex required for transcriptional enhancement. *Proceedings of the National Academy of Sciences of the USA*, **91**, 2905–8.

Piña, B., Barettino, D., Truss, M., and Beato, M. (1990*a*). Structural features of a regulatory nucleosome. *Journal of Molecular Biology*, **216**, 975–90.

Piña, B., Brüggemeier, U., and Beato, M. (1990*b*). Nucleosome positionining modulates accessibility of regulatory proteins to the mouse mammary tumor virus promoter. *Cell*, **60**, 719–31.

Piña, B., Truss, M., Ohlenbusch, H., Postma, J., and Beato, M. (1990*c*). DNA rotational positioning in a regulatory nucleosome is determined by base sequence. An algorithm to model the preferred superhelix. *Nucleic Acids Research*, **18**, 8981–7.

Piña, B., Barettino, D., and Beato, M. (1991). Nucleosome positioning and regulated gene expression. In *Nucleosome positioning and regulated gene expression* (ed. N. Maclean), pp. 83–117. Oxford University Press, Oxford.

Richard-Foy, H. and Hager, G.L. (1987). Sequence-specific positioning of nucleosomes over the steroid-inducible MMTV promoter. *EMBO Journal*, **6**, 2321–8.

Ringold, G.M., Yamamoto, K.R., Tomkins, G.M., Bishop, J.M., and Varmus, H.E. (1975). Dexamethasone-mediated induction of mouse mammary tumor virus RNA: a system for studying glucocorticoid action. *Cell*, **6**, 299–305.

Ringold, G.M., Yamamoto, K.R., Bishop, J.M., and Varmus, H.E. (1977). Glucocorticoid-stimulated accumulation of mouse mammary tumor virus RNA: increase rate of synthesis of viral RNA. *Proceedings of the National Academy of Sciences of the USA*, **74**, 2879–83.

Rollini, P., Billotte, J., Kolb, E., and Diggelmann, H. (1992). Expression pattern of Mouse Mammary Tumor Virus in transgenic mice carrying exogenous proviruses of different origins. *Journal of Virology*, **66**, 4580–6.

Ross, S.R. and Sloter, D. (1985). Glucocorticoid regulation of mouse mammary tumor virus sequences in transgenic mice. *Proceedings of the National Academy of Sciences of the USA*, **82**, 5880–4.

Rupp, R.A.W., Kruse, U., Multhaupt, G., Göbel, U., Beyreuther, K., and Sippel, A.E. (1990). Chicken NFI/TGGCA proteins are encoded by at least three independent genes: NFI-A, NFI-B and NFI-C with homologous in mammalian genomes. *Nucleic Acids Research*, **18**, 2607–16.

Sadovsky, Y., Webb, P., Lopez, G., Baxter, J.D., Fitzpatrick, P.M., Gizang-Ginsberg, E., *et al.* (1995). Transcriptional activators differ in their responses to overexpression of TATA-box-binding protein. *Molecular and Cellular Biology*, **15**, 1554–63.

Schatt, M.D., Rusconi, S., and Schaffner, W. (1990). A single DNA-binding transcription factor is sufficient for activation from a distant enhancer and/or from a promoter position. *EMBO Journal*, **9**, 481–7.

Scheidereit, C., Geisse, S., Westphal, H.M., and Beato, M. (1983). The glucocorticoid receptor binds to defined nucleotide sequences near the promoter of mouse mammary tumour. *Nature*, **304**, 749–52.

Schmid, W., Strähle, U., Schütz, G., Schmitt, J., and Stunnenberg, H. (1989). Glucocorticoid receptor binds cooperatively to adjacent recognition sites. *EMBO Journal*, **8**, 2257–63.

Schüle, R., Muller, M., Otsuka-Murakami, H., and Renkawitz, R. (1988*a*). Many transcription factors interact synergistically with steroid receptors. *Science*, **242**, 1418–20.

Schüle, R., Müller, M., Otsuka-Murakami, H., and Renkawitz, R. (1988*b*). Cooperativity of the glucocorticoid receptor and the CACCC-box binding factor. *Nature*, **332**, 87–90.

Strähle, U., Klock, G., and Schütz, G. (1987). A DNA sequence of 15 base pairs is sufficient to mediate both glucocorticoid and progesterone induction of gene expression. *Proceedings of the National Academy of Sciences of the USA*, **84**, 7871–5.

Strähle, U., Schmid, W., and Schütz, G. (1988). Synergistic action of the glucocorticoid receptor with transcription factors. *EMBO Journal*, **7**, 3389–95.

Tanaka, H., Dong, Y., Li, Q., Okret, S., and Gustafsson, J.-A. (1991). Identification and characterization of a *cis*-acting element that interferes with glucocorticoid-inducible activation of the mouse mammary tumor virus promoter. *Proceedings of the National Academy of Sciences of the USA*, **88**, 5393–7.

Toohey, M.G., Lee, J.W., Huang, M., and Peterson, D.O. (1990). Functional elements of the steroid hormone-responsive promoter of mouse mammary tumor virus. *Journal of Virology*, **64**, 4477–88.

Truss, M. and Beato, M. (1993). Steroid hormone receptors. Interaction with DNA and transcription factors. *Endocrine Reviews*, **14**, 459–79.

Truss, M., Chalepakis, G., and Beato, M. (1992). Interplay of steroid hormone receptors and transcription factors on the mouse mammary tumor virus promoter. *Journal of Steroid Biochemistry and Molecular Biology*, **43**, 365–78.

Truss, M., Bartsch, J., and Beato, M. (1994). Antiprogestins prevent progesterone receptor binding to hormone responsive elements *in vivo*. *Proceedings of the National Academy of Sciences of the USA*, **91**, 11333–7.

Truss, M., Bartsch, J., Schelbert, A., Haché, R.J.G., and Beato, M. (1995). Hormone induces binding of receptors and transcription factors to a rearranged nucleosome on the MMTV promoter *in vivo*. *EMBO Journal*, **14**, 1737–51.

Ucker, D.S. and Yamamoto, K.R. (1984). Early events in the stimulation of mammary tumor virus RNA synthesis by glucocorticoids. Novel assays of transcription rates. *Journal of Biological Chemistry*, **259**, 7416–20.

Van Holde, K.E. (1993). The omnipotent nucleosome. *Nature*, **362**, 111–12.

Vettese-Dadey, M., Walter, P., Chen, H., Juan, L.-J., and Workman, J.L. (1994). Role of the histone amino termini in facilitated binding of a transcription factor, GAL4-AH, to nulceosomal cores. *Molecular and Cellular Biology*, **14**, 970–81.

Vom Baur, E., Zechel, C., Heery, D., Heine, M., Garnier, J.M., Vivat, V., *et al.* (1995). Differential ligand-dependent interactions between the AF-2 activation domain of nuclear receptors and the putaive transcriptional intermediary factors mSUG1 and TIF1. *EMBO Journal*, **15**, 119–24.

Wan, J.S., Mann, R.K., and Grunstein, M. (1995). Yeast histone H3 and H4 N termini function through different GAL1 regulatory elements to repress and activate transcription. *Proceedings of the National Academy of Sciences of the USA*, **92**, 5664–8.

Wendler, W., Altmann, H., and Winnacker, E.L. (1994). Transcriptional activation of NFI/CTF1 depends on a sequence motif strongly related to the carboxyterminal domain of RNA polymerase II. *Nucleic Acids Research*, **22**, 2601–3.

Winston, F. and Carlson, M. (1992). Yeast SNF/SWI transcriptional activators and the SPT/SIN chromatin connection. *Trends in Genetics*, **8**, 387–91.

Wolffe, A.P. (1994). Nucleosome positioning and modification: chromatin structures that potentiate transcription. *Trends in Biochemical Sciences*, **19**, 240–4.

Yoshida, M., Kijima, M., Akita, M., and Beppu, T. (1990). Potent and specific inhibition of mammalian histone deacetylase both *in vivo* and *in vitro* by trichostatin A. *Journal of Biological Chemistry*, **265**, 17174–9.

Yoshinaga, S.K., Peterson, C.L., Herskowitz, I., and Yamamoto, K.R. (1992). Roles of SWI1, SWI2, and SWI3 proteins for transcriptional enhancement by steroid receptors. *Science*, **258**, 1598–604.

Young, H.A., Shih, T.Y., Scolnick, E.M., and Parks, W.P. (1977). Steroid induction of mouse mammary tumor virus: effect upon synthesis and degradation of viral RNA. *Journal of Virology*, **21**, 139–46.

Zaret, K.S. and Yamamoto, K.R. (1984). Reversible and persistent changes in chromatin structure accompanying activation of a glucocorticoid-dependent enhancer element. *Cell*, **38**, 29–38.

3

Histone modification and transcriptional competence

Alan P. Wolffe

Introduction

Chromosomes are highly differentiated and dynamic structures. Transcriptional activity and chromosome morphology vary throughout the cell cycle. These events can be correlated with global changes in the post-translational modification of the histone proteins that occur in all chromosomes. At a more local level individual chromosomes, chromatin domains, or genes can have their transcription modulated through developmental or signal transduction pathways. These local transitions in gene activity can also be related to changes in chromatin structure and histone modification. The focus of this chapter is to discuss the nature of histone modifications and their potential structural and functional consequences.

Histone modifications are diverse: they include acetylation, methylation, phosphorylation, ADP ribosylation and ubiquitination. Most of the early studies on the significance of these modifications have concerned their interrelationship with transcriptional activity and the cell cycle (van Holde 1988; Bradbury 1992). More recent work has focused on the modification of histones associated with particular chromosomal domains or promoters and the impact of these targeted alterations on transcription (Wolffe 1995).

Nucleosome structure and sites of histone modification

Each metazoan nucleosome contains 180–200 bp of DNA, two molecules of each of the four core histones (H2A, H2B, H3, and H4), and a single molecule of a linker histone. The core histones each contain a structured histone fold domain and a basic amino (N-) terminal tail domain (Arents et al. 1991). Histone H2A has a carboxyl (C-) terminal tail as well. The histone fold domains are involved in histone–histone interactions and in wrapping DNA (Fig. 3.1). The core histone tails are the sites of post-translational modification (summarized in Fig. 3.2). These tail domains lie on the outside of the nucleosome (Fig. 3.1), where they are accessible both to the enzymes that carry out the modification and to specific trans-acting factors that recognize the tail domains (see below).

Linker histones interact both with the core histones and nucleosomal DNA (Fig. 3.1; Pruss et al. 1995). A typical metazoan linker histone (e.g. H1) has a

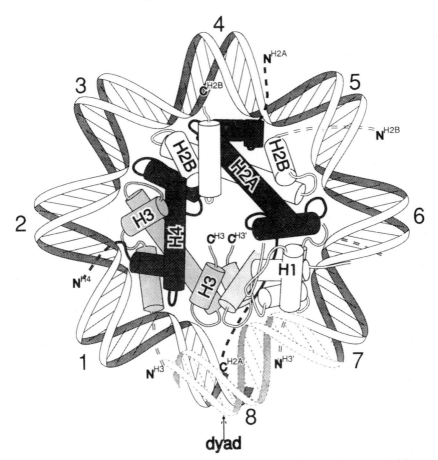

Fig. 3.1 A model for the nucleosome. The approximate structure of the histone (H3,H4)$_2$ tetramer, the histone H2A,H2B dimer and histone H1 bound to DNA is shown. The dyad axis of the nucleosome is indicated. Numbers refer to integral turns of DNA away from the dyad axis. The approximate positions of the amino (N-) and carboxyl (C-) terminal tails of the core histones are indicated (dashed lines).

central structured domain flanked by basic N- and C-terminal tails. The central domain has a similar structure to that of the 'winged-helix' family of transcriptional regulators. The terminal tail has the potential to form α-helical structures (Clark *et al.* 1988). It is also the site of extensive post-transcriptional phosphorylation (Hohmann 1983; Fig. 3.2). Phosphorylation of histone H1 has been shown directly to weaken the interaction of the basic tails of the protein with DNA. Surprisingly, these changes influence the binding of the protein to chromatin even more than to DNA (Hill *et al.* 1991).

The structural consequences of histone modification for the nucleosome have been determined in greatest detail for acetylation of the core histones. However,

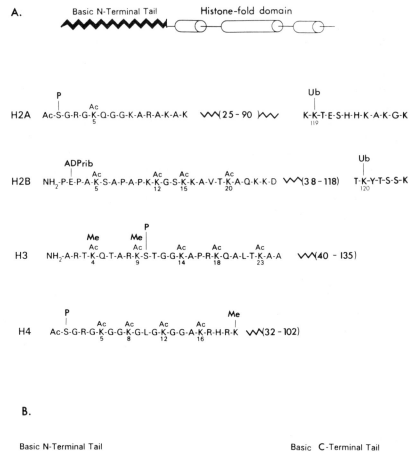

A.

Basic N-Terminal Tail

Histone-fold domain

H2A   Ac-S-G-R-G-K-Q-G-G-K-A-R-A-K-A-K   ～(25 – 90)～   K-K-T-E-S-H-H-K-A-K-G-K

H2B   NH₂-P-E-P-A-K-S-A-P-A-P-K-K-G-S-K-K-A-V-T-K-A-Q-K-K-D ～(38 – 118)   T-K-Y-T-S-S-K

H3   NH₂-A-R-T-K-Q-T-A-R-K-S-T-G-G-K-A-P-R-K-Q-A-L-T-K-A-A ～(40 – 135)

H4   Ac-S-G-R-G-K-G-G-K-G-L-G-K-G-G-A-K-R-H-R-K ～(32 – 102)

B.

Basic N-Terminal Tail   Basic C-Terminal Tail

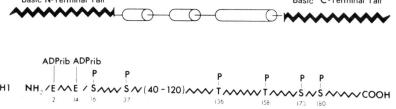

H1   NH₂ /E～E / S～～S～(40–120)～～T～～～T～～S～S～～COOH

Fig. 3.2   Histones and their modifications. (a) A cartoon of a core histone is shown. The sequence of calf thymus core histones is shown together with sites of acetylation (Ac), phosphorylation (P), methylation (Me), ADP ribosylation (ADPrib) and ubiquitination (Ub). The positions of lysine residues (K) have been indicated. (b) A cartoon of a linker histone is shown. The sequence of calf thymus histone H1 is given together with sites of phosphorylation (P) and ADP ribosylation (ADPrib). The positions of serine (S) and threonine (T) residues have been indicated.

it might be expected that both acetylation and phosphorylation of the N-terminal tail domains will exert similar effects since both lead to a reduction in net positive charge under physiological conditions. The hyperacetylation of the core histone tails (i.e. modification of the majority of lysine residues) leads to subtle changes in nucleosome conformation (Bode *et al.* 1983; Imai *et al.* 1986; Ausio and van Holde 1986; Oliva *et al.* 1990). The electrophoretic mobility of nucleosome cores decreases as the average degree of acetylation increases, potentially due to a change in shape (Bode *et al.* 1983). Likewise, the sedimentation coefficient of core particles increases with acetylation (Ausio and van Holde 1986). The helical repeat of nucleosome-bound DNA is unchanged upon acetylation, and the extent of DNA–protein contacts is unaltered (Bauer *et al.* 1994). Detailed analysis of the supercoiling of DNA around the core histones suggests a small decrease in the number of times DNA winds around the nucleosome central axis as the histones are acetylated (Norton *et al.* 1989, 1990; Bauer *et al.* 1994).

Chemical crosslinking experiments have shown that weak interactions can occur between the N-terminal tail of histone H2B and linker DNA, although the particular example examined was a sea urchin sperm H2B variant with a particularly long N-terminal tail domain (Bavykin *et al.* 1990; Hill and Thomas 1990). The N-terminal tails of somatic H2B and H2A contact DNA at the periphery of the nucleosome, as far as 80 bp from the dyad axis (Pruss and Wolffe 1993). The C-terminal tail of H2A binds to DNA around the dyad axis (Guschin *et al.* 1991; Usachenko *et al.* 1994). The N- terminal tail of histone H4 can be crosslinked to DNA at about 1.5 turns from the dyad axis, coincident with the sites of strong DNA deformation (Ebralidse *et al.* 1988). The N-terminal tail of histone H3 may interact with the ends of DNA in the nucleosome and perhaps also with histone H1 (Glotov *et al.* 1978). High resolution protein nuclear magnetic resonance data indicate the stable association of the N-terminal tails of histones H3 and H4 with DNA in the nucleosome core particle at physiological ionic strength (less than 0.3 M NaCl). In contrast, the tails of the normal somatic variants of histones H2A and H2B are mobile at all ionic strengths examined (Cary *et al.* 1982). The weak interaction of the core histone tails with DNA in the nucleosome is reflected in the lack of structural change in the organization of DNA and in the integrity of the nucleosome following their proteolytic removal (Ausio *et al.* 1989; Hayes *et al.* 1991*b*). Removal of the histone tails from arrays of nucleosomes deficient in linker histones does impede the capacity of these arrays to compact into higher order structures (Garcia-Ramirez *et al.* 1992). This observation suggests that a major effect of the histone tails on chromatin structure may be through their influence on the integrity of higher order structures.

An important consequence of acetylation or removal of the core histone tails is that nucleosomal DNA becomes more accessible to *trans*-acting factors (Lee *et al.* 1993; Vettesse-Dadey *et al.* 1994; Godde *et al.* 1995). This result suggests that histone acetylation might have a major regulatory role in the transcription process. This view is strengthened by the recent observation that known transcriptional regulators in yeast have the capacity to acetylate histones (Brownell *et al.* 1996; Wolffe and Pruss, 1996). Histone hyperacetylation increases the general accessibility of DNA in chromatin to DNAase I. This implies that higher order chromatin structures containing acetylated nucleosomes are less stable (Perry and Annunziato 1989). Maintenance of histone acetylation in nascent chromatin might also contribute to a reduction in the stable sequestration of linker histones (Perry and Annunziato 1989). Histone hyperacetylation does not directly influence the association of histone H1 with nucleosomes (Ura *et al.* 1994). Constitutive hyperacetylation of the core histones does not reduce the association of histone H1 with bulk chromatin (Dimitrov *et al.* 1993; Almouzni *et al.* 1994), however it inhibits the complete condensation of interphase chromatin (Annunziato *et al.* 1988).

ADP ribosylation of the tails of histone H2B and linker histones will introduce a negatively charged branched molecule into chromatin that will superficially resemble a single-stranded nucleic acid. Thus, such a structure might be expected to locally disrupt higher order chromatin structure and potentially displace the more weakly bound histones from nucleosomes, such as histone H1 and the histone dimer H2A/H2B. Ubiquitination of H2A and H2B also introduces a bulky adduct into chromatin (West and Bonner 1980). Ubiquitination is the attachment of a 76 amino acid peptide to the C-terminal tail of histone H2A and H2B. Ubiquitinated H2A is known to be incorporated into nucleosomes (Levinger and Varshavsky 1980). However, there are no major changes in the organization of nucleosome cores (Kleinschmidt and Martinson 1981). Since the C-terminus of histone H2A contacts nucleosomal DNA at the dyad axis of the nucleosome (Guschin *et al.* 1991; Usachenko *et al.* 1994), ubiquitination of this tail domain might be anticipated to disrupt the interaction of linker histones with nucleosomal DNA. The bulky ubiquitin adduct might also be anticipated to disrupt higher order chromatin structures by impeding internucleosomal interactions. Methylation of the core histone tails is without any known structural consequences.

Functional consequences of histone modification

A causal relationship between histone modification and transcriptional activity has not yet been established for any gene. However, so many correlations have accumulated over the past three decades that an interrelationship seems certain (Allfrey *et al.* 1964). The cell cycle provides a useful context in which to consider all the histone modifications and transcription (Fig. 3.3). RNA polymerase I- and II-mediated transcription is severely inhibited during mitosis (Johnson and Holland 1965; Morcillo *et al.* 1976). This inhibition involves the post-translational modification of components of the basal transcriptional machinery, as well as chromatin structural components (Hartl *et al.* 1993).

During S phase, all of the core histones are acetylated, but the predominant

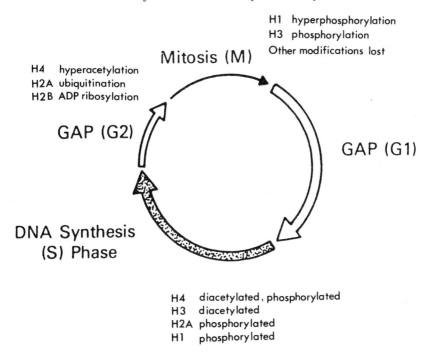

Fig. 3.3 A cartoon of the cell cycle showing the major changes in histone modification associated with each stage.

modifications are mono- and diacetylation of histones H3 and H4 (Waterborg and Matthews 1982, 1984). In G2 the histones H3 and H4 become hyperacetylated and in mitosis all four core histones are deacetylated. With respect to phosphorylation, histone H2A is phosphorylated throughout the cell cycle (Gurley *et al.* 1978), histone H3 is phosphorylated during mitosis (Paulson and Taylor 1982), and histone H1 phosphorylation occurs throughout S phase, increases during G2 and becomes maximal at metaphase with 22–24 phosphates per H1 molecule (Mueller *et al.* 1985). ADP ribosylation and ubiquitination are present throughout S phase, becoming maximal in G2 (Kidwell and Mage 1976; Mueller *et al.* 1985). These modifications also decline during mitosis. Taken together these data suggest that with the exception of the phosphorylation of H2A, H3 and most notably H1, all histone modifications are significantly reduced or eliminated during mitosis. This indicates the dynamic nature of these modifications and the necessity of reconfiguring the chromosome following mitosis. A current hypothesis is that histone H1 phosphorylation releases the protein from contacts with linker DNA, thereby allowing local chromatin structure to become destabilized and hence deformable (Roth and Allis 1992). The reduction of histone acetylation, ubiquitination, and ADP ribosylation correlates with transcriptional inactivation at mitosis.

³²P-labeled

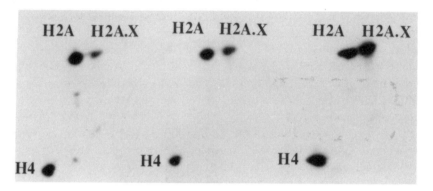

Markers

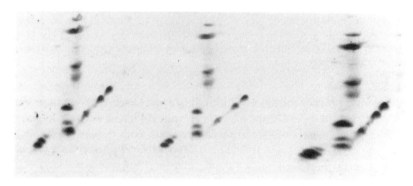

Fig. 3.4   Phosphorylation of proteins in *Xenopus* sperm chromatin during remodelling into a paternal pronucleus. The upper panels show proteins (histones H2A, H2AX and H4) that are radiolabelled by [γ³²P]ATP, the lower panels show marker proteins. The efficiency of decondensation increases from left to right, as does the efficiency with which histone H2AX is phosphorylated (see Dimitrov *et al.* (1994) for details of a similar experiment).

Nascent chromatin assembled at the replication fork during S phase contains diacetylated histone H4 and phosphorylated H2A and H4 (Ruiz-Carillo *et al.* 1975). These modifications are particularly prevalent in the early embryonic chromatin of several organisms exhibiting rapid cell cycles and in the remodelled chromatin of egg pronuclei (Poccia and Green 1992; Dimitrov *et al.* 1993, 1994; Fig. 3.4). It is possible that modification of the core histones is necessary for their targeting to the molecular chaperones engaged in chromatin assembly. Modified histones might generally facilitate *trans*-acting factor access to DNA in nascent chromatin facilitating the duplication of functional chromosomal structures (Lee *et al.* 1993).

The generation of antibodies against acetylated core histones has allowed a number of general correlations to be made with the transcriptional activity of chromatin (Allfrey *et al.* 1964; Gorovsky *et al.* 1973; Mathis *et al.* 1978; Wolffe 1995). In *Saccharomyces cerevisiae* most of the genome is transcriptionally active and contains hyperacetylated core histones (Clark *et al.* 1993). However, there are also transcriptionally inactive domains in yeast, such as the silent mating type cassettes and telomeric sequences. These contain histone H4 that is hypoacetylated (Braunstein *et al.* 1993). In higher eukaryotes acetylation of histone H4 increases during the reactivation of transcription in the initially inactive chicken erythrocyte nucleus, following the fusion of the erythrocyte with a transcriptionally active cultured cell to form a heterokaryon (Turner 1991). Histone acetylation has been shown to be particularly prevalent over the chromatin domain containing the chicken β-globin gene family; these genes are potentially transcribed in erythrocytes (Hebbes *et al.* 1988, 1992, 1994). This result is indicative of a very specific targeting of histone acetyltransferase activity by largely unknown mechanisms. The fact that transcription factors can direct histone acetylation partially explains the localization of acetylated histones (Wolffe and Pruss 1996). Immunolabelling of polytene chromosomes in *Chironomus* and *Drosophila* also reveals a non-random distribution of histone H4 acetylation correlating with transcriptional activity (Turner *et al.* 1992; Bone *et al.* 1994). Within female mammals the transcriptionally inactive X chromosome is distinguished by a lack of histone H4 acetylation (Jeppesen and Turner 1993). Therefore, several independent experimental approaches have shown that actively transcribed and potentially active chromatin domains are selectively enriched in hyperacetylated histones, whereas transcriptionally inactive chromatin contains hypoacetylated histones.

The second type of core histone modification to receive extensive experimental study with respect to transcription is phosphorylation (Wolffe 1995). Histone H3 is rapidly phosphorylated on serine residues within the N-terminal tail when extracellular signals, such as growth factors or phorbol esters, stimulate quiescent cells to proliferate (Mahadevan *et al.* 1991). Several studies have suggested a change in nucleosomal conformation or higher order structure within the chromatin of the proto-oncogenes c-*fos* and c-*jun*, following their rapid induction to higher levels of transcriptional activity by phorbol esters (Chen and Allfrey 1987; Chen *et al.* 1990). DNAase I sensitivity of chromatin rapidly increases and proteins with exposed sulphydryl groups accumulate on the proto-oncogene chromatin. The proteins containing exposed sulphydryl groups include both non-histone proteins, such as RNA polymerase, and molecules of histone

H3. The histone H3 cysteine residues, the only ones in the nucleosome, are normally buried within the particle. Exposure of the sulphydryl groups might imply a major disruption of nucleosome structure. For example, the dissociation of histone H1 and of a H2A/H2B dimer might allow access from solution to this region of histone H3. Phosphorylation and acetylation of histone H3 might act in concert to cause these changes.

Ubiquitination of histone H2A is also associated with transcriptional activity. Only one nucleosome in 25 contains ubiquitinated histone H2A in non-transcribed chromatin. This increases to one nucleosome in two for the transcriptionally active hsp70 genes (Levinger and Varshavsky 1982). Enrichment in ubiquitinated H2A is especially prevalent at the 5' end of transcriptionally active genes (Varshavsky *et al.* 1982). Since ubiquitination directs many proteins towards degradation, Varshavsky and co-workers have suggested that nucleosomal structure might be lost on actively transcribed genes following ubiquitination of H2A (Levinger and Varshavsky 1982).

The phosphorylation of histone H1 at mitosis is believed to weaken the interaction of the linker histone with chromatin, thereby destabilizing local higher order structures (Roth and Allis 1992). Similar effects might be expected to facilitate transcription. *Tetrahymena* histone H1 is phosphorylated as transcriptional activity increases (Roth *et al.* 1988). More significantly, during the *Tetrahymena* sexual cycle transcription is entirely repressed and histone H1 is completely dephosphorylated (Lin *et al.* 1991). A similar process occurs during the final stages of chicken erythrocyte development, when the erythrocyte nucleus becomes transcriptionally repressed (Hentschel and Tata 1978). The newly synthesized chicken erythrocyte linker histone variant H5 is highly phosphorylated, but when the erythrocyte chromatin becomes condensed and transcription is inactivated histone H5 is quantitatively dephosphorylated (Aubert *et al.* 1991). That these events are directly linked receives further support from experiments in which the gene for histone H5 is expressed in fibroblasts. This specialized linker histone would not normally be found in these cells. The accumulation of histone H5 in the fibroblasts inhibits cell growth concomitant with chromatin compaction (Sun *et al.* 1989; Aubert *et al.* 1991). Under these circumstances histone H5 is not phosphorylated. Introduction of the protein into transformed cells leads to phosphorylation of histone H5. In this case nuclear condensation does not occur and the cells continue to grow and divide. These results clearly illustrate the impact of linker histone variants on nuclear activities.

In general, the presence of histone modifications within chromatin correlates with increased transcriptional activity, whereas the absence of any modifications is a feature of transcriptionally inert chromatin. The exception to this rule is the presence of hyperphosphorylated histones H1 and H3 in transcriptionally inactive mitotic chromosomes.

Developmental and epigenetic effects of histone modification

Histone modifications have been proposed to have an important role in establishing heritable chromatin structures during early development (Wolffe 1995). Such modifications might contribute to the maintenance of an epigenetic imprint, whereby particular genes are maintained in stable active or inactive states.

The level of acetylation of the core histones is controlled by an equilibrium between histone acetyltransferases and deacetylases (Turner 1993). It is normally possible to induce histone hyperacetylation using inhibitors of deacetylases such as sodium butyrate (Candido *et al.* 1978) or Trichostatin A (Yoshida *et al.* 1995). However, in *Xenopus* hyperacetylated histones only accumulate in the presence of these inhibitors for the first time during gastrulation (Dimitrov *et al.* 1993). This result implies that histone acetyltransferase activity is developmentally regulated. Hyperacetylation of the core histones correlates with the capacity to selectively induce the transcriptional activity of particular genes (Khochbin and Wolffe 1993; Fig. 3.5), and leads to a delay in the completion of gastrulation (Almouzni *et al.* 1994). *Xenopus* embryos maintained in Trichostatin A have defects in mesoderm formation. In the starfish *Asterina pectinifera* the inhibition of histone deacetylation leads to the arrest of development at the early gastrula stage before mesenchyme formation (Ikegami *et al.* 1993). Thus, the capacity to modulate histone acetylation levels appears important in establishing stable states of differential gene activity. In the mouse embryo core histone acetylation changes dramatically during the first few cell divisions following fertilization. Chromatin containing acetylated histone H4 becomes enriched at the nuclear periphery when the zygotic genome becomes strongly activated at the two-cell stage (Worrad *et al.* 1995). Inhibition of histone deacetylation using Trichostatin A increases the overall efficiency of gene expression. Acetylated chromatin co-localizes with RNA polymerase II suggesting that it is the site of active transcription. This localization to the nuclear periphery is lost in the four-cell embryo and during subsequent development.

The promoters of exogenous DNA templates injected into two-cell mouse embryos are normally repressed (Wiekowski *et al.* 1991). However, incubation with sodium butyrate, leading to histone hyperacetylation, relieves this repression (Wiekowski *et al.* 1993). Similar results are obtained if these templates are injected into the maternal, but not the paternal pronucleus of the egg. These results strongly implicate histone acetylation as a regulatory factor during early mammalian embryogenesis.

The capacity of chromatin structures to function in maintaining an epigenetic imprint follows from the fact that pre-existing chromatin structural proteins will be utilized to reassemble daughter chromatids following replication (Wolffe 1995). For example, histone H4 that is hyperacetylated or histone H1, which is hyperphosphorylated, will segregate to daughter chromatids within the newly assembled nascent chromatin (Perry *et al.* 1993). This may well facilitate the

A   **B**

H1°- CAT gene   Endogenous gene

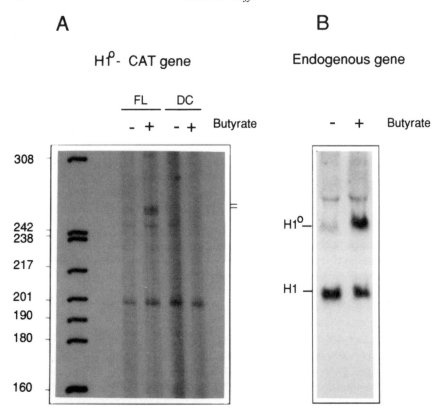

Fig. 3.5   Regulation of *Xenopus* histone H1° gene expression by inhibitors of histone deacetylation. (a) *Xenopus* A6 cells were transfected with either full length (FL) or a deletion construct (DC) of the histone H1° promoter before incubation (+) or not (−) with 5 mM sodium butyrate for 6 h before lysis. RNA was extracted and H1° transcripts quantitated by primer extension. Double lines at the right margin of the panel indicate the position of the specifically initiated transcripts from the H1° promoter. (b) The accumulation of the endogenous H1° mRNA under the conditions described above is shown. A Northern blot of RNA resolved on a 1 per cent agarose gel is shown. The positions of H1° and H1 mRNA are indicated.

assembly of a transcriptionally active chromatin environment. The converse is true for hypoacetylated histone H4 and hypophosphorylated histone H1. Aside from any direct effects, the segregation of histone acetyltransferases, deacetylases, kinases, and phosphatases together with any other tail binding proteins might influence the structural and functional properties of chromatin. A useful example to test these hypotheses would be the two X chromosomes of females. Nucleosomes on the active X chromosome contain predominantly acetylated histones whereas those on the inactive X chromosome are not acetylated (Jeppesen and Turner 1993). Since these states are differentially maintained within the same nucleus, it is possible to speculate that the differential packaging with modified histones not only modulates transcriptional activity (see also Bone *et al.* 1994), but also propagates chromatin structures that are permissive or restrictive for transcription.

Future directions

A major increase in our understanding of the role of histone modification in chromatin function will follow from the molecular characterization of the enzymes that modify the histones. These enzymes will allow the exact manipulation of histone modification by genetic or biochemical approaches and thus establish causal relationships much more effectively than inhibitor studies. It is also possible that considerable insights will follow concerning how states of modification are regulated in all chromosomes through the cell cycle, yet potentially targeted to individual chromosomes, chromosomal domains, and genes in the context of regulated gene expression. It might be anticipated that a large number of enzymes will exist for modifications such as acetylation and phosphorylation. This follows from existing biochemical evidence concerning the specificity of histone modifications (for example see Turner *et al.* 1992).

Our appreciation of the structural and biochemical properties of nucleosomes and chromatin containing histone modifications is still rudimentary. In this regard the development of defined dinucleosomes or nucleosomal arrays competent to transcribe will be very useful (Hansen and Wolffe 1992; Schild *et al.* 1993; Ura *et al.* 1995). The capacity to both manipulate structure and to determine the functional consequences on a defined template is essential. Efforts to understand the influence of histone modification on long-range chromatin structure and function will await the development of suitable assays. Reconstituted synthetic chromatin and nuclei in *Xenopus* oocyte, egg and embryo extracts are likely to prove useful for this type of analysis (Almouzni and Wolffe 1993*a*, *b*).

Although the link between histone modification and transcriptional activity is a strong one, there is still a need to obtain more examples. We do not know the exact sequence of events leading histone hyperacetylation to promote *Xenopus* histone H1° promoter activity (Khochbin and Wolffe 1993) or ubiquitinated H2A to accumulate on the *Drosophila* hsp70 promoter (Levinger and Varshavsky 1982). There is a need not only to consider the precise assembly of nucleoprotein architectures including histones on the regulatory DNA of eukaryotic genes, but also to consider the structural genes and functional consequences of modifying the histones on particular nucleoprotein complexes.

Summary

The large number of reversible covalent modifications of the histones (acetylation, phosphorylation, ubiquitination, ADP ribosylation, methylation) reflects the dynamic nature of the chromosome. The targeting of these modifications directs the differentiation of the chromosome into highly specialized functional domains. This differentiation is important for metazoan development. In general, core histone acetylation, phosphorylation, and ubiquitination increase with transcriptional activity, as does linker histone phosphorylation. These relationships are not invariant; in particular during mitosis, histones H1 and H3 are phosphorylated and transcription is repressed. Cause and effect for the role of histone modification in transcription is yet to be conclusively established. Nevertheless it is clear that modified histones will have an essential function in establishing and maintaining stable states of gene activity.

References

Allfrey, V., Faulkner, R.M., and Mirsky, A.E. (1964). Acetylation and methylation of histones and their possible role in the regulation of RNA synthesis. *Proceedings of the National Academy of Sciences of the USA*, **51**, 786–94.

Almouzni, G. and Wolffe, A.P. (1993*a*). Replication coupled chromatin assembly is required for the repression of basal transcription *in vivo*. *Genes and Development*, **7**, 2033–47.

Almouzni, G. and Wolffe, A.P. (1993*b*). Nuclear assembly, structure and function: the use of *Xenopus in vitro* systems. *Experimental Cell Research*, **205**, 1–15.

Almouzni, G., Khochbin, S., Dimitrov, S., and Wolffe, A.P. (1994). Histone acetylation influences both gene expression and development of *Xenopus laevis*. *Developmental Biology*, **165**, 654–69.

Annunziato, A.T., Frado, L.L.-Y., Seale, R.L., and Woodcock, C.L.F. (1988). Treatment with sodium butyrate inhibits the complete condensation of interphase chromatin. *Chromosoma*, **96**, 132–8.

Arents, G., Burlingame, R.W., Wang, B.W., Love, W.E., and Moudrianakis, E.N. (1991). The nucleosomal core histone octamer at 3.1Å resolution: a tripartite protein assembly and a left-handed superhelix. *Proceedings of the National Academy of Sciences of the USA*, **88**, 10148–52.

Aubert, D., Garcia, M., Benchaibi, M., Poncet, D., Chebloune, Y., Verdier, G., *et al.* (1991). Inhibition of proliferation of primary avian fibroblasts through expression of histone H5 depends on the degree of phosphorylation of the protein. *Journal of Cell Biology*, **11**, 497–506.

Ausio, J. and van Holde, K.E. (1986). Histone hyperacetylation: its effects on nucleosome conformation and stability. *Biochemistry*, **25**, 1421–28.

Ausio, J., Dong, F., and van Holde, K.E. (1989). Use of selectively trypsinized nucleosome core particles to analyze the role of the histone tails in the stabilization of the nucleosome. *Journal of Molecular Biology*, **206**, 451–63.

Bauer, W.R., Hayes, J.J., White, J.H., and Wolffe, A.P. (1994). Nucleosome structural changes due to acetylation. *Journal of Molecular Biology*, **236**, 685–90.

Bavykin, S.G., Usachenko, S.I., Zalensky, A.O., and Mirzabekov, A.D. (1990). Structure of nucleosomes and organization of internucleosomal DNA in chromatin. *Journal of Molecular Biology*, **212**, 495–511.

Bode, J., Gomez-Lira, M.J., and Schroter, H. (1983). Nucleosomal particles open as the histone core becomes hyperacetylated. *European Journal of Biochemistry*, **130**, 437–45.

Bone, J.R., Lavender, J., Richman, R., Palmer, M.J., Turner, M.B., *et al.* (1994).

Acetylated histone H4 on the male X chromosome is associated with dosage compensation in *Drosophila. Genes and Development*, **8**, 96–104.

Bradbury, E.M. (1992). Reversible histone modification and the cell cycle. *BioEssays*, **14**, 9–16.

Braunstein, M., Rose, A.B., Holmes, S.G., Allis, C.D., and Broach, J.R. (1993). Transcriptional silencing in yeast in associated with reduced histone acetylation. *Genes and Development*, **7**, 592–604.

Brownell, J.E., Zhou, J., Ranalli, T., Kobayashi, R., Edmondson, D.G., Roth, S.Y., *et al.* (1996). Tetrahymena histone acetyltransferase A: a homolog to yeast Gcn5p linking histone acetylation to gene activation. *Cell*, **84**, 843–51.

Candido, E.P.M., Reeves, R., and Davie, J.R. (1978). Sodium butyrate inhibits histone deacetylation in cultured cells. *Cell*, **14**, 105–15.

Cary, P.D., Crane-Robinson, C., Bradbury, E.M., and Dixon, G.H. (1982). Effect of acetylation on the binding of N-terminal peptides of histone H4 to DNA. *European Journal of Biochemistry*, **127**, 137–43.

Chen, T.A. and Allfrey, V.G. (1987). Rapid and reversible changes in nucleosome structure accompany the activation, repression and super induction of the murine proto-oncogenes c-fos and c-myc. *Proceedings of the National Academy of Sciences of the USA*, **84**, 5252–6.

Chen, T.A., Sterner, R., Cozzolino, A., and Allfrey, V.G. (1990). Reversible and irreversible changes in nucleosome structure along the c-fos and c-myc oncogenes following inhibition of transcription. *Journal of Molecular Biology*, **212**, 481–93.

Clark, D.J., Hill, C.S., Martin, S.R., and Thomas, J.O. (1988) α-Helix in the carboxy-terminal domains of histones H1 and H5. *EMBO Journal*, **7**, 69–75.

Clark, D.J., O'Neill, L.P., and Turner, B.M. (1993). Selective use of H4 acetylation sites in the yeast *Saccharomyces cerevisiae. Biochemical Journal*, **294**, 557–61.

Dimitrov, S., Almouzni, G., Dasso, M., and Wolffe, A.P. (1993). Chromatin transitions during early *Xenopus* embryogenesis: changes in histone H4 acetylation and in linker histone type. *Developmental Biology*, **160**, 214–27.

Dimitrov, S., Dasso, M.C., and Wolffe, A.P. (1994). Remodeling sperm chromatin in *Xenopus laevis* egg extracts: the role of core histone phosphorylation and linker histone B4 in chromatin assembly. *Journal of Cell Biology*, **126**, 591–601.

Ebralidse, K.K., Grachev, S.A., and Mirzabekov, A.D. (1988). A highly basic histone H4 domain bound to the sharply bent region of nucleosomal DNA. *Nature*, **331**, 365–7.

Garcia-Ramirez, M., Dong, F., and Ausio, J. (1992). Role of the histone 'tails' in the folding of oligonucleosomes depleted of histone H1. *Journal of Biological Chemistry*, **267**, 19587–95.

Glotov, B.O., Itkes, A.V., Nikolaev, L.G., and Severin, E.S. (1978). Evidence for close proximity between histones H1 and H3 in chromatin of intact nuclei. *FEBS Letters*, **91**, 149–52.

Godde, J.S., Nakatani, Y., and Wolffe, A.P. (1995). The amino-terminal tails of the core histones and the translational position of the TATA box determine TBP/TFIIA association with nucleosomal DNA. *Nucleic Acids Research*, **23**, 4557–64.

Gorovsky, M.A., Pleger, G.L., Keevert, J.B., and Johmann, C.A. (1973). Studies on histone fraction F2A1 in macro and micronuclei of *Tetrahymena pyriformis. Journal of Cell Biology*, **57**, 773–81.

Gurley, L.R., D'Anna, J.A., Barham, S.S., Deavan, L.L., and Tobey, R.A. (1978). Histone phosphorylation and chromatin structure during mitosis in chinese hamster cells. *European Journal of Biochemistry*, **84**, 1–15.

Gushchin, D.Y., Ebralidse, K.K., and Mirzabekov, A.D. (1991). Structure of the nucleosome: localization of the segments of the H2A histones that interact with DNA by DNA–protein crosslinking. *Molecular Biology*, **25**, 1400–11.

Hansen, J.C. and Wolffe, A.P. (1992). Chromatin folding inhibits both transcription initiation and elongation by RNA polymerase III. *Biochemistry*, **31**, 7977–88.

Hartl, P., Gottesfeld, J., and Forbes, D.J. (1993) Mitotic repression of transcription *in vitro*. *Journal of Cell Biology*, **120**, 613–24.

Hayes, J.J., Clark, D.J., and Wolffe, A.P. (1991*a*). Histone contributions to the structure of DNA in the nucleosome. *Proceedings of the National Academy of Sciences of the USA*, **88**, 6829–33.

Hayes, J.J., Bashkin, J., Tullius, T.D., and Wolffe, A.P. (1991*b*). The histone core exerts a dominant constraint on the structure of DNA in a nucleosome. *Biochemistry*, **30**, 8434–40.

Hebbes, T.R., Thorne, A.W., and Crane-Robinson, C. (1988). A direct link between core histone acetylation and transcriptionally active chromatin. *EMBO Journal*, **7**, 1395–402.

Hebbes, T.R., Thorne, A.W., Clayton, A.L., and Crane-Robinson, C. (1992). Histone acetylation and globin gene switching. *Nucleic Acids Research*, **20**, 1017–22.

Hebbes, T.R., Clayton, A.L., Thorne, A.W., and Crane-Robinson, C. (1994). Core histone hyperacetylation co-maps with generalized DNase I sensitivity in the chicken β-globin chromosomal domain. *EMBO Journal*, **13**, 1823–30.

Hentschel, C.C. and Tata, J.R. (1978). Template-engaged and free RNA polymerases during *Xenopus* erythroid cell maturation. *Developmental Biology*, **65**, 496–507.

Hill, C.S. and Thomas, J.O. (1990). Core histone–DNA interactions in sea urchin sperm chromatin: the N-terminal tail of H2B interacts with linker DNA. *European Journal of Biochemistry*, **187**, 145–53.

Hill, C.S., Rimmer, J.M., Green, B.N., Finch, J.T., and Thomas, J.O. (1991). Histone–DNA interactions and their modulation by phosphorylation of Ser-Pro-X-Lys/Arg-motifs. *EMBO Journal*, **10**, 1939–48.

Hohmann, P. (1983). Phosphorylation of H1 histones. *Molecular and Cellular Biochemistry*, **57**, 81–95.

Ikegami, S., Ooe, Y., Shimizu, T., Kasahara, T., Tsuruta, T., *et al.* (1993). Accumulation of multiacetylated forms of histones by trichostatin A and its developmental consequences in early starfish embryos. *Roux's Archives of Developmental Biology*, **202**, 144–51.

Imai, B.S., Yau, P., Baldwin, J.P., Ibel, K., May, R.P., and Bradbury, E.M. (1986). Hyperacetylation of core histones does not cause unfolding of nucleosomes: neutron scatter data accords with disc structure of the nucleosome. *Journal of Biological Chemistry*, **261**, 8784–92.

Jeppesen, P. and Turner, B.M. (1993). The inactive X chromosome in female mammals is distinguished by a lack of histone H4 acetylation, a cytogenetic marker for gene expression. *Cell*, **74**, 281–91.

Johnson, T.C. and Holland, J.J. (1965). Ribonucleic acid and protein synthesis in mitotic HeLa cells. *Journal of Cell Biology*, **27**, 565–74.

Khochbin, S. and Wolffe, A.P. (1993). Developmental regulation and butyrate inducible transcription of the *Xenopus* histone H1° promoter. *Gene*, **128**, 173–80.

Kidwell, W.R. and Mage, M.G. (1976). Change in poly(adenosine diphosphate ribose) and poly(adenosine diphosphate ribose) polymerase in synchronized HeLa cells. *Biochemistry*, **15**, 1213–17.

Kleinschmidt, A.M. and Martinson, H.G. (1981). Structure of nucleosome core particles containing uH2A. *Nucleic Acids Research*, **9**, 2423–31.

Lee, D.Y., Hayes, J.J., Pruss, D., and Wolffe, A.P. (1993). A positive role for histone acetylation in transcription factor binding to nucleosomal DNA. *Cell*, **72**, 73–84.

Levinger, L. and Varshavsky, A. (1980). High-resolution fractionation of nucleosomes: minor particles 'whiskers' and separation of nucleosomes containing and lacking A24 semihistone. *Proceedings of the National Academy of Sciences of the USA*, **77**, 3244–8.

Levinger, L. and Varshavsky, A. (1982). Selective arrangement of ubiquitinated and D1 protein containing nucleosomes within the *Drosophila* genome. *Cell*, **28**, 375–85.

Lin, R., Cook, R.G., and Allis, C.D. (1991). Proteolytic removal of core histone amino termini and dephosphorylation of histone H1 correlate with the formation of condensed chromatin and transcriptional silencing during *Tetrahymena* macronuclear development. *Genes and Development*, **5**, 1601–10.

Mahadevan, L.C., Willis, A.C., and Barrah, M.J. (1991). Rapid histone H3 phosphorylation in response to growth factors, phorbol esters, okadaic acid and protein synthesis inhibitors. *Cell*, **65**, 775–83.

Mathis, D.J., Oudet, P., Waslyk, B., and Chambon, P. (1978). Effect of histone acetylation on structure and *in vitro* transcription of chromatin. *Nucleic Acids Research*, **5**, 3523–47.

Morcillo, G., de La Torre, C., and Gimenez-Martin, G. (1976). Nucleolar transcription during plant mitosis. *Experimental Cell Research*, **102**, 311–16.

Mueller, R.D., Yasuda, H., Hatch, C.L., Bonner, W.M., and Bradbury, E.M. (1985). Identification of ubiquitinated histones H2A and H2B in *Physarum polycelphalum*. Disappearance of these proteins at metaphase and reappearance at anaphase. *Journal of Biological Chemistry*, **260**, 5147–53.

Norton, V.G., Imai, B.S., Yau, P., and Bradbury, E.M. (1989). Histone acetylation reduces nucleosome core particle linking number change. *Cell*, **57**, 449–57.

Norton, V.G., Marvin, K.W., Yau, P., and Bradbury, E.M. (1990). Nucleosome linking number change controlled by acetylation of histones H3 and H4. *Journal of Biological Chemistry*, **265**, 19848–52.

Oliva, R., Bazett-Jones, D.P., Locklear, L., and Dixon, G.H. (1990). Histone hyperacetylation can induce unfolding of the nucleosome core particle. *Nucleic Acids Research*, **18**, 2739–47.

Paulson, J.R. and Taylor, S.S. (1982). Phosphorylation of histones 1 and 3 and non histone high mobility group 14 by an endogenous kinase in HeLa metaphase chromosomes. *Journal of Biological Chemistry*, **257**, 6064–72.

Perry, C.A. and Annunziato, A.T. (1989). Influence of histone acetylation on the solubility, H1 content and DNaseI sensitivity of newly replicated chromatin. *Nucleic Acids Research*, **17**, 4275–91.

Perry, C.A., Allis, C.D., and Annunziato, A.T. (1993). Parental nucleosomes segregated to newly replicated chromatin are underacetylated relative to those assembled *de novo*. *Biochemistry*, **32**, 13615–23.

Poccia, D.L. and Green, G.R. (1992). Packaging and unpackaging the sea urchin genome. *Trends in Biochemical Sciences*, **17**, 223–7.

Pruss, D. and Wolffe, A.P. (1993). Histone–DNA contacts in a nucleosome core containing a *Xenopus* 5S rRNA gene. *Biochemistry*, **32**, 6810–14.

Pruss, D., Hayes, J.J., and Wolffe, A.P. (1995). Nucleosomal anatomy—where are the histones? *BioEssays*, **17**, 161–70.

Roth, S.Y. and Allis, C.D. (1992). Chromatin decondensation: does histone H1 dephosphorylation play a role? *Trends in Biochemical Sciences*, **17**, 93–8.

Roth, S.Y., Schulman, I.G., Richman, R., Cook, R.G., and Allis, C.D. (1988). Characterization of phosphorylation sites in histones H1 in the amitotic macronucleus of *Tetrahymena* during different physiological states. *Journal of Cell Biology*, **107**, 2473–82.

Ruiz-Carrillo, A., Wangh, L.J., and Allfrey, V.G. (1975). Processing of newly synthesized histone molecules. *Science*, **190**, 117–28.

Schild, C., Claret, F-X., Wahli, W., and Wolffe, A.P. (1993). A nucleosome-dependent static loop potentiates estrogen-regulated transcription from the *Xenopus* vitellogenin B1 promoter *in vitro*. *EMBO Journal*, **12**, 423–33.

Sun, J-M., Wiaderkiewicz, R., and Ruiz-Carrillo, A. (1989). Histone H5 in the control of DNA synthesis and cell proliferation. *Science*, **245**, 68–71.

Turner, B.M. (1991). Histone acetylation and control of gene expression. *Journal of Cell Science*, **99**, 13–20.

Turner, B.M. (1993). Decoding the nucleosome. *Cell*, **75**, 5–8.

Turner, B.M., Birley, A.J., and Lavender, J. (1992). Histone H4 isoforms acetylated at specific lysine residues define individual chromosomes and chromatin domains in *Drosophila* polytene nuclei. *Cell*, **69**, 375–84.

Ura, K., Wolffe, A.P., and Hayes, J.J. (1994). Core histone acetylation does not block linker histone binding to a nucleosome including a *Xenopus borealis* 5S rRNA gene. *Journal of Biological Chemistry*, **269**, 27171–4.

Ura, K., Hayes, J.J., and Wolffe, A.P. (1995). A positive role for nucleosome mobility in the transcriptional activity of chromatin templates: restriction by linker histones. *EMBO Journal*, **14**, 3752–65.

Usachenko, S.I., Bavykin, S.G., Gavin, I.M., and Bradbury, E.M. (1994). Rearrangement of the histone H2A C-terminal domain in the nucleosome. *Proceedings of the National Academy of Sciences of the USA*, **91**, 6845–9.

van Holde, K.E. (1988). *Chromatin*. Springer-Verlag, New York.

Varshavsky, A., Levinger, L., Sundin, O., Barsoum, J., Ozkaynak, E., Swerdlow, P., and Finley, D. (1982). Cellular and SV40 chromatin: replication, segregation ubiquitination, nuclease hypersensitive sites, HMG-containing nucleosomes and heterochromatin-specific protein. *Cold Spring Harbor Symposia on Quantitative Biology*, **47**, 511–28.

Vettesse-Dadey, M., Walter, P., Chen, H., Juan, L-J., and Workman, J.L. (1994). Role of the histone amino termini in facilitated binding of a transcription factor, GAL4-AH to nucleosome cores. *Molecular and Cellular Biology*, **14**, 970–81.

Waterborg, J.H. and Matthews, H.R. (1982). Control of histone acetylation. Cell cycle dependence of deacetylase activity in *Physarum* nuclei. *Experimental Cell Research*, **138**, 462–6.

Waterborg, J.H. and Matthews, H.R. (1984). Patterns of histone acetylation in *Physarum polycephalum*. H2A and H2B acetylation and H3 and H4 acetylation. *European Journal of Biochemistry*, **142**, 329–33.

West, M.H.P. and Bonner, W.M. (1980). Histone H2B can be modified by the attachment of ubiquitin. *Nucleic Acids Research*, **8**, 4671–80.

Wiekowski, M., Miranda, M., and De Pamphilis, M.L. (1991). Regulation of gene expression in preimplantation mouse embryos: effects of zygotic gene expression and the first mitosis on promoter and enhancer activities. *Developmental Biology*, **147**, 403–14.

Wiekowski, M., Miranda, M., and De Pamphilis, M.L. (1993). Requirements for promoter activity in mouse oocytes and embryos distinguish paternal pronuclei from maternal and zygotic nuclei. *Developmental Biology*, **159**, 366–78.

Wolffe, A.P. (1995). *Chromatin: structure and function*. Academic Press, London.

Wolffe, A.P. and Pruss, D. (1996). Targeting chromatin disruption: transcription regulators that acetylate histones. *Cell*, **84**, 817–19.

Worrad, D.M., Turner, B.M., and Schultz, R.M. (1995). Temporally restricted spatial localization of acetylated isoforms of histone H4 and RNA polymerase II in the 2-cell mouse embryo. *Development*, **121**, 2949–59.

Yoshida, M., Horinouoshi, S., and Beppu, T. (1995). Trichostatin A and trapoxin: novel chemical probes for the role of histone acetylation in chromatin structure and function. *BioEssays*, **17**, 423–30.

4

Mechanisms for alleviating chromatin-mediated repression of transcription

Janet Quinn and Craig L. Peterson

Introduction

There is now a wealth of data derived from both biochemical and genetic studies that confirm the importance of chromatin in regulating gene expression (for recent reviews see Felsenfeld 1992; Grunstein 1992; Lewin 1994; Paranjape *et al.* 1994; Peterson 1994). In all eukaryotes the basic building block of chromatin is the nucleosome, which is comprised of approximately 145 base pairs (bp) of DNA wrapped about twice around an octamer of histone proteins. The histone octamer is organized as a central tetramer of histones H3 and H4 flanked by two heterodimers of histones H2A and H2B. In higher eukaryotes a single molecule of histone H1 binds to both the outer surface of DNA that circles the octamer and to the DNA that links neighbouring nucleosomes.

In vivo and *in vitro* studies indicate that nucleosomes repress transcription by blocking the access of both the general transcription machinery and regulatory factors to their DNA sites. It is highly plausible that transcription factors that recognize specific sequences in B-form DNA may encounter several problems following the incorporation of this DNA into a nucleosome (reviewed in Hayes and Wolffe 1992; see Chapters 1, 3, and 4). For instance, one face of the DNA helix is completely concealed upon association with the histone core and the structure of the DNA is changed dramatically upon wrapping around the histone proteins. Concordantly, there are several examples of *trans*-acting factors that cannot bind to their preferred sequences when presented in a nucleosomal context, including human heat shock factor (HSF) and the yeast positive regulator PHO4 (Taylor *et al.* 1991; Venter *et al.* 1994). Nonetheless, regulatory factors must often gain access to their respective binding sites, even when present in a nucleosomal context, and activate the formation of transcription complexes at core promoter elements. Evidence is beginning to emerge that the eukaryotic cell employs both replication-dependent and replication-independent pathways to contend with nucleosome-mediated repression of transcription. In the replication-dependent mechanism transcription factors bind to the DNA template immediately following passage of the replication fork and prior to the assembly of the freshly replicated DNA into chromatin, thus preventing the formation of repressive chromatin structures. In the replication-independent pathway promoter regions are assembled into organized nucleosomal arrays following replication. Such repressive chromatin structures must then be directly disrupted and there is now growing evidence that a distinct class of transcriptional co-activators exist, whose primary role is to reconfigure chromatin structure. The major emphasis of this chapter is to review the recent advances that have been made in understanding how the cell contends with chromatin-mediated repression and the identification of proteins that possess chromatin remodelling properties. First of all however, we will describe earlier work illustrating the repressive nature of nucleosomes and higher order chromatin structures on transcription initiation.

Nucleosomes as repressors of transcription

It is now over 20 years since the discovery of the nucleosome (Kornberg 1974). Suggestions that chromatin may play a role in transcriptional regulation were quick to follow when it was noted that transcriptional activation in eukaryotes often correlated with changes in chromatin structure (for early reviews see Elgin 1981; Igokemenes *et al.* 1982). Such inducible changes in chromatin structure have been studied vigorously in yeast and mammalian systems. One of the best examples in yeast is the highly regulated *PHO5* gene which encodes an acid phosphatase. Transcription of *PHO5* is repressed by growth in high phosphate medium and induced upon phosphate starvation. As illustrated in Fig. 1(a), the promoter region of the yeast *PHO5* gene is packaged into six precisely positioned nucleosomes in the uninduced state (Almer and Hörz 1986). Both *in vitro* and *in vivo* footprinting analyses have demonstrated that the upstream regulatory region of *PHO5* contains two binding sites (UASp1 and UASp2) for the positive regulator PHO4 (Vogel *et al.* 1989; Venter *et al.* 1994). UASp1 is contained in a short DNAase I-hypersensitive region (HS) in the inactive promoter, whereas UASp2 is located in the adjacent nucleosome (Fig. 4.1(a), nucleosome –2). Induction of *PHO5* gene transcription is accompanied by striking changes in nucleosome position or structure; four nucleosomes surrounding the HS are lost from the promoter region or their structure is altered so dramatically that the nuclease accessibility of this region is close to that observed with naked DNA (Almer *et al.* 1986). Recent *in vivo* footprinting analysis of the *PHO5* promoter has illustrated that binding of PHO4 to UASp2 (which is critical for *PHO5* activation) is prevented by the presence of the nucleosome at position –2 and is only facilitated after disruption of this nucleosome in a process that is dependent on the UASp1 element (Venter *et al.* 1994). Thus, there is good evidence that the positioned nucleosomes at the *PHO5* promoter contribute directly to repression of *PHO5* transcription.

In mammalian systems, the chromatin structure surrounding the steroid-responsive mouse mammary tumour virus (MMTV) promoter has been similarly well characterized (for a detailed review see Chapter 2). In the uninduced state,

A. *PHO5*

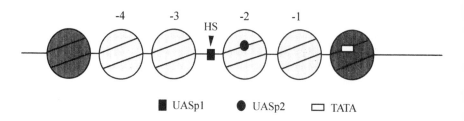

B. *MMTV*

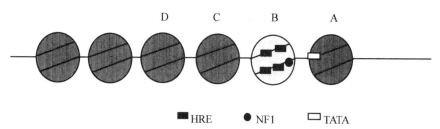

Fig. 4.1   Chromatin structure of the *PHO5* (a) and *MMTV* (b) promoters. The unshaded large circles represent nucleosomes that undergo remodelling upon gene induction; the shaded large circles represent stable nucleosomes. *Cis*-acting regulatory elements are shown as black boxes or circles, and the TATA elements are represented by an open box. The position of the DNAase I-hypersensitive site (HS) that appears over the *PHO5* promoter upon gene induction is also shown.

six positioned nucleosomes cover the length of the promoter (Richard-Foy and Hager 1987). One of these (Fig. 4.1(b), nucleosome B) encompasses the four hormone receptor binding sites (HREs), in addition to binding sites for the NF1 and OTF1 positive regulators (Perlmann and Wrange 1988; Piña *et al.* 1990). Immediately upon hormone induction, a DNAase I-hypersensitive site becomes apparent over the promoter region and only then can NF1, an abundant nuclear protein, bind to its site and stimulate MMTV transcription (Piña *et al.* 1990; Archer *et al.* 1991). The formation of this hypersensitive site was originally thought to indicate nucleosome loss or, at the very least, severe nucleosome disruption. However, it has recently been demonstrated that nucleosome B is neither removed nor shifted upon hormone induction (Truss *et al.* 1995). Instead, hormone induction leads to an unknown change in chromatin structure, which then facilitates binding of all the relevant transcription factors to the rearranged nucleosome. Thus the nucleosome at the MMTV promoter appears to play two very different roles. In the first instance it prevents transcription factor binding prior to hormone induction. Subsequently, however, maintenance of a remodelled nucleosome may be required to allow consecutive binding of transcription factors. This is further evidenced by the fact that these factors cannot bind simultaneously to naked DNA templates (Brüggemeier *et al.* 1990).

In vitro experiments have similarly illustrated that nucleosomes can inhibit transcription when positioned over a promoter (Lorch *et al.* 1987; 1992; Knezetic *et al.* 1988; Laybourn and Kadonaga 1991) or upstream regulatory elements (Taylor *et al.* 1991). Such observations suggested a simple model in which the transcription complex and *trans*-acting factors compete with histones for binding to DNA. Further studies have demonstrated that formation of pre-initiation complexes prior to chromatin assembly (Knezetic *et al.* 1988) and the presence of TFIID (Workman and Roeder 1987) or activator proteins (Workman *et al.* 1991) during chromatin assembly allows for subsequent transcription.

Finally, genetic studies with the yeast *Saccharomyces cerevisiae* have provided some of the most convincing evidence of a functional relationship between chromatin structure and regulation of transcription. Such studies were facilitated by the ease of genetic manipulation of *S. cerevisiae* and the presence of only two copies of each of the core histone genes per haploid organism (compare with *Drosophila melaganster* in which there are approximately 110 of each!). To examine the function of histone H4 *in vivo* Grunstein and colleagues constructed a yeast strain in which the two endogenous H4 genes were deleted and an episomal copy of the histone H4 gene was introduced to support cell viability. This episomal copy of histone H4 was placed under the transcriptional control of the *GAL1* promoter, which is induced by galactose and repressed by glucose. Glucose-treated cells lose approximately half of their nucleosomes as a round of replication occurs in the absence of H4 synthesis. It was shown that nucleosome loss resulted in activation of *PHO5*, *CYC1*, *GAL1*, *CUP1*, and *HIS3* gene promoters in the absence of inducers (Han and Grunstein 1988; Han *et al.* 1988; Durrin *et al.* 1992). This gene activation did not require the function of the UAS elements, but was dependent on the presence of TATA and initiation sequences. Since nucleosome depletion precedes transcriptional initiation, these results indicate that changes in chromatin can alter transcription activation rather than vice versa. This is consistent with earlier studies by Clark-Adams *et al.* (1988), showing that changes in the stoichiometry of histones (achieved by over-expression of individual histones or histone pairs) could also alter transcription initiation *in vivo*.

It was also found that nucleosome loss does not result in a global increase in mRNA accumulation, but only appears to effect the transcription of inducible genes. For instance, expression of the constitutive *TRP1*, *ARG4*, and *HIS4* genes is not increased upon nucleosome depletion (Kim *et al.* 1988). This is likely to be a consequence of the fact that the regulatory regions of constitutive genes are generally found to be nucleosome free.

In concluding this section it is important to note that there is increasing evidence that nucleosomes do not function exclusively as repressive structures.

The exact positioning of nucleosomes has also been shown, in some instances, to make a significant contribution to transcriptional activation. As mentioned previously, the presence of a nucleosome at the MMTV promoter allows contiguous binding of transcription factors that are unable to do so on naked DNA (Truss *et al.* 1995). Nucleosome positioning can also facilitate gene activation by bringing together widely separated regulatory elements. One such example is the case of the *Drosophila hsp26* gene, in which HSF and GAGA factor bound approximately 200 bp apart are thought to be brought into proximity with each other and the initiation complex by the wrapping of intervening DNA around a specifically positioned nucleosome (Thomas *et al.* 1988; Lu *et al.* 1993). Similarly, a nucleosome positioned between the promoter and enhancer elements of the *Drosophila adh* gene has been suggested to be involved in the formation of a tertiary nucleoprotein structure that potentiates transcription (Jackson and Benyajati 1993). Although a functional role for these nucleosomes has yet to be established, reconstruction of a positioned nucleosome *in vitro* between an enhancer and promoter in the *Xenopus vitellogenin* gene was found to facilitate enhancer activity approximately 10-fold (Schild *et al.* 1993).

Higher order chromatin structures

Arrays of nucleosomes represent only the first level of DNA compaction in the nucleus. This 'beads on a string' structure is further condensed six- to seven-fold to form the 30 nm fibre, which represents the majority of chromatin present during interphase. This increase in the compaction of nucleosomal DNA is expected to inhibit transcription initiation further, due to an increase in the inaccessibility of transcription factor binding sites. This has recently been demonstrated by Hansen and Wolffe (1992). In this study nucleosomes were assembled onto either closed circular plasmids containing a single *Xenopus* 5S RNA gene, or a linear tandemly repeated array of *Lytechinus* 5S RNA genes. Compaction of these chromatin templates by increasing the concentration of divalent cations led to a significant increase in the inhibition of transcription initiation by RNA polymerase III. Interestingly, such compaction occurred under ionic conditions (5 mM $MgCl_2$) commonly used for *in vitro* transcription assays. This suggests that a substantial fraction of the transcriptional repression that was previously noted upon incorporation of promoter elements and regulatory sequences into nucleosomes may have in fact been due to compaction of the nucleosomal templates.

How is the compaction state of DNA regulated? The 30 nm fibre appears to be a very dynamic structure that either folds or unfolds depending upon the level of transcriptional activity. Traditionally, histone H1 is thought to play a primary role in the stabilization of the 30 nm fibre structure. Currently however, the relationship between the presence of linker histones in chromatin, stabilization of the chromatin fibre, and further condensation of the chromosome is controversial.

Shen and colleagues (1995) report that disruption of *Tetrahymena thermophila* histone H1 genes results in a detectable reduction in chromatin condensation *in vivo*. Nonetheless, chromatin condensation occurs in *Xenopus* mitotic extracts depleted in histone H1 (Ohsumi *et al.* 1993). The basis for this difference remains unclear, but provides some evidence that histone H1 may play a role in processes other than DNA compaction. In addition, several observations indicate that linker histones are non-essential. For instance, linker histones are absent in *S. cerevisiae* (Van Holde 1989), *in vivo* knockouts of linker histones in *Tetrahymena* are viable (Shen *et al.* 1995), and immunodepletion of linker histones from *Xenopus* egg extracts has no effect on the formation of functional nuclei (Dasso *et al.* 1994).

There is now increasing evidence that the core histone components themselves can influence the formation of higher order structures. Hansen and colleagues have illustrated, using homogeneous nucleosomal arrays, that core histones can direct DNA compaction, comparable to that within the chromatin fibre, at physiological ionic strengths (Hansen *et al.* 1989; Hansen and Wolffe 1992, 1994; Fletcher *et al.* 1994). Furthermore, both a reduction in the extent of folding and an enhancement in the initiation of transcription are observed for arrays that are subsaturated, i.e. partially depleted of histone octamers (Hansen and Lohr 1993), or arrays that are deficient in H2A and H2B dimers (Hansen and Wolffe 1994). Tryptic removal of the histone N-termini has also been found to decrease the extent of chromatin compaction (Fletcher and Hansen 1995). These findings have led to the hypothesis that changes in histone octamer content or modification may influence gene expression in part by modulating the extent of chromatin folding (Hansen and Wolffe 1994).

Some chromosomal regions have much more condensed chromatin structures than that exhibited by the 30 nm chromatin fibre. Heterochromatin, for example, which is often found adjacent to telomeres or centromeres, remains very highly condensed during interphase. Unfortunately, there is little information regarding how such condensed chromatin structures are resolved to facilitate transcription (for more information see Chapters 9, 13, and 14). Hence, we would now like to review the significant advances that have recently been made in understanding the mechanisms employed by the cell to contend with the primary level of chromatin structure: the nucleosomal fibre.

Mechanisms for alteration of chromatin structure

During the past few years there has been an explosion of genetic and biochemical research on the mechanisms employed by the eukaryotic cell to alleviate chromatin-mediated repression of transcription. Although we do not yet know the precise molecular mechanisms behind chromatin disruption, evidence is beginning to emerge suggesting that there are two basic strategies employed by the eukaryotic cell to contend with repressive chromatin structures (reviewed in Workman and Buchman 1993; Paranjape *et al.* 1994; Wallrath *et al.* 1994).

The first strategy is replication dependent and involves binding of transcription factors to the DNA template immediately after passage of the replication fork and prior to the assembly of chromatin. Therefore, in this strategy it is the disruption of the chromatin structure that occurs upon DNA replication that allows binding of factors and the formation of active chromatin configurations. This strategy has been shown to operate in synthetic nuclei at the β^A-globin gene of the chicken β-globin locus (Barton and Emerson 1994). Chromatin structural studies have demonstrated that the β^A-enhancer is nucleosome-free throughout erythroid development, whereas the promoter only acquires an accessible conformation at the onset of β^A-globin expression in definitive red blood cells. Two erythroid proteins, GATA-1 and NF-E4, have been implicated in the formation of accessible chromatin structures at both promoter and enhancer regions of the β^A-globin gene. Barton and Emerson (1995) found that transcription of the β^A-globin gene requires the interaction of specific erythroid activators with the promoter and enhancer regions prior to chromatin assembly. Templates assembled into chromatin in the absence of the erythroid factors remain transcriptionally repressed even if these proteins are added later. However, transcription is activated after a single round of DNA replication in the presence of appropriate amounts of erythroid proteins. Thus the transient disruption of nucleosomal structure that occurs during DNA replication provides an opportunity for the erythroid proteins to bind to the β-globin DNA and prevent the formation of repressive chromatin structures. The formation of active structures in this way may be regulated by controlling the levels of the DNA binding factors present at the time of replication.

Permanently active, nucleosome-free structures are also found over promoter regions of constitutively transcribed 'housekeeping' genes. One possibility is that these nucleosome-free regions are created by the binding of constitutive transcription factors immediately following DNA replication. Many 'housekeeping' genes have multiple binding sites for the Sp1 transcription factor and thus it may be a member of this proposed class of factors (Saffer *et al.* 1991). Previous work has illustrated that repeated Sp1 sites in the SV40 early promoter are important for the formation of nucleosome-free regions in SV40 minichromosomes *in vivo* (Gross and Garrard 1988) and that Sp1 binding to multiple sites prevents nucleosome repression of transcription *in vitro* (Laybourn and Kadonaga 1991).

The replication-dependent pathway may also allow for the formation of 'preset' promoters in which constitutive factors that bind following DNA replication serve to maintain the accessibility of promoter elements to inducible factors. This serves to generate a chromatin structure that is repressed but poised for transcription. This is exemplified by the 'inducible preset' *Drosophila* heat shock protein (hsp) promoters. At the hsp promoters binding of the constitutively expressed GAGA factor is believed to be essential in maintaining the accessibility of upstream elements for heat shock factors (reviewed in Becker 1994; Wallrath *et al.* 1994). A group of yeast proteins, known collectively as general regulatory factors or GRFs (Buchman *et al.* 1988; Chasman *et al.* 1990), are also thought to play a similar role. One such factor, GRF2, binds to the UAS of the bidirectional *GAL1/GAL10* genes and is thought to maintain the accessibility of this region to the *trans*-activator protein GAL4 (Fedor *et al.* 1988).

The second strategy employed by the cell to reconfigure chromatin structure is a replication-independent mechanism that involves induced nucleosome remodelling or nucleosome displacement. This occurs at genes whose upstream and promoter elements are packaged into a continuous array of nucleosomes (Workman and Buchman 1993; Wallrath *et al.* 1994). In this case there is evidence that co-activators alter the chromatin structure at *cis*-regulatory elements or TATA boxes to allow transcription factor access. In the case of the *PHO5* and MMTV genes (Fig. 4.1) changes in chromatin structure and transcriptional activation are known to occur in the absence of DNA replication (Richard-Foy and Hager 1987; Schmid *et al.* 1992).

These two different strategies employed by the cell are not, however, mutually exclusive. There are examples of genes that utilize features of both replication-dependent and -independent strategies. As described above, a nucleosome-free region is found at the UAS of the bidirectional *GAL1/GAL10* genes in *S. cerevisiae* that is attributed to the replication-dependent binding of GRF2. However, the TATA boxes of both *GAL1* and *GAL10* genes are contained in nucleosomes, which must be remodelled during induction (Fedor *et al.* 1988). Thus, the upstream regulatory region makes use of the replication-dependent strategy, whereas the chromatin remodelling at the TATA elements operates through a replication-independent pathway.

Investigation into the mechanisms that underlie the replication-independent pathway of nucleosome remodelling has led to the identification of cofactors that play a role in antagonizing repressive chromatin structures. The most significant discovery to date is the presence of an evolutionarily conserved multimeric protein assembly, the SWI/SNF complex, that uses the energy of ATP hydrolysis to facilitate binding of transcription factors to nucleosomal sites. Currently, the SWI/SNF complex is the only well-characterized factor known to remodel chromatin structure (Peterson and Tamkun 1995). Recently, however, the identification of *Drosophila* factors that also appear to mediate an ATP-dependent remodelling of chromatin structure has been reported (Pazin *et al.* 1994; Tsukiyama *et al.* 1994; Varga-Weisz *et al.* 1995; Wall *et al.* 1995). Although the relationship between these factors and the SWI/SNF complex remains to be determined, initial data suggest that these factors may in fact represent novel chromatin remodelling activities.

The SWI/SNF complex: a chromatin remodelling machine

Five subunits of the yeast SWI/SNF complex, encoded by the *SWI1* (*ADR6*), *SWI2* (*SNF2*), *SWI3*, *SNF5*, and *SNF6* genes, were initially identified as positive regulators of two yeast genes: *HO* and *SUC2* (reviewed in Winston and Carlson 1992; Peterson and Tamkun 1995). These SWI/SNF products are now known to be required for the expression of a large number of yeast genes, including *INO1*, *ADH1*, *ADH2*, *GAL1*, and *GAL10*. The observation that many SWI/SNF-dependent genes are highly regulated prompted the suggestion that SWI/SNF proteins may be required to facilitate activator protein function (Peterson and Herskowitz 1992). Subsequent studies with simple reporter genes illustrated that this is indeed the case, with SWI/SNF proteins being required for the functioning of many different activators, including GAL4, INO1/INO4, HSF, and GCN4 (Peterson and Herskowitz 1992; C.L. Peterson, unpublished data). The effect of *swi/snf* mutations on activator protein function is dramatic. For example, the activity of GAL4 is reduced at least 10- to 30-fold in the absence of SWI or SNF proteins. In contrast, SWI/SNF proteins are not required for basal transcription *in vivo* (Peterson and Herskowitz 1992) or *in vitro* (Cairns *et al.* 1994).

Recent biochemical studies have confirmed earlier suggestions (Laurent *et al.* 1991; Peterson and Herskowitz 1992) that the yeast SWI and SNF proteins function as a multi-subunit complex. The SWI1, SWI2, SWI3, SNF5, and SNF6 polypeptides co-elute during gel filtration with an apparent native molecular mass of 2 MDa (Côté *et al.* 1994; Peterson *et al.* 1994), co-purify through several chromatographic steps, and are co-immunoprecipitated by antibodies to either SNF6, SWI3, or an epitope-tagged version of SWI2 (Cairns *et al.* 1994; Côté *et al.* 1994). Surprisingly, five additional polypeptides were found to co-purify and co-immunoprecipitate with the known SWI/SNF subunits. The cloning of genes for these proteins has yet to be reported and is awaited with interest. In the meantime, however, an eleventh component of the yeast SWI/SNF complex has been identified. Using a two-hybrid screen to identify proteins that interact with the *N*-terminal region of SWI2, Treich *et al.* (1995) isolated a gene, designated *SNF11*, that encodes a previously uncharacterized 19 kDa subunit of the SWI/SNF complex.

The chromatin connection

A connection between the function of the SWI/SNF complex and chromatin was initially established through genetic studies in *S. cerevisiae* in which mutations in several chromatin components were found to alleviate *swi/snf* phenotypes. For example, deletion of one of the two gene clusters that encode histones H2A and H2B, which is believed to result in a reduced level of these histones, was found to suppress defects in *SUC2* expression resulting from mutations in *SWI2*, *SNF5*, and *SNF6* (Hirschhorn *et al.* 1992). Mutations that inactivate the *SIN1* gene, which encodes a putative HMG-like component of chromatin, were also found to restore activator function in the absence of SWI or SNF gene products (Kruger and Herskowitz 1991). Furthermore, specific mutations in histones H3 and H4 were identified that bypass the need for a functional SWI/SNF complex (Prelich and Winston 1993; Kruger *et al*. 1995). This impressive collection of genetic data supports the notion that the SWI/SNF complex facilitates activator function by antagonizing the repressive actions of chromatin. Consistent with this hypothesis, Hirschhorn *et al*. (1992) found that SWI2 and SNF5 are required for changes in chromatin structure surrounding the *SUC2* promoter that occur during derepression of *SUC2* transcription.

The SWI/SNF complex is conserved in higher eukaryotes

The search for SWI/SNF homologues in higher eukaryotes was stimulated by the finding that the heterologous activators, *Drosophila* Ftz (Peterson and Herskowitz 1992), mammalian steroid receptors (Yoshinaga *et al*. 1992), and a LexA–Bicoid fusion protein (Laurent and Carlson 1992), require SWI/SNF proteins to facilitate transcription in yeast. One of the first relatives of SWI/SNF proteins to be discovered was a *Drosophila* homologue of SWI2 (Tamkun *et al*. 1992). This protein, designated *brahma* or *brm*, was initially identified in screens for dominant suppressors of mutations in Polycomb, a heterochromatin-like protein that represses homeotic gene transcription (Kennison and Tamkun 1988; see Chapter 14); thus a role for *brm* in regulating chromatin structure had previously been established. Subsequently, potential homologues of SWI2 have been identified in mice (Randazzo *et al*. 1994) and humans (Khavari *et al*. 1993; Muchardt and Yaniv 1993; Chiba *et al*. 1994). Interestingly, humans appear to have at least two genes which show close similarity to SWI2/*brm*: BRG1 (*brm*-related gene 1) and h*brm* (human *brm*). In addition, a human homologue of the yeast SNF5 gene has also been isolated (Kalpana *et al*. 1994). Intriguingly, this protein was cloned by its ability to bind tightly to HIV-1 integrase in a two-hybrid screen. Designated INI1 (integrase interactor 1), this protein shows considerable sequence homology to the yeast SNF5 polypeptide.

Perhaps the most compelling evidence that SWI/SNF complexes exist in higher eukaryotes has come from biochemical studies (Kwon *et al*. 1994). Two chromatographically separable complexes were isolated from HeLa cells with native molecular weights similar to that of the yeast complex. In addition, a separate study demonstrated that the human SNF5 homologue INI1 co-fractionated with BRG1 through five different chromatography columns and was also retained on a BRG1 immunoaffinity column (Kalpana *et al*. 1994). Collectively these data provide strong evidence for the presence of functional SWI/SNF complexes in higher eukaryotes.

The SWI2 subunit is a DNA-stimulated ATPase

The sequences of all known SWI/SNF subunits have been closely examined as a means of providing clues towards elucidating the function of the SWI/SNF complex. This search led to the discovery that SWI2 contains seven sequence motifs similar to those found in DNA-stimulated ATPases and DNA helicases (for an excellent review see Carlson and Laurent 1994). These sequence similarities suggested a provocative model in which the SWI/SNF complex functions as a helicase machine that reorganizes large chromatin domains via its processive DNA unwinding activity (Travers 1992). However, no discernible helicase activity could be detected in the purified yeast SWI/SNF complex (Côté *et al.* 1994). Furthermore, a reinspection of the SWI2 motifs suggested that these sequences bear more similarity to DNA-dependent ATPases than to DNA helicases (Henikoff 1993).

In contrast to the lack of helicase activity, SWI2 does exhibit DNA-stimulated ATPase activity. This was illustrated initially using a recombinant SWI2–maltose binding protein fusion purified from *Escherichia coli* (Laurent *et al.* 1993). Subsequently, the intact yeast and human SWI/SNF complexes have been demonstrated to have potent DNA-stimulated ATPase activities (Cairns *et al.* 1994; Côté *et al.* 1994; Kwon *et al.* 1994). Importantly, these ATPase motifs are crucial for SWI/SNF complex function. Single amino acid changes in several motifs eliminate SWI/SNF complex function *in vivo* (Laurent *et al.* 1993; Khavari *et al.* 1993; T.J. Richmond and C.L. Peterson, unpublished data) and *in vitro* (Côté *et al.* 1994).

Function of the SWI/SNF complex

Recent biochemical studies with purified yeast and partially purified human SWI/SNF complexes have provided the first direct biochemical link between the function of the complex and chromatin. The purified yeast complex caused a striking 10- to 30-fold stimulation in the binding of GAL4 derivatives to mononucleosomal DNA templates in a reaction that required ATP hydrolysis (Côté *et al.* 1994). Similarly, both of the partially purified human SWI/SNF complexes facilitated the ATP-dependent binding of TBP and GAL4 derivatives to nucleosomal DNA (Kwon *et al.* 1994; Imbalanzo *et al.* 1994). These studies suggest that a primary function of the SWI/SNF complex is to use the energy of ATP hydrolysis to drive transcription factors onto nucleosomal sites.

How does the SWI/SNF complex facilitate transcription factor binding to nucleosomal sites? From *in vitro* studies using both human and yeast complexes it is evident that the SWI/SNF complex does not lead to nucleosomal displacement, but promotes the formation of a ternary complex comprised of the transcription factor, DNA probe, and histone proteins (Côté *et al.* 1994; Imbalanzo *et al.* 1994; Kwon *et al.* 1994). Moreover, both human and yeast SWI/SNF complexes appear to interact directly with nucleosomal DNA and disrupt histone–DNA contacts. On DNA probes where the histone octamer is rotationally positioned, DNAase I digestion yields a diagnostic 10 bp repeat, reflecting the periodic accessibility of the minor groove of the DNA. However, addition of yeast or human SWI/SNF complex in the presence of ATP results in a severe disruption of this 10 bp repeat, which is indicative of a loss of helical positioning.

Genetic data also suggest that the SWI/SNF complex disrupts DNA–histone interactions, as point mutations in histones H3 and H4 that are believed to affect surface residues that interact with DNA alleviate the need for a functional SWI/SNF complex (Kruger *et al.* 1995).

In vitro studies with the yeast SWI/SNF complex found that low concentrations of nucleoplasmin, a histone chaperone with *in vivo* specificity for histones H2A and H2B, facilitated the SWI/SNF reaction, doubling the amount of GAL4-AH binding to the nucleosome (Côté *et al.* 1994). This observation suggested to Côté and colleagues that the yeast SWI/SNF complex may facilitate activator binding by directly or indirectly promoting loss of one or both H2A/H2B dimers from the GAL4 ternary complex. This model is consistent with earlier genetic data that illustrated that depletion of H2A/H2B dimers *in vivo* alleviates the transcriptional defects in *swi/snf* mutants (Hirschhorn *et al.* 1992).

A recent study, which examined the chromatin remodelling due to glucocorticoid induction of the MMTV promoter *in vivo*, suggests that H2A/H2B dimers are not lost from the regulatory nucleosome (Truss *et al.* 1995). This remodelling is believed to be due to SWI/SNF function (Yoshinaga *et al.* 1992; Muchardt and Yaniv 1993; Truss *et al.* 1995). In this study it was found that upon steroid induction the centre of the regulatory nucleosome ((Fig. 4.1(b), nucleosome B) becomes more accessible to DNAase I and restriction enzymes (indicative of disruption of DNA–histone contacts), but the 145 bp core region remains protected against micrococcal nuclease (MNase) digestion. Dissociation of H2A/H2B dimers has previously been reported to (i) make the dyad axis significantly more accessible to MNase, resulting in the formation of protected 73 bp fragments (Dong and van Holde 1991), or (ii) result in a decrease in the size of protected fragments from 145 to 120 bp when using hydroxy-radical cleavage reagents (Hayes *et al.* 1991). These observations suggest that at the regulatory nucleosome of the MMTV promoter the H2A/H2B dimers remain associated with the H3/H4 tetramer. Although not conclusive, this study by Truss *et al.* (1995) illustrates that further investigation is needed to directly test the 'dimer dissociation' model for SWI/SNF function.

How does the SWI/SNF complex utilize the energy of ATP hydrolysis to disrupt histone–DNA and histone–histone contacts? Our recent work has identified some novel properties of the yeast SWI/SNF complex, which may play a significant role in the disruption of chromatin structure (Quinn *et al.* 1996). We found that the SWI/SNF complex has a high affinity for DNA and that the DNA binding properties of the complex are highly reminiscent of those exhibited by HMG box domains. One of the most intriguing of these properties is the ability of the complex to introduce positive supercoils into plasmid DNA in the presence of topoisomerase I. One possibility is that the complex uses the energy of DNA binding to directly change the helical twist, resulting in an overwinding of the DNA. Such a change in helical twist would be incompatible with nucleosome stability, thus it is tempting to speculate that this property of the SWI/SNF complex is responsible for the disruption of DNA–histone contacts. Consistent with this view, the human SWI/SNF complex was shown to alter the linking number of a circular mini-chromosome (Kwon *et al.* 1994). In this case as well, the change in linking number was consistent with overwinding of DNA. However, the ability of the SWI/SNF complex to introduce topological changes on naked DNA templates does not require ATP hydrolysis (Quinn *et al.* 1996), whereas SWI/SNF-induced disruption of chromatin structure and changes in linking number of a mini-chromosome template are ATP dependent. These data suggest that the SWI/SNF complex requires the energy of ATP hydrolysis to specifically introduce structural changes into chromatin templates.

Targeting of the SWI/SNF complex

The yeast SWI/SNF complex is required for the expression of only a subset of genes and it has been estimated on the basis of yields from several purifications that there are only 50–100 copies of the SWI/SNF complex per yeast cell (Côté *et al.* 1994). These findings suggest that the SWI/SNF complex is not a general component of chromatin, and thus one outstanding question is how the complex is targeted to specific nucleosomal positions. One hypothesis is that the SWI/SNF complex associates with activator proteins, which then target SWI/SNF function to specific genes. This possibility is supported by a previous study in which it was shown that immunoprecipitation of the glucocorticoid receptor (GR) from wild-type *SWI*+ extracts co-precipitated a fraction of the SWI3 protein. Such receptor/SWI complexes were not detected in *swi1*− and *swi2*− strains, thus illustrating that GR is associated with the complex (Yoshinaga *et al.* 1992). However, understanding how the complex could interact with all SWI/SNF-dependent activators is an intriguing, and thus far unanswered question.

Alternatively, the complex may be targeted by directly binding to specific DNA sequences. As mentioned above, the yeast SWI/SNF complex has an intrinsic high affinity for DNA (Quinn *et al.* 1996). However, it is unclear whether this DNA binding activity plays a role in targeting of the SWI/SNF complex since the complex appears rather promiscuous on naked DNA templates. Nonetheless, the complex was also found to bind to structured DNAs. Thus it is possible that the complex exhibits a more marked specificity for specific DNA structures that are present at SWI/SNF-dependent genes in a chromatin context.

Identification of a novel factor from *Drosophila* embryos that mediates nucleosome disruption

Although the SWI/SNF complex is the most characterized chromatin remodelling activity to date, there is evidence that other molecular machines operate to remodel chromatin structure. For example, the recent development of *in vitro* systems designed to study nucleosome disruption has allowed for the identification of a potentially novel factor from *Drosophila* embryos that hydrolyses ATP to mediate nucleosome disruption. This system was first described by Tsukiyama *et al.* (1994), who developed an *in vitro* assay for nucleosome remodelling using a chromatin assembly extract from *Drosophila* embryos, plasmid DNA, and purified transcription factors. As a model system the *Drosophila hsp70* promoter was analysed. This promoter contains binding sites for the heat shock transcription factor, HSF, and GAGA factor, which is a constitutively expressed factor that binds to GA/CT-rich sites present in many *Drosophila* genes. Tsukiyama and colleagues found that introduction of recombinant GAGA factor during or after nucleosome assembly *in vitro* resulted in a disruption of nucleosome structure at the *hsp70* promoter (Tsukiyama *et al.* 1994). This disruption was characterized by DNAase I hypersensitivity at TATA and heat shock elements and a rearrangement of adjacent nucleosomes. Addition of polyclonal antibodies raised against bacterially expressed GAGA factor abolished the disruption of nucleosomes over the *hsp70* promoter, confirming that GAGA factor was critical for the observed remodelling activity. To investigate the energy requirements of this nucleosome disruption, the ATP-hydrolysing enzyme apyrase was added after completion of nucleosome assembly (an energy-requiring process; Becker and Wu 1992), but before introduction of GAGA factor. This treatment was found to suppress the changes in chromatin structure, but had no effect on GAGA factor DNA binding properties. Addition of fresh ATP restored the chromatin remodelling observed in the presence of GAGA factor, suggesting that this is an energy-dependent process.

One possibility is that GAGA factor itself possesses intrinsic chromatin remodelling activities. This fits in well with other previously characterized features of GAGA factor, such as its ability to act as a transcriptional antirepressor (Kerrigan *et al.* 1991), to establish the correct chromatin configuration at the *hsp26* promoter (Lu *et al.* 1993), and to function as an enhancer of position effect variegation (Farkas *et al.* 1994). However, GAGA factor has no recognizable ATP binding motif and no detectable ATP binding activity (Tsukiyama *et al.* 1994). Further doubt that GAGA factor can itself remodel chromatin in an ATP-dependent manner came from a related study on the *Drosophila hsp26* promoter (Wall *et al.* 1995). Using the same cell-free chromatin reconstitution system derived from *Drosophila* embryos, it was found that the transactivator HSF could also trigger energy-dependent nucleosome rearrangements in the same manner as that observed for GAGA factor. This observation may indicate that both of these transcription factors share a common function, however the lack of sequence similarity between these proteins argues against this. Alternatively, the increased access of both these factors to nucleosomal DNA may be due to a previously unrecognized chromatin remodelling activity present in the crude *Drosophila* assembly extract.

What is the nature of the observed chromatin disruption? When the *hsp70* plasmid DNA is reconstituted into chromatin, the MNase digestion pattern indicates that the DNA surrounding the *hsp70* promoter is arranged in a regularly spaced array of nucleosomes with a characteristic repeat length of 180 bp.

Addition of GAGA factor resulted in a dramatically different pattern of digestion surrounding the GAGA factor binding sites. The abundance of the 146 bp nucleosomal core fragment is significantly reduced and subnucleosomal fragments shorter than 146 bp are evident. Tsukiyama and colleagues propose that the generation of small fragments is indicative of an invasion and cleavage of the DNA within the nucleosome core particle by MNase, illustrating that the nucleosomal structure had been disrupted at the promoter region. The precise nature of the changes in nucleosomal structure, however, remains unclear (Tsukiyama *et al.* 1994).

Upon studying the *hsp26* promoter, Wall and colleagues (1995) found that the GAGA factor-induced changes in chromatin at the *hsp26* promoter appear identical in nature to that reported in analogous experiments at the *hsp70* promoter. They suggest, however, that the short nucleosomal fragments are due to protection of the DNA by the bound transcription factor. This notion was supported by MNase footprinting analysis, which revealed protection of GAGA factor-bound DNA from MNase digestion and an enhanced cutting at either side of the GAGA factor binding sites, effectively releasing short fragments. Wall *et al.* (1995) propose that the mechanism of chromatin remodelling may involve not nucleosome disruption but nucleosome sliding. They found that binding of HSF and GAGA factor to chromatin templates in the presence of ATP results in removal of nucleosomes from factor binding sites to a new location next to the bound factor. Furthermore, Varga-Weisz *et al.* (1995) have shown that ATP-dependent nucleosome sliding can occur in the absence of any transcription factor. In all studies nucleosome movements occurred even in the presence of histone H1. On the basis of these findings, Becker and colleagues propose that transcription factors, such as GAGA factor and HSF, take advantage of factor-mediated nucleosome movements to bind to the transient nucleosome-free regions formed and, once bound, function as boundaries that redirect nucleosome positions (Becker 1994). Importantly, the mobility of nucleosomes in this model does not depend on the interacting transcription factor. In this respect it is interesting that interaction of a third transcription factor, GAL4, with chromatin reconstituted in a *Drosophila* embryo 'S190' extract, also leads to an ATP-dependent rearrangement of nucleosome positions (Pazin *et al.* 1994). As seen with the GAGA factor and HSF proteins, this remodelling was not inhibited by histone H1. These similarities suggest that the same nucleosome remodelling activity is also present in 'S190' extracts.

The nature of the activity present in the *Drosophila* extracts that hydrolyses ATP to render nucleosomes mobile is presently unclear. The activity appears to be chromatin associated, since it is also present in chromatin immobilized on paramagnetic beads (Sandaltzopoulos *et al.* 1994). One possible candidate for this activity is the *Drosophila* SWI/SNF complex. The relationship between *Drosophila* SWI/SNF and this sliding activity is not known. Recent studies, however, with the yeast SWI/SNF complex indicate that it does not alter nucleosome positioning (J. Côté and J.L. Workman, personal communication). It is more likely that this is a distinct chromatin remodelling activity. Purification and further characterization of this novel activity is necessary to ascertain the role it plays in chromatin remodelling.

Future directions

We anticipate that one of the future directions in understanding how the eukaryotic cell alleviates chromatin-mediated repression of transcription will involve the identification of additional novel chromatin remodelling activities. Already there is growing evidence for multiple SWI/SNF-like complexes in various organisms. For example, there are four genes in the yeast data base that possess significant homology to *SWI2*. Multiple SWI2 homologues have also been identified in higher eukaryotes, including *Drosophila*, mice, and humans. In addition, biochemical analysis of the human SWI/SNF complex led to the identification of two chromatographically separable SWI/SNF complexes from HeLa cells (Kwon *et al.* 1994). We speculate that such multiple SWI/SNF-like complexes may either function to remodel chromatin using different mechanisms, or may respond to different regulatory signals. It is possible that other chromatin remodelling activities will be identified that are distinct from the SWI/SNF proteins. In this respect we eagerly await the purification and characterization of the recently identified activity in *Drosophila* extracts that functions to increase the mobility of nucleosomes.

Summary

One of the most exciting advances that has been made in understanding how the cell contends with chromatin repression has been the identification of activities that possess chromatin remodelling activities. One such activity, the SWI/SNF complex, uses the energy of ATP hydrolysis to distort histone–DNA contacts, thus driving transcription factors onto nucleosomal transcription factor binding sites. The huge size of the SWI/SNF complex (2 MDa, about half the size of a ribosome) suggests that it is truly a protein machine that plays several roles in chromatin remodelling and transcription factor function. In addition, recent studies have identified at least one distinct chromatin remodelling activity that uses the energy of ATP hydrolysis to increase the mobility of nucleosomes. Although this activity has yet to be purified, it is already evident that the eukaryotic cell employs different mechanisms to actively contend with repressive chromatin structures.

References

Almer, A. and Hörz , W. (1986). Nuclease hypersensitive region with adjacent positioned nucleosomes mark the gene boundaries of the PHO5/PHO3 locus in yeast. *EMBO Journal*, **5**, 2681–7.

Almer, A., Rudolph, H., Hinnen, A., and Hörz, W. (1986). Removal of positioned nucleosomes from the yeast PHO5 promoter upon PHO5 induction releases additional upstream activating DNA elements. *EMBO Journal*, **5**, 2689–96.

Archer, T.K., Cordingley, M.G., Wolford, R.G., and Hager, G.L. (1991). Transcription factor access is mediated by accurately positioned nucleosomes on the mouse mammary tumor virus promoter. *Molecular and Cellular Biology*, **11**, 688–98.

Barton, M.C. and Emerson, B.M. (1994). Regulated expression of the β-globin gene locus in synthetic nuclei. *Genes and Development*, **8**, 2453–65.

Barton, M.C. and Emerson, B.M. (1995). Regulated expression of the beta-globin locus in synthetic nuclei. *Genes and Development* **8**, 2453–65.

Becker, P.B. (1994). The establishment of active promoters in chromatin. *BioEssays*, **16**, 541–7.

Becker, P.B. and Wu, C. (1992). Cell-free system for assembly of transcriptionally repressed chromatin from *Drosophila* embryos. *Molecular and Cellular Biology*, **12**, 2241–9.

Brüggemeier, U., Rogge, L., Winnacker, E.L., and Beato, M. (1990). Nuclear Factor I acts as a transcription factor on the MMTV promoter but competes with steroid hormone receptors for DNA binding. *EMBO Journal*, **9**, 2233–9.

Buchman, A.R., Lue, N.F., and Kornberg, R.D. (1988). Connections between transcriptional activators, silencers and telomeres as revealed by functional analysis of a yeast DNA-binding protein. *Molecular and Cellular Biology*, **8**, 5086–99.

Cairns, B.R., Kim, Y-J., Sayre, M.H., Laurent, B.C., and Kornberg, R.D. (1994). A multisubunit complex containing the *SWI1/ADR6*, *SWI2/SNF2*, *SWI3*, *SNF5*, and *SNF6* gene products isolated from yeast. *Proceedings of the National Academy of Sciences of the USA*, **91**, 1950–4.

Carlson, M. and Laurent, B.C. (1994). The SNF/SWI family of global transcriptional activators. *Current Opinion in Cell Biology*, **6**, 396–402.

Chasman, D.I., Lue, N.F., Buchman, A.R., LaPointe, J.W., Lorch, Y., and Kornberg, R.D. (1990). A yeast protein that influences the chromatin structure of UAS$_G$ and functions as a powerful auxiliary gene activator. *Genes and Development*, **4**, 503–14.

Chiba, H., Muramatsu, M., Nomoto, A., and Kato, H. (1994). Two human homologues of *Saccharomyces cerevisiae SWI2/SNF2* and *Drosophila brahma* are transcriptional coactivators cooperating with the estrogen receptor and the retinoic acid receptor. *Nucleic Acids Research*, **22**, 1815–20.

Clark-Adams, C.D., Norris, D., Osley, M.A., Fassler, J.S., and Winston, F. (1988). Changes in histone gene dosage alter transcription in yeast. *Genes and Development*, **2**, 150–9.

Côté, J., Quinn, J., Workman, J.L., and Peterson, C.L. (1994). Stimulation of GAL4 derivative binding to nucleosomal DNA by the yeast SWI/SNF complex. *Science*, **265**, 53–60.

Dasso, M., Dimitrov, S., and Wolffe, A.P. (1994). Nuclear assembly is independent of linker histones. *Proceedings of the National Academy of Sciences of the USA*, **91**, 12477–81.

Dong, F. and van Holde, K.E. (1991). Nucleosome positioning is determined by the (H3-H4)$_2$ tetramer. *Proceedings of the National Academy of Sciences of the USA*, **88**, 10596–600.

Durrin, L.K., Mann, R.K., and Grunstein, M. (1992). Nucleosome loss activates *CUP1* and *HIS3* promoters to fully induced levels in the yeast *Saccharomyces cerevisiae*. *Molecular and Cellular Biology*, **12**, 1621–9.

Elgin, S.C.R. (1981). DNase I hypersensitive sites of chromatin. *Cell*, **27**, 413–15.

Farkas, G., Gausz, J. Galloni, M., Reuter, G., Gyurkovics, H., and Karch, F.(1994). The *Trithorax-like* gene encodes the *Drosophila* GAGA factor. *Nature*, **371**, 806–8.

Fedor, M.J., Lue, N.F., and Kornberg, R.D. (1988). Statistical positioning of nucleosomes by specific protein-binding to an upstream activating sequence in yeast. *Journal of Molecular Biology*, **204**, 109–27.

Felsenfeld, G. (1992). Chromatin as an essential part of the transcription mechanism. *Nature*, **355**, 219–24.

Fletcher, T.M. and Hansen, J.C. (1995). Core histone tail domains mediate oligonucleosome folding and nucleosomal DNA organization through distinct mechanisms. *Journal of Biological Chemistry*, **270**, 25359–62

Fletcher, T.M., Serwer, P., and Hansen, J.C. (1994). Quantitative analysis of macromolecular conformational changes using agarose gel electrophoresis: application to chromatin folding. *Biochemistry*, **33**, 10859–63.

Gross, D.S. and Garrard, W.T. (1988). Nuclease hypersensitive sites in chromatin. *Annual Reviews in Biochemistry*, **57**, 159–97.

Grunstein, M. (1992). Histones as regulators of genes. *Scientific American*, **267**, 68–74.

Han, M. and Grunstein, M. (1988). Nucleosome loss activates yeast downstream promoters *in vivo*. *Cell*, **55**, 1137–45.

Han, M., Kim, U-J., Kayne, P., and Grunstein, M. (1988). Depletion of histone H4 and nucleosomes activates the *PHO5* gene in *Saccharomyces cerevisiae*. *EMBO Journal*, **7**, 2221–8.

Hansen, J.C. and Wolffe, A.P. (1992). Influence of chromatin folding on transcription initiation and elongation by RNA polymerase III. *Biochemistry*, **31**, 7977–88.

Hansen, J.C. and Wolffe, A.P. (1994). A role for histones H2A/H2B in chromatin folding and transcriptional repression. *Proceedings of the National Academy of Sciences of the USA*, **91**, 2339–43.

Hansen, J.C. and Lohr, D. (1993). Assembly and structural properties of subsaturated chromatin arrays. *Journal of Biological Chemistry*, **268**, 5840–8.

Hansen, J.C., Ausio, J., Stanik, V.H., and van Holde, K.E. (1989). Homogeneous reconstituted oligonucleosomes, evidence for salt-dependent folding in the absence of histone H1. *Biochemistry*, **28**, 9129–36.

Hayes, J.J. and Wolffe, A.P. (1992). The interaction of transcription factors with nucleosomal DNA. *BioEssays*, **14**, 597–603.

Hayes, J.J., Clark, D.J., and Wolffe, A.P. (1991). Histone contributions to the structure of DNA in the nucleosome. *Proceedings of the National Academy of Sciences of the USA*, **88**, 6829–33.

Henikoff, S. (1993). Transcriptional activator components and poxvirus DNA-dependent ATPases comprise a single family. *Trends in Biochemical Sciences*, **18**, 291–2.

Hirschhorn, J.N., Brown, S.A., Clark, C.D., and Winston, F. (1992). Evidence that SNF2/SWI2 and SNF5 activate transcription in yeast by altering chromatin structure. *Genes and Development*, **6**, 2288–98.

Igokemenes, T., Hörz, W., and Zachau, H.G. (1982). Chromatin. *Annual Reviews in Biochemistry*, **51**, 89–121.

Imbalzano, A.N., Kwon, H., Green, M. R., and Kingston, R.E. (1994). Facilitated binding of TATA-binding protein to nucleosomal DNA. *Nature*, **370**, 481–5.

Jackson, J.R. and Benyajati, C. (1993). DNA–histone interactions are sufficient to position a single nucleosome juxtaposing *Drosophila Adh* adult enhancer and distal promoter. *Nucleic Acids Research*, **21**, 957–67.

Kalpana, G.V., Marmon, S., Wang, W., Crabtree, G.R., and Goff, S.P. (1994). Binding and stimulation of HIV-1 integrase by a human homolog of yeast transcription factor SNF5. *Science*, **266**, 2002–6.

Kennison J.A. and Tamkun, J.W. (1988). Dosage-dependent modifiers of homeotic mutations in *Drosophila melanogaster. Genetics*, **116**, 75–86.

Kerrigan, L.A., Croston, G.E., Lira, L.M., and Kadonaga, J.T. (1991). Sequence-specific transcriptional antirepression of the *Drosophila Krüppel* gene by the GAGA factor. *Journal of Biological Chemistry*, **266**, 574–82.

Khavari, P.A., Peterson, C.L., Tamkun, J.W., Mendel, D.B., and Crabtree, G.R. (1993). BRG1 contains a conserved domain of the *SWI2/SNF2* family necessary for normal mitotic growth and transcription. *Nature*, **366**, 170–4.

Kim, U-J., Han, M., Kayne, P., and Grunstein, M. (1988). Effects of histone H4 depletion on the cell cycle and transcription of *Saccharomyces cerevisiae. EMBO Journal*, **7**, 2211–19.

Knezetic, J.A., Jacob, G.A., and Luse, D.S. (1988). Assembly of RNA polymerase II initiation complexes before assembly of nucleosomes allows efficient initiation of transcription on nucleosomal templates. *Molecular and Cellular Biology*, **8**, 3114–21.

Kornberg, R. (1974). Chromatin structure: a repeating unit of histones and DNA. *Science*, **184**, 868–71.

Kruger, W. and Herskowitz, I. (1991). A negative regulator of *HO* transcription, SIN1 (SPT2), is a nonspecific DNA-binding protein related to HMG1. *Molecular and Cellular Biology*, **11**, 4135–46.

Kruger, W., Peterson, C., Sil, A., Coburn, C., Arents, G., Moudrianakis, E., *et al.* (1995). Residues in the globular domains of histones H3 and H4 are necessary for proper transcriptional regulation by the yeast SWI/SNF complex. *Genes and Development*, **9**, 2770–9.

Kwon, H., Imbalzano, A.N., Khavari, P.A., Kingston, R.E., and Green, M.R. (1994). Nucleosome disruption and enhancement of activator binding by a human SWI/SNF complex. *Nature*, **370**, 477–81.

Laurent, B.C. and Carlson, M. (1992). Yeast SNF2/SWI2, SNF5, and SNF6 proteins function co-ordinately with the gene-specific functional activators GAL4 and Bicoid. *Genes and Development*, **6**, 1707–15.

Laurent, B.C., Treitel, M.A., and Carlson, M. (1991). Functional interdependence of the yeast SNF2, SNF5 and SNF6 proteins in transcriptional activation. *Proceedings of the National Academy of Sciences of the USA*, **88**, 2687–91.

Laurent, B.C., Treich, I., and Carlson, M. (1993). The yeast SNF2/SWI2 protein has DNA-stimulated ATPase activity required for transcriptional activation. *Genes and Development*, **7**, 583–91.

Laybourn P.J. and Kadonaga, J.T. (1991). Role of nucleosomal cores and histone H1 in regulation of transcription by RNA polymerase II. *Science*, **254**, 238–45.

Lewin, B. (1994). Chromatin and gene expression: constant questions but changing answers. *Cell*, **79**, 397–406.

Lorch, Y., LaPointe, J.W., and Kornberg, R.D. (1987). Nucleosomes inhibit the initiation of transcription but allow chain elongation with the displacement of histones. *Cell*, **49**, 203–10.

Lorch, Y., LaPointe, J.W., and Kornberg, R.D. (1992). Initiation on chromatin templates in a yeast RNA polymerase II transcription system. *Genes and Development*, **6**, 2282–7.

Lu, Q., Wallrath, L.L., Granok, H., and Elgin, S.C.R. (1993). (CT)n·(GA)n repeats and heat shock elements have distinct roles in chromatin structure and transcriptional activation of the *Drosophila hsp26* gene. *Molecular and Cellular Biology*, **13**, 2802–14.

Muchardt, C. and Yaniv, M. (1993). A human homologue of *Saccharomyces cerevisiae SNF2/SWI2* and *Drosophila brm* genes potentiates transcriptional activation by the glucocorticoid receptor. *EMBO Journal*, **12**, 4279–90.

Ohsumi, K., Katagiri, C., and Kishimoto, T. (1993). Chromosome condensation in *Xenopus* mitotic extracts without histone H1. *Science*, **262**, 2033–5.

Paranjape, S.M., Kamakaka, R.T., and Kadonaga, J.T. (1994). Role of chromatin structure in the regulation of transcription by RNA polymerase II. *Annual Reviews in Biochemistry*, **63**, 265–97.

Pazin, M.J., Kamakaka, R.T., and Kadonaga, J.T. (1994). ATP-dependent nucleosome reconfiguration and transcriptional activation from preassembled chromatin templates. *Science*, **266**, 2007–11.

Perlmann, T. and Wrange, Ö. (1988). Specific glucocorticoid receptor binding to DNA reconstituted in a nucleosome. *EMBO Journal*, **7**, 3073–9.

Peterson, C.L. (1994). Contending with chromatin-mediated repression. Genetic approaches. In *Transcription mechanisms and regulation* (ed. R.C. Conway and J.W. Conway), pp. 535–55. Raven Press, New York.

Peterson, C.L. and Herskowitz, I. (1992). Characterisation of the yeast *SWI1*, *SWI2*, and *SWI3* genes, which encode a global activator of transcription. *Cell*, **68**, 573–83.

Peterson, C.L. and Tamkun, J.W. (1995). The SWI–SNF complex: a chromatin remodeling machine? *Trends in Biochemical Sciences*, **20**, 143–6.

Peterson, C.L., Dingwall, A., and Scott, M.P. (1994). Five *SWI/SNF* gene products are components of a large multisubunit complex required for transcriptional enhancement. *Proceedings of the National Academy of Sciences of the USA*, **91**, 2905–8.

Piña, B., Brüggemeier, U., and Beato, M. (1990). Nucleosome positioning modulates accessibility of regulatory proteins to the mouse mammary tumor virus promoter. *Cell*, **60**, 719–31.

Prelich, G. and Winston, F. (1993). Mutations that suppress the deletion of an upstream activating sequence in yeast: involvement of a protein kinase and histone H3 in repressing transcription *in vivo*. *Genetics*, **135**, 665–76.

Quinn, J., Fyrberg, A.M., Ganster, R.W., Schmidt, M. C., and Peterson, C.L. (1996). DNA binding properties of the yeast SWI/SNF complex. *Nature*, **379**, 844–7.

Randazzo, F.M., Khavari, P., Crabtree, G., Tamkun, J., and Rossant, J. (1994). *brg1*: a putative murine homologue of the *Drosophila brahma* gene, a homeotic gene regulator. *Developmental Biology*, **161**, 229–42.

Richard-Foy, H. and Hager, G.L. (1987). Sequence-specific positioning of nucleosomes over the steroid-inducible MMTV promoter. *EMBO Journal*, **6**, 2321–8.

Saffer, J.D., Jackson, S.P., and Annarella, M.B. (1991). Developmental expression of Sp1 in the mouse. *Molecular and Cellular Biology*, **11**, 2189–99.

Sandaltzopoulos, R., Blank, T., and Becker, P.B. (1994). Transcriptional repression by nucleosomes but not H1 in reconstituted preblastoderm *Drosophila* chromatin. *EMBO Journal*, **13**, 373–9.

Schild, C., Claret, F-X., Wahli, W., and Wolffe, A.P. (1993). A nucleosome dependent static loop potentiates estrogen-regulated transcription from the *Xenopus* vitellogenin B1 promoter *in vitro*. *EMBO Journal*, **12**, 423–33.

Schmid, A., Fascher, K.D., and Hörz, W. (1992). Nucleosome disruption at the yeast *PHO5* promoter upon *PHO5* induction occurs in the absence of DNA replication. *Cell*, **71**, 853–64.

Shen, X., Yu, L., Weir, J.W., and Gorovsky, M.A. (1995). Linker histones are not essential and affect chromatin condensation *in vivo*. *Cell*, **82**, 47–56.

Tamkun, J.W., Deuring, R., Scott, M.P., Kissinger, M., Pattatucci, A.M., Kaufman, T.C., *et al.* (1992). brahma: a regulator of *Drosophila* homeotic genes structurally related to the yeast transcriptional activator SNF2/SWI2. *Cell*, **68**, 561–72.

Taylor, I.C.A., Workman, J.L., Schuetz, T.J,. and Kingston, R.E. (1991). Facilitated binding of GAL4 and heat shock factor to nucleosomal templates: differential function of DNA binding domains. *Genes and Development*, **5**, 1285–98.

Thomas, G.H. and Elgin, S.C.R. (1988). Protein/DNA architecture of the DNaseI hypersensitive region of the *Drosophila hsp26* promoter. *EMBO Journal*, **7**, 7291–301.

Travers, A.A. (1992). The reprogramming of transcriptional competence. *Cell*, **69**, 573–5.

Treich, I., Cairns, B.R., Santos, T., Brewster, E., and Carlson, M. (1995). SNF11, a new component of the yeast SWI/SNF complex that interacts with a conserved region of SWI2. *Molecular and Cellular Biology*, **15**, 4240–8.

Truss, M., Bartsch, J., Schelbert, A., Haché, R.J.G., and Beato, M. (1995). Hormone induces binding of receptors and transcription factors to a rearranged nucleosome on the MMTV promoter *in vivo*. *EMBO Journal*, **14**, 1737–51.

Tsukiyama, T., Becker, P.B., and Wu, C. (1994). ATP-dependent nucleosome disruption at a heat-shock promoter mediated by binding of GAGA transcription factor. *Nature*, **367**, 525–32.

van Holde, K.E. (1989). *Chromatin*. Springer-Verlag, New York.

Varga-Weisz, P.D., Blank, T.A., and Becker, P.B. (1995). Energy-dependent chromatin accessibility and nucleosome mobility in a cell-free system. *EMBO Journal*, **14**, 2209–16.

Venter, U., Svaren, J., Schmitz, J. Schmid, A., and Hörz, W. (1994). A nucleosome precludes binding of the transcription factor Pho4 *in vivo* to a critical target site in the *PHO5* promoter. *EMBO Journal*, **13**, 4848–55.

Vogel, K., Hörz, W., and Hinnen, A. (1989). The two positively acting regulatory proteins PHO2 and PHO4 physically interact with the *PHO5* upstream activation regions. *Molecular and Cellular Biology*, **9**, 2050–7.

Wall, G., Varga-Weisz, P.D., Sandaltzopoulos, R., and Becker, P.B. (1995). Chromatin remodeling by GAGA factor and heat shock factor at the hypersensitive *Drosophila* hsp26 promoter *in vitro*. *EMBO Journal*, **14**, 1727–36.

Wallrath, L.L., Lu, Q., Granok, H., and Elgin, S.C.R. (1994). Architectural variations of inducible promoters: preset and remodeling chromatin structures. *BioEssays*, **16**, 165–70.

Winston, F. and Carlson, M. (1992). Yeast SNF/SWI transcriptional activators and the SPT/SIN chromatin connection. *Trends in Genetics*, **8**, 387–91.

Workman, J.L. and Buchman, A.R. (1993). Multiple functions of nucleosomes and regulatory factors in transcription. *Trends in Biochemical Sciences*, **18**, 90–5.

Workman, J.L. and Roeder, R.G. (1987). Binding of transcription factor TFIID to the major late promoter during *in vitro* nucleosome assembly potentiates subsequent initiation by RNA polymerase II. *Cell*, **51**, 613–22.

Workman, J.L., Taylor, I.C.A., and Kingston, R.E. (1991). Activation domains of stably bound GAL4 derivatives alleviate repression of promoters by nucleosomes. *Cell*, **64**, 533–44.

Yoshinaga, S.K., Peterson, C.L., Herskowitz, I., and Yamamoto, K.R. (1992). Roles of

SWI1, SWI2, and SWI3 proteins for transcriptional enhancement by steroid receptors. *Science*, **258**, 1598–604.

Part II

Higher order structure of the genome
and the nucleus

5

Chromatin domains and boundary elements

Victor G. Corces and Tatiana I. Gerasimova

Introduction

The temporal and spatial expression of eukaryotic genes is regulated by transcription factors through their interaction with enhancer elements. After this initial interaction, enhancer-bound transcription factors cause changes in the transcription complex assembled on the promoter to activate transcription (Thompson and McKnight 1992). The mechanisms of this activation are not well understood and several alternative mechanisms have been put forward to explain this process. Transcriptional activation might involve direct protein–protein interactions through the looping of intervening sequences that separate enhancers and promoters, tracking of the transcription factors from the enhancers to the promoter, changes in chromatin structure that take place upon factor binding to its target sequence and can be transmitted to the promoter to activate transcription, or localization to a particular nuclear compartment that is permissive for gene expression. Independent of the particular details of the process, the intrinsic nature of enhancer–promoter interactions lacks the necessary specificity required for the precise temporal and spatial patterns of gene expression that underlies the development of eukaryotic organisms. To account for the observed specificity in enhancer–promoter interactions it has been proposed that the eukaryotic genome has organizational properties that rely on the ability of chromatin to establish autonomous functional units delimiting levels and patterns of gene expression. Functional specificity of transcriptional enhancers might then be imposed by the existence of chromosomal domains, discrete and topologically independent units of gene expression in which transcription from a specific promoter results only from activation by enhancers located within the same domain, whereas enhancers in a second domain are unable to act on a promoter located in the first one.

Identifying boundary elements that define chromatin domains

The possible existence of structural compartments charged with the function of organizing the DNA within chromosomes into higher order domains was originally inferred from cytological studies on insect polytene chromosomes and the lampbrush chromosomes of amphibian oocytes. The characteristic and reproducible banding pattern of polytene chromosomes is suggestive of an underlying structural organization perhaps imposed by the DNA sequence on the higher order organization of chromatin. This specific structural layout appears to form the foundation for the functional organization of the chromosome as suggested by the correlation between transcriptional activation and decondensation of particular polytene bands (Tissières *et al.* 1974; Lewis *et al.* 1975). Similarly, the finding of active genes in the loops of lampbrush chromosomes was taken as an early indication of a direct relationship between activation of gene expression and location within a specific structural chromosomal domain (Callan 1986). Since these early observations, results from experiments designed to identify regions of functional importance in the control of gene expression have given further support to the idea of a domain organization of eukaryotic chromosomes. Analyses of enhancer activity in assays that involve integration of the gene into the host chromosome have given results that depend on the integration site and thus support the idea that the organization of the chromatin in the region surrounding the transgene determines its ability to be transcribed. Germ line transformation of exogenous DNA into cultured cells and transgenic animals has then been used as a functional assay to identify *cis*-acting sequences that can provide transcriptional activation independent of the integration site within the genome (Grosveld *et al.* 1987). Sequences thus identified have been termed locus control regions (LCRs) and boundary or insulator elements. LCRs define active domains of gene expression in a dominant fashion and are required for proper tissue-specific regulated transcription of many vertebrate genes (see Chapter 8). The second type of domain indicator, a boundary or insulator, is defined as a sequence that ensures position-independent transcription by insulating the expression of a gene from the effect of neighbouring sequences. These buffering properties of chromatin insulators presumably depend on their ability to separate two regions with distinct chromatin structures: the chromatin is compacted and inaccessible to nucleases on one side of the boundary, whereas the other side presents an open chromatin conformation permissive to transcriptional activation. This property of boundary elements is shown for a generic gene in Fig. 5.1.

Based on this functional property, the standard assay to determine boundary function for a specific sequence is to analyse whether this sequence confers position-independent expression after integration into a chromosome. The *white* gene of *Drosophila* has been used routinely as a sensitive reporter transgene because alterations in the level of expression caused by repressive position effects of flanking sequences translate into easily measurable changes in eye pigmentation that vary between red (wild type) and white (null mutant) (Kellum and Schedl 1991). When boundary elements are placed flanking this gene, expression is normal and transformed flies show red eye coloration. The sensitivity of this gene to position effects that enhance gene expression can also be measured in transgenes that carry a minimal *white* gene lacking most introns

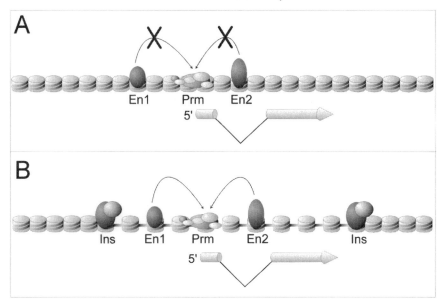

Fig. 5.1 Insulator elements buffer gene expression from repressive effects of adjacent chromatin. En1 and En2 represent two different enhancers and their associated transcription factors bound to nucleosomal DNA. Prm is the promoter of the gene where the different components of the transcription complex are present. Ins is an insulator element with its associated proteins. Arrows indicate a positive activation of transcription by the enhancer element; a cross over the arrow indicates a repression of this effect. (a) A transgene integrated in the chromosome in a region of condensed chromatin is not properly expressed. The repressive chromatin structure of the surrounding region presumably spreads into transgene sequences inhibiting enhancer–promoter interactions. (b) If the transgene is flanked by insulator elements, these sequences inhibit the spreading of the repressive chromatin conformation, allowing normal transcription of the gene.

and containing only approximately 300 bp of 5' flanking sequences. Expression of this mini-*white* gene varies between white (when repressed by adjacent sequences) and red (when activated by positive regulatory elements). When boundary elements or insulator sequences are placed flanking this mini-*white* gene, transgenic flies show the orange eye coloration expected if the *white* gene was buffered from position effects (Kellum and Schedl 1991, 1992; Roseman *et al.* 1993). A second property of boundary and insulator elements is their ability to interfere with the activation of a promoter by a specific enhancer when the insulator is located in between (Fig. 5.2(a) and (b)). This property of insulators conceptually separates boundary elements from classical repressors, since an insulator has no effect on transcription when placed more distally from the promoter than the enhancers that control expression of the gene (Fig. 5.2(c)). In the following sections we will review the functional properties of three well-characterized boundary elements: the scs and scs' sequences of *Drosophila*, a

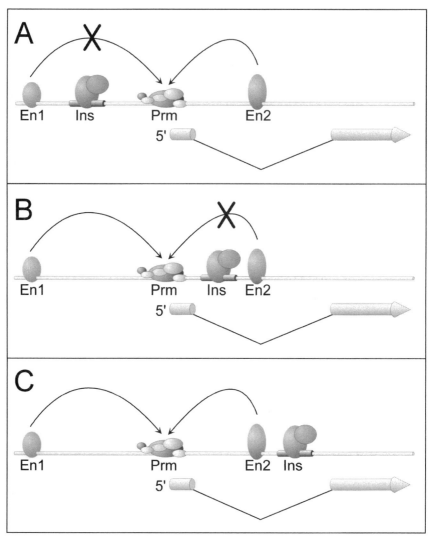

Fig. 5.2   Polar effect of an insulator on enhancer–promoter interactions. Symbols are as in Fig. 5.1. (a) An insulator located in the 5' region of the gene inhibits its transcriptional activation by an upstream enhancer (En1) without affecting the function of a second enhancer (En2) located in the intron of the gene. (b) When the insulator is located in the intron, expression from the downstream enhancer (En2) is blocked, whereas the upstream enhancer (En1) is still active. (c) When the insulator is located in the intron but distal to the En2 enhancer, both enhancers are active and transcription of the gene is normal. This property distinguishes an insulator from a typical repressor.

boundary element present in the chicken β-globin gene, and the insulator element of the *Drosophila* gypsy retrotransposon.

The scs and scs' boundary elements of the *Drosophila* hsp70 locus

The *Drosophila* heat shock locus at 87A contains two divergently transcribed hsp70 genes separated by short intergenic sequences and it is flanked by two unusual chromatin structures that display properties expected of domain boundaries. These specialized chromatin structures, termed scs and scs', are located at the edges of the puff that arises by induction of gene expression by heat shock, suggesting that they might demarcate the extent of chromatin that decondenses upon induction of transcription by temperature elevation (Udvardy *et al.* 1985). This idea is also in line with the chromatin organization observed in these sequences: each of these elements is defined by two strong nuclease-hypersensitive sites surrounding a nuclease-resistant core and this central structure is flanked by additional weaker nuclease cleavage sites present at approximately nucleosome-length intervals. This specific chromatin organization is altered by transcriptional activation of the heat shock genes (Udvardy *et al.* 1985).

The boundary properties of the scs and scs' sequences have been established by measuring their ability to insulate the *Drosophila* hsp70 promoter from the effect of the yolk protein *yp1* enhancer in a hybrid hsp70–lacZ reporter gene construct. This chimeric gene expresses β-galactosidase in the female fat body under the influence of the *yp1* enhancer, but this effect is abolished in transformed flies when scs sequences are interposed between the enhancer and the promoter (Kellum and Schedl 1991, 1992). Considerable progress has been made in dissecting the components of these boundary elements. The role of specific sequences and particular chromatin structures, contained within the originally defined scs element, has been determined by testing the ability of specific subfragments to repress the activation by the eye and testis enhancers of the *white* gene on a downstream promoter (Vazquez and Schedl 1994). Results from this analysis have led to the conclusion that sequences associated with the DNAase I-hypersensitive sites are essential for complete blocking activity of enhancer function, whereas the central nuclease-resistant AT-rich region is dispensable for this effect. Deletion of sequences associated with some hypersensitive sites leads to a reduction in enhancer blocking, whereas multimerization of subfragments with partial activity restores full boundary function (Vazquez and Schedl 1994). This result suggests that the effect of boundary elements on transcription might not be the result of the presence of specific proteins bound to insulator sequences, but rather a critical number of relatively non-specific proteins that could cause chromatin alterations as a consequence of their interaction with DNA.

The nature of the protein components present on boundary elements has been determined in the case of the scs' sequences (Zhao *et al.* 1995). A 32 kDa protein termed BEAF-32 (for boundary element-associated factor of 32 kDa) has been purified from nuclei of a *Drosophila* cell line. This protein binds to a palindromic sequence that flanks the two hypersensitive regions in the scs' sequence. This sequence acts as a typical boundary element in an enhancer-blocking assay involving stable transformation into cultured cells of a reporter gene containing

CAT under the control of ecdysone response and heat shock regulatory elements. The BEAF-32 binding site blocks the activity of both heat shock and ecdysone responsive enhancers in stably but not transiently transfected cells, although the effect on the heat shock activation is less dramatic (Zhao *et al.* 1995). Immunolocalization of BEAF-32 shows the presence of this protein in specific subnuclear regions and its exclusion from the nucleolus. BEAF-32 is present in the interband region that separates the highly reproducible and characteristic polytene bands of *Drosophila* third instar larval chromosomes. Interbands contain lower amounts of DNA than bands and are presumed to be regions of partial unfolding of the 30 nm chromatin fibre. As expected, BEAF-32 is present at the scs'-containing border of the 87A7 chromomere and is also found at one of the edges of many developmental puffs typically seen in polytene chromosomes at this stage of larval development (Zhao *et al.* 1995). This observation suggests that BEAF-32 might have general structural and functional roles in defining many boundary elements throughout the *Drosophila* genome. The specific one-side staining of these domains is suggestive of a bipartite polar organization of boundary elements in which a domain might be established through interactions between proteins of varied nature present on different chromosomal boundaries. This characteristic of boundary elements opens the possibility for a great versatility in the nature of chromosomal domains that can be formed in various tissues and at different times during the development of an organism, allowing highly dynamic patterns of gene expression.

Boundary elements in the chicken β-globin locus

The enhancer-blocking assay has also been used to analyse the insulating properties of the 5' constitutive hypersensitive site present in the chicken β-globin locus adjacent to the LCR (Chung *et al.* 1993). This LCR acts as an activator of expression of all globin genes present in the domain by decondensing the higher order chromatin structure of this region in erythroid tissues (Elder *et al.* 1990; see Chapter 8). This decondensation extends for several hundred kilobase pairs in the 3' direction, but is limited in the 5' direction by the location of a strong constitutive hypersensitive site, suggesting that this site might be the natural boundary of the β-globin domain. In agreement with this hypothesis, a 1.2 kb DNA fragment containing the 5' constitutive hypersensitive site can insulate a reporter gene from the effects of the LCR. When a reporter gene containing the β-globin promoter and the coding region of the bacterial neomycin gene is stably transformed into human erythroleukaemia cells, the presence of the LCR results in an increase in reporter gene expression, as well as a 30- to 100-fold increase in the number of resistant colonies obtained (Chung *et al.* 1993). Flanking the reporter gene with the chicken 5' element caused a 10-fold reduction in the number of resistant colonies and this reduction was even more dramatic when additional copies of the 5' element were used, leading to an almost complete insulation of LCR activation. This effect is observed only when the 5' element is placed between the LCR and the reporter gene, but not when located outside, supporting the premise that this sequence behaves as a typical insulator.

Insights into the mechanisms by which the chicken globin insulator represses LCR function have come from comparisons of restriction enzyme site accessibility in the promoters of the reporter and endogenous β-globin gene. The effect of the LCR on the expression of the genes in the β-globin domain is mediated by changes in chromatin structure that create an altered nucleosome structure or nucleosome-free region over the promoters of these genes. In particular, an *Apa*I restriction site present in the β-globin promoter is inaccessible to the enzyme in HeLa cells, where the gene is inactive presumably due to the presence of a nucleosome over these sequences. In cells in which the β-globin gene is active, the restriction site becomes accessible due to the displacement of this nucleosome. A similar analysis carried out on the promoter of a transformed reporter gene under the control of the LCR shows a decrease in *Apa*I accessibility when the reporter gene is separated from the LCR by insulator sequences. These results indicate that the interference of the chicken globin insulator with LCR–promoter interactions involves alterations of chromatin structure that block nucleosome displacement over the promoter region (Chung *et al.* 1993).

Interestingly, the chicken globin insulator can also act as a boundary in *Drosophila*. When a *white* mini-gene lacking most of its regulatory sequences is stably integrated into the germ line by P element-mediated transformation, transformed flies show a variable eye coloration that depends on the integration site. If this *white* gene is flanked by copies of the chicken globin insulator, transformants have an uniform orange eye colour, suggesting that the chicken sequences can buffer *white* gene expression from repressive effects of adjacent chromatin (Chung *et al.* 1993). This finding is highly significant, since it suggests a dramatic conservation in the protein components of chromatin insulators, as well as their mechanism of action across species as divergent as *Drosophila* and chickens.

The gypsy insulator of *Drosophila*

Sequences present in the 5' transcribed untranslated region of gypsy can repress transcription from specific enhancers and this property is intimately related to the mechanism by which this retrotransposon causes mutations. Insertion of gypsy into different *Drosophila* genes gives rise to a variety of phenotypes that affect only specific tissues at particular times of development. These mutant phenotypes arise as a consequence of the presence within gypsy of a sequence that is homologous to the octamer motif present in enhancer and promoter elements of many vertebrate genes (Geyer *et al.* 1988; Pfeifer and Bender 1988). This sequence acts as a typical boundary element and interferes with the action of specific enhancers, which are responsible for the expression of the mutant gene in the affected tissues, on the promoter of the gene (Jack *et al.* 1991; Holdridge and Dorsett 1991; Geyer and Corces 1992). This effect displays a polarity expected of an insulator element: only enhancers located distally, either upstream or downstream, from the promoter with respect to the insertion site of the gypsy insulator are affected (Fig. 5.2). The *Drosophila* genome contains approximately 200 of these boundary elements, randomly distributed throughout all chromosomes (Spana *et al.* 1988). These sequences are not associated with the presence of gypsy elements and presumably play a role in assuring normal transcription of a great variety of *Drosophila* genes by creating domains of independent gene expression.

In the case of the gypsy insulator the nature of at least one of its components has long been intuited based on genetic evidence indicating that mutations in the *su(Hw)* gene reverse the phenotype of gypsy-induced mutations, suggesting that this protein is probably responsible for the effect of the insulator in the repression of distal enhancers (Corces and Geyer 1991). The gypsy insulator is composed of 12 binding sites for the su(Hw) protein; the core of this recognition sequence is homologous to the octamer motif (Spana *et al.* 1988; Mazo *et al.* 1989; Dorsett 1990), and it is flanked by AT tracts that impose a bend in the DNA necessary for proper protein–DNA recognition (Spana and Corces 1990). The su(Hw) protein contains 12 zinc fingers involved in DNA binding. Interaction of the su(Hw) protein with its target sequence through the zinc fingers is a prerequisite for proper insulator function (Harrison *et al.* 1993). In addition, su(Hw) protein contains two acidic domains, located in the amino- and carboxy-terminal ends of the protein, which are dispensable for its effect on enhancer function and its ability to buffer against chromosomal position effects. An α-helical region, homologous to the second helix coiled coil region of basic HLH–bzip proteins, is absolutely required for insulator function (Harrison *et al.* 1993). Since leucine zipper regions usually mediate protein–protein interactions and su(Hw) does not interact with itself, this observation suggests the involvement of other proteins in addition to su(Hw) in the formation of the boundary element.

Classical genetic analyses have identified a second component of the su(Hw) insulator by searching for other mutations that alter gypsy-induced phenotypes (Georgiev and Gerasimova 1989). Mutations in the modifier of *mdg4* (*mod(mdg4)*) gene reverse the polar effect of the insulator on enhancer function by causing a bidirectional repression of all enhancers, independent of whether they are located distal or proximal to the promoter with respect to the location of the boundary element. The *mod(mdg4)* gene encodes at least three different proteins, arising from alternatively spliced RNAs that contain a BTB domain (BTB is named for three of the proteins that contain this domain: the broad complex, tramtrack, and bric-á-brac (Dorn *et al.* 1993; Gerasimova *et al.* 1995)). Genetic and molecular analyses suggest that mod(mdg4) proteins interact directly with su(Hw) and therefore constitute a second component of the gypsy insulator. In the absence of mod(mdg4) the su(Hw) protein transforms the insulator into a silencer, causing bidirectional repression of transcription (Gerasimova *et al.* 1995).

The BTB domain is present in various transcription factors (Zollman *et al.* 1994) including GAGA, a transcriptional activator encoded by the *Trithorax-like* (*Trl*) gene that functions by influencing chromatin structure (Farkas *et al.* 1994). Mutations in *Trl* result in lower levels of expression of homeotic genes by a mechanism involving chromatin packaging (Farkas *et al.* 1994) and the GAGA factor itself has been shown to remodel chromatin *in vitro* (Tsukiyama *et al.* 1994; Pazin *et al.* 1994). Further evidence for a role of *Trl* in chromatin structure comes from the observation that mutations in this gene act as enhancers of position-effect variegation. Interestingly, mutations in *mod(mdg4)* display a variegated phenotype affecting the gene adjacent to the insulator sequences. In addition, *mod(mdg4)* mutations act as enhancers of position-effect variegation for genes brought close to heterochromatin by chromosomal rearrangements (Dorn *et al.* 1993; Gerasimova *et al.* 1995). These results suggest that the bidirectional repression of enhancer function caused by the gypsy insulator lacking the mod(mdg4) protein is the result of heterochromatization of sequences surrounding the insulator insertion site. Since changes in gene expression caused by position-effect variegation are concomitant with alterations in chromatin structure (Wallrath and Elgin 1995), the enhancer silencing observed in *mod(mdg4)* mutants might also be caused by changes in the packaging of nucleosomes. It is important to emphasize that this effect is caused by the presence of su(Hw) protein bound to insulator sequences in the absence of mod(mdg4) protein. Both the leucine zipper and acidic domains of su(Hw) are involved in this effect, and the heterochromatization of adjacent chromatin might be caused by the interaction of su(Hw) with other proteins in the absence of mod(mdg4) (Gerasimova *et al.* 1995).

Insulator function involves alterations in chromatin structure

The finding of the BEAF-32, su(Hw), and mod(mdg4) proteins and their involvement in various chromatin- and chromosome-related processes suggests that boundary elements and chromatin insulators are multiprotein component complexes that regulate gene expression by effecting changes in the adjacent chromatin structure. The similarities between dominant position effects in the *brown-Dominant* (*bw*[D]) mutation of *Drosophila* caused by adjacent heterochromatic sequences (Dreesen *et al.* 1991) and *trans*-effects caused by *mod(mdg4)* mutations further support this hypothesis. The silencing effect of su(Hw) protein in the absence of mod(mdg4) can be transmitted to the paired homologous chromosome, causing repression of specific enhancers (Georgiev and Corces 1995). The tissue specificity suggests an effect on enhancer function rather than a general effect on the promoter. A female *Drosophila* that is heterozygous for the gypsy-induced *yellow* (*y*[2]) mutation in an otherwise wild-type fly shows normal coloration, since expression from one of the copies of the *yellow* gene is normal. Nevertheless, in the background of a mutation in *mod(mdg4)* the flies have abnormal bristle pigmentation, suggesting that the bristle enhancer is not functional. This effect is not observed when one of the copies of the *yellow* gene is moved by a chromosomal rearrangement that disrupts pairing between the two homologues. The only explanation for this observation is that the silencing effect of su(Hw) in the absence of the mod(mdg4) protein can be transmitted to the homologous paired chromosome and interfere with the function of the bristle enhancer (Georgiev and Corces 1995). This *trans*-effect is similar to the dominant position-effect variegation observed at the *Drosophila brown* (*bw*) locus in the *bw*ᴰ allele (Dreesen *et al.* 1991). In this case heterochromatic sequences adjacent to the *bw* gene in one chromosome cause *trans*-inactivation of the normal gene located in the paired homologue, suggesting that the inhibitory effect of heterochromatin can be transmitted between chromosomes.

Further evidence suggesting a relationship between the function of the gypsy insulator and chromatin structure comes from the observation that boundary elements can affect dosage compensation in *Drosophila*. In this organism equal expression of X-linked genes in both sexes is due to a two-fold increase in the transcription of genes on the male X chromosome. This is accomplished by the assembly of multimeric male-specific lethal (MSL) protein complexes on the X chromosome of the male that remodel nucleosome structure through the specific acetylation of Lys 16 on histone H4 (Bone *et al.* 1994). Transgenes that contain X chromosome loci and are inserted in autosomal regions fail to dosage compensate completely, suggesting an inhibitory effect of autosomal chromatin on the compensation process. Nevertheless, when the copy of the transgene is flanked on both sides by the gypsy insulator almost 90 per cent of autosomal insertions show proper dosage compensation (Roseman *et al.* 1995). The mechanism by which the gypsy insulator causes this effect is not clear, but it might act by stabilizing the interaction of the MSL complex with the transgene, or it might prevent access of histone deacetylases to increase the degree of nucleosome modification in the inserted sequences (Roseman *et al.* 1995)

This modulation of chromatin conformation must affect the interaction between an enhancer and a promoter without affecting the functionality of the enhancer itself. This hypothesis has now been elegantly confirmed (Cai and Levine 1995; Scott and Geyer 1996). When the gypsy insulator or the scs element are interposed between the two enhancers that control expression of the even-skipped gene in stripes 2 and 3 during *Drosophila* embryogenesis, transcription from a promoter located on the other side of the insulator is repressed, whereas expression from a different promoter located on the same side of the enhancer is still active (Cai and Levine 1995). The same effect has been observed for the shared enhancers present in the intergenic region of the *Drosophila* yolk protein (*yp*) genes that control transcription of the divergently expressed *yp1* and *yp2* genes in the fat body of the fly. When the gypsy insulator is interposed between the yp1 promoter and the enhancer this gene is not expressed, but the *yp2* gene is transcribed normally (Scott and Geyer 1996). This conclusion suggests that the presence of an insulator does not disrupt the interaction of transcription factors with enhancer sequences.

A second constraint imposed by the functional properties of an insulator on its mechanism of action is that putative changes induced by an insulator on the chromatin structure of adjacent sequences do not interfere with the actual process of transcription by RNA polymerase. When the gypsy insulator is inserted in the intron of the *yellow* gene, only enhancers located downstream are affected by its presence, whereas transcription of *yellow* in other tissues is otherwise normal (Geyer and Corces 1992). These properties of chromatin insulators shape the types of models we can put forward to explain the mechanisms underlying insulator function.

Can SARs and MARs act as functional boundary or insulator elements?

Chromatin insulators and boundary elements have been defined by functional analyses that measure the effect of these sequences on enhancer–promoter interactions. In addition, structural studies of chromosome organization have led to the discovery of scaffold or matrix attachment regions (SARs, MARs, or S/MARs) (Mirkovitch *et al.* 1984; Cockerill and Garrard 1986; see Chapter 6). These are AT-rich DNA sequences, often containing topoisomerase II cleavage sites, which mediate the anchoring of the chromatin fibre to the chromosome scaffold or nuclear matrix and delimit the boundaries of discrete and topologically independent higher order domains (Laemli *et al.* 1992). Proteins bound to these sequences tether the chromatin fibre into structural loops held together by the components of the chromosome scaffold. It is easy to visualize how these structural domains that organize the chromosome could also serve a functional purpose by organizing genes into specific hierarchies of regulated transcription by acting as boundary or insulator elements. Although some S/MARs have failed to confer position-independent expression on reporter genes (Poljak *et al.* 1992), S/MARs from genes such as the chicken lysozyme and human β-interferon can ensure elevated expression of reporter genes when stably integrated in the chromosome independent of their location (see Chapter 7). S/MARs flanking the human apolipoprotein B (apoB) gene have also been shown to have similar properties (Kalos and Fournier 1995). The nuclear attachment sites in the apoB gene are located 5 kb upstream and 43 kb downstream from the transcription start site. These sequences contain DNAase I-hypersensitive sites at the junctions between regions of closed versus open chromatin. These properties suggest that the 5' and 3' S/MARs are the physical boundaries of the apoB chromatin domain. Recent results indicate that they also have a functional role in the control of apoB expression unrelated to classical enhancer activity. The apoB S/MARs can direct proper expression of a reporter gene in stably transformed, but not transiently transfected human hepatoma cells. Transformed cells containing a single copy of the reporter gene support low and variable transcription. In contrast, expression of the reporter flanked by 5' and 3' S/MARs was 200-fold higher and independent of its integration site (Kalos and Fournier 1995). Interestingly, the insulating effect of S/MARs might not be constitutive and might depend on the cell type and time of development of the organism. For example, S/MARs isolated from the human β-interferon gene, when flanking the murine HSP70.1 gene, display a functional boundary activity in a tissue-specific fashion (Thompson *et al.* 1994). This gene is expressed at the time of activation of the zygotic genome in the two-cell embryo and later shows ubiquitous constitutive expression in a number of differentiated tissues. When flanked by interferon S/MARs, expression of an HSP70.1 transgene is regulated in a copy-dependent manner in pre-implantation mouse embryos. In contrast, correlation between copy number and levels of expression was lost and position effects were observed in differentiated tissues of newborn and adult transgenic mice (Thompson *et al.* 1994). These results suggest that topological chromosomal domains established by S/MARs are not static entities, but rather dynamic ones that can change during development to adapt to new patterns of gene expression required by changes in the state of cell differentiation.

Future directions

The mechanisms by which insulators affect enhancer function in a polar fashion are constrained by how enhancers affect activation of transcription from specific promoters. Simple models that view insulators as roadblocks that interfere with tracking, or as sinks that sequester enhancer-bound transcription factors, are easy to visualize, but fail to account for the observed connection between insulator function and chromatin structure. The role of chromatin in mediating insulator function was first established by the fact that boundary elements separate regions of chromatin with different accessibility to nucleases and that they work only in stably integrated transgenes. This is now further supported by the finding of insulator binding proteins throughout the chromosome in regions where they could play structural functions in chromosome domain organization (Zhao *et al.* 1995). Furthermore, genes encoding insulator components act as enhancers of position-effect variegation, suggesting that these proteins might also function to repress formation of heterochromatin (Gerasimova *et al.* 1995). Finally, a direct relationship between insulator function and nucleosome positioning has been clearly demonstrated in the case of the chicken β-globin insulator (Chung *et al.* 1993). Based on these observations one attractive argument to integrate mechanisms of enhancer and insulator function is that activation of transcription by enhancer elements correlates with subtle changes in chromatin, such as changes in nucleosomal phasing. Boundary elements or insulators could then interfere with this remodelling of chromatin conformation, thus explaining the uni-directionality of the effect. This certainly seems to be the case for the chicken β-globin insulator (Chung *et al.* 1993) and might well be true for other boundary elements. These effects on chromatin must be compatible with the observation that an insulated enhancer is still able to function on a promoter located within the same chromatin domain, and with the finding that RNA polymerase is able to transcribe through an insulator.

Many questions still remain to be answered before we understand the structure and role of chromatin insulators, but their unique properties lend themselves to crucial roles in the control of eukaryotic gene expression. Although boundary sequences identified so far do not show homology to each other, hybrid chromatin domains could be formed by the interaction between two different insulators if the proteins that constitute these entities are promiscuous and able to interact with each other. This conclusion is supported by the observation that scs and scs' might interact *in vivo* to form a domain: although the BEAF-32 protein does not bind to scs, DNA immunoprecipitated with BEAF-32 antibodies contains scs sequences, suggesting that both elements might interact (Zhao *et al.* 1995). The presence of tissue-specific boundary proteins could then allow for a greater flexibility in the establishment of higher order levels of chromatin organization with the formation of fluid chromatin domains that could change in response to specific developmental programmes. Differential affinities among protein components of the insulators could afford the establishment of a hierarchy of chromatin domains with different abilities to affect enhancer–promoter interactions, depending on the strength of these two transcription signals. Only a detailed analysis of their structure and mechanisms of action will tell whether insulators play an important role in the complex world of eukaryotic gene regulation.

Summary

Boundary or insulator elements are DNA sequences defined functionally by two characteristic effects on gene expression: they confer position-independent transcription to transgenes stably integrated into the chromosome and protect a promoter from activation by a specific enhancer when the insulator is located in between. Characterization of proteins that interact with these sequences suggests that insulators are multicomponent nucleoprotein complexes that affect transcription by modulating changes in chromatin structure. The properties of these novel regulatory sequences suggest that they might play a general role in gene regulation by organizing the genome into functional domains of gene expression that prevent promiscuous interactions between enhancers and promoters.

Acknowledgement

Work carried out in the authors' laboratory was supported by United States Public Health Service award GM35463 from the National Institutes of Health.

References

Bone, J.R., Lavender, J., Richman, R., Palmer, M.J., Turner, B.M., *et al.* (1994). Acetylated histone H4 on the male X chromosome is associated with dosage compensation in *Drosophila*. *Genes and Development*, **8**, 96–104.

Cai, H. and Levine, M. (1995). Modulation of enhancer–promoter interactions by insulators in the *Drosophila* embryo. *Nature*, **376**, 533–6.

Callan, H.G. (1986). *Lampbrush chromosomes*. Springer-Verlag, Berlin.

Chung, J.H., Whiteley, M., and Felsenfeld, G. (1993). A 5' element of the chicken β-globin domain serves as an insulator in human erythroid cells and protects against position effects in *Drosophila*. *Cell*, **74**, 505–14.

Cockerill, P.N. and Garrard, W.T. (1986). Chromosomal loop anchorage of the kappa immunoglobulin gene occurs next to the enhancer in a region containing topoisomerase II sites. *Cell*, **44**, 273–82.

Corces, V.G. and Geyer, P.K. (1991). Interactions of retrotransposons with the host genome: the case of the gypsy element of *Drosophila*. *Trends in Genetics*, **7**, 86–90.

Dorn, R., Krauss, V., Reuter, G., and Saumweber, H. (1993). The enhancer of position-effect variegation of *Drosophila* E(var)3-93D codes for a chromatin protein containing a conserved domain common to several transcriptional regulators. *Proceedings of the National Academy of Sciences of the USA*, **90**, 11376–80.

Dorsett, D. (1990). Potentiation of a polyadenylation site by a downstream protein–DNA interaction. *Proceedings of the National Academy of Sciences of the USA*, **87**, 4373–7.

Dreesen, T.D., Henikoff, S., and Loughney, K. (1991). A pairing-sensitive element that mediates trans-inactivation is associated with the *Drosophila* brown gene. *Genes and Development*, **5**, 331–40.

Elder, J.T., Forrester, W.C., Thompson, C., Mager, D., Henthorn, P., Peretz, M., *et al.* (1990). Translocation of an erythroid-specific hypersensitive site in deletion-type hereditary persistence of fetal hemoglobin. *Molecular and Cellular Biology*, **10**, 1382–9.

Farkas, G., Gausz, J., Galloni, M., Reuter, G., Gyurkovics, H., and Karch, F. (1994). The *Trithorax-like* gene encodes the *Drosophila* GAGA factor. *Nature*, **371**, 806–8.

Georgiev, P.G. and Corces, V.G. (1995). The su(Hw) protein bound to gypsy sequences in one chromosome can repress enhancer–promoter interactions in the paired gene located in the other homolog. *Proceedings of the National Academy of Sciences of the USA*, **92**, 5184–8.

Georgiev, P.G. and Gerasimova, T.I. (1989). Novel genes influencing the expression of the yellow locus and mdg4 (gypsy) in *Drosophila melanogaster*. *Molecular and General Genetics*, **220**, 121–6.

Gerasimova, T.I., Gdula, D.A., Gerasimov, D.V., Simonova, O., and Corces, V.G. (1995). A *Drosophila* protein that imparts directionality on a chromatin insulator is an enhancer of position-effect variegation. *Cell*, **82**, 587–97.

Geyer, P.K. and Corces, V.G. (1992). DNA position-specific repression of transcription by a *Drosophila* Zn finger protein. *Genes and Development*, **6**, 1865–73.

Geyer, P.K., Green, M.M., and Corces, V.G. (1988). Mutant gene phenotypes mediated by a *Drosophila melanogaster* retrotransposon require sequences homologous to mammalian enhancers. *Proceedings of the National Academy of Sciences of the USA*, **85**, 8593–7.

Grosveld, F., van Assendelft, G.B., Greaves, D.R., and Kollias, G. (1987). Position-independent, high level expression of the human β-globin gene in transgenic mice. *Cell*, **51**, 975–85.

Harrison, D.A., Gdula, D.A., Coyne, R.S., and Corces, V.G. (1993). A leucine zipper domain of the suppressor of Hairy-wing protein mediates its repressive effect on enhancer function. *Genes and Development*, **7**, 1966–78.

Holdridge, C. and Dorsett, D. (1991). Repression of hsp70 heat shock gene transcription by the suppressor of Hairy-wing protein of *Drosophila melanogaster*. *Molecular and Cellular Biology*, **11**, 1894–900.

Jack, J., Dorsett, D., Delotto, Y., and Liu, S. (1991). Expression of the cut locus in the *Drosophila* wing margin is required for cell type specification and is regulated by a distal enhancer. *Development*, **113**, 735–48.

Kalos, M. and Fournier, R.E.K. (1995). Position-independent transgene expression mediated by boundary elements from the apolipoprotein B chromatin domain. *Molecular and Cellular Biology*, **15**, 198–207.

Kellum, R. and Schedl, P. (1991). A position-effect assay for boundaries of higher order chromatin domains. *Cell*, **64**, 941–50.

Kellum, R. and Schedl, P. (1992). A group of scs elements function as domain boundaries in an enhancer-blocking assay. *Molecular and Cellular Biology*, **12**, 2424–31.

Laemli, U.K., Kös, E., Poljak, L., and Adachi, Y. (1992). Scaffold-associated regions: cis-acting determinants of chromatin structural loops and functional domains. *Current Opinion in Genetics and Development*, **2**, 275–85.

Lewis, M., Helmsing, P.J., and Ashburner, M. (1975). Parallel changes in puffing activity and patterns of protein synthesis in salivary glands of *Drosophila*. *Proceedings of the National Academy of Sciences of the USA*, **72**, 3604–8.

Mazo, A.M., Mizrokhi, L.J., Karavanov, A.A., Sedkov, Y.A., Krichevskaja, A.A., and Ilyin, Y.V. (1989). Suppression in *Drosophila*: su(Hw) and su(f) gene products interact with a region of gypsy (mdg4) regulating its transcriptional activity. *EMBO Journal*, **8**, 903–11.

Mirkovitch, J., Mirault, M.-E., and Laemmli, U.K. (1984). Organization of the higher-order chromatin loop: specific DNA attachment sites on nuclear scaffold. *Cell*, **39**, 223–32.

Pazin, M.J., Kamakaka, R.T., and Kadonaga, J.T. (1994). ATP-dependent nucleosome reconfiguration and transcriptional activation from preassembled chromatin templates. *Science*, **266**, 2007–11.

Pfeifer, M., and Bender, W. (1988). Sequences of the gypsy transposon of *Drosophila* necessary for its effects on adjacent genes. *Proceedings of the National Academy of Sciences of the USA*, **85**, 9650–4.

Poljak, L., Seum, C., Mattioni, T., and Laemmli, U.K. (1994). SARs stimulate but do not confer position independent gene expression. *Nucleic Acids Research*, **22**, 4386–94.

Roseman, R.R., Pirrotta, V., and Geyer, P.K. (1993). The su(Hw) protein insulates expression of the *Drosophila melanogaster white* gene from chromosomal position-effects. *EMBO Journal*, **12**, 435–42.

Roseman, R.R., Swan, J.M., and Geyer, P.K. (1995). A *Drosophila* insulator protein facilitates dosage compensation of the X chromosome mini-white gene located at autosomal insertion sites. *Development*, **121**, 3573–82.

Scott, K.S. and Geyer, P.M. (1996). Effects of the *Drosophila* su(Hw) insulator protein on the expression of the divergently transcribed yolk protein genes. *EMBO Journal*, **14**, 6258–79.

Spana, C. and Corces, V.G. (1990). DNA bending is a determinant of binding specificity for a *Drosophila* zinc finger protein. *Genes and Development*, **4**, 1505–15.

Spana, C., Harrison, D.A., and Corces, V.G. (1988). The *Drosophila melanogaster* suppressor of Hairy-wing protein binds to specific sequences of the gypsy retrotransposon. *Genes and Development*, **2**, 1414–23.

Thompson, C.C. and McKnight, S.L. (1992). Anatomy of an enhancer. *Trends in Genetics*, **8**, 232–6.

Thompson, E.M., Christians, E., Stinnakre, M.-G., and Renard, J.-P. (1994). Scaffold attachment regions stimulate HSP70.1 expression in mouse preimplantation embryos but not in differentiated tissues. *Molecular and Cellular Biology*, **14**, 4694–703.

Tissières, A., Mitchell, H.K., and Tracy, U.M. (1974). Protein synthesis in salivary glands of *Drosophila melanogaster*: relation to chromosome puffs. *Journal of Molecular Biology*, **84**, 389–98.

Tsukiyama, T., Becker, P.B., and Wu, C. (1994). ATP-dependent nucleosome disruption at a heat shock promoter mediated by binding of GAGA transcription factor. *Nature*, **367**, 525–32.

Udvardy, A., Maine, E., and Schedl, P. (1985). The 87A7 chromomere: identification of novel chromatin structures flanking the heat shock locus that may define the boundaries of higher order domains. *Journal of Molecular Biology*, **185**, 341–58.

Vazquez, J. and Schedl, P. (1994). Sequences required for enhancer blocking activity of scs are located within two nuclease-hypersensitive regions. *EMBO Journal*, **13**, 5984–93.

Wallrath, L.L. and Elgin, S.C. (1995). Position effect variegation in *Drosophila* is associated with an altered chromatin structure. *Genes and Development*, **9**, 1263–77.

Zhao, K., Hart, C.M., and Laemmli, U.K. (1995). Visualization of chromosomal domains with boundary element-associated factor BEAF-32. *Cell*, **81**, 879–89.

Zollman, S., Godt, D., Prive, G.G., Couderc, J.L., and Laski, F.A. (1994). The BTB domain, found primarily in zinc finger proteins, defines an evolutionarily conserved family that includes several developmentally regulated genes in *Drosophila*. *Proceedings of the National Academy of Sciences of the USA*, **91**, 10717–21.

6

Nuclear matrix-associated regions

M. E. Eva Ludérus and Roel van Driel

Introduction

Eukaryotic interphase chromosomes are thought to be organized in topologically independent loop domains. Such loops can be visualized emanating from permeabilized nuclei after extraction of histone proteins (Vogelstein *et al.* 1980; Jackson *et al.* 1984; Gerdes *et al.* 1994). At their bases the loops are fastened to the residual nuclear structure, which is known as nuclear matrix or nuclear scaffold. It is believed that the loop organization of chromatin is important not only for the compaction and spatial organization of the chromatin, but also for the regulation of gene expression. Each loop domain may represent an independent unit of chromatin structure and gene activity.

A specific group of nuclear matrix-bound DNA sequences, termed scaffold/ matrix-attached regions (S/MARs), have been identified as candidates to form the bases of the loops. In this review we will evaluate their role in chromatin organization and gene expression. Three independently testable aspects of the presumed domain boundary function of S/MARs will be discussed: their role (i) as physical anchors of chromatin loops at the nuclear matrix; (ii) as boundaries of chromatin structure and topology; and (iii) as regulatory elements in transcription.

Structure of S/MARs

S/MARs have been identified in many different eukaryotic species. They contain about 70 per cent A+T base pairs, and their size varies between several hundreds of base pairs and several kilobases. Although there is no consensus sequence for S/MARs, they share several characteristic primary and secondary sequence features (reviewed in Boulikas 1995). Many AT base pairs in S/MARs are found in arrays of oligo(dA)/oligo(dT) sequences, so-called A-tracts. In addition, S/MARs are generally enriched in (A)ATATTT(TT) elements and sequences related to the *in vitro* cleavage consensus for topoisomerase II. Secondary sequence features include a narrow minor groove and DNA bending. Under superhelical strain S/MARs have a strong potential for extensive base unpairing, a property which is not merely due to their general AT richness, but for which specific nucleation sites have been identified (Bode *et al.* 1992; reviewed by Bode *et al.* 1995).

S/MARs as anchors of chromatin loops at the nuclear matrix

Interaction of S/MARs with the nuclear matrix

S/MARs were originally identified as DNA elements with a high binding affinity for the nuclear matrix (Mirkovitch *et al.* 1984; Cockerill and Garrard 1986a). Their interaction with the matrix is primarily based on the recognition of general structural features of S/MAR DNA, rather than a precise base sequence. The involvement of A-tracts has been deduced from binding experiments using synthetic AT-rich DNA polymers and the drug distamycin, a DNA binding ligand with a high specificity for the narrow minor groove of oligo(dA)/oligo(dT) sequences (Käs *et al.* 1989). Bode *et al.* (1992) have shown that the base unpairing property of S/MARs plays a role in matrix binding. Specific point mutations that abolish the unwinding capacity of artificial S/MARs were found to inhibit their interaction with the nuclear matrix. In line with this notion, single-stranded random sequence DNA has been shown to compete with double-stranded S/MARs for binding to the matrix (Kay and Bode 1994; Ludérus *et al.* 1994). S/MAR–matrix binding is thus largely mediated by interactions with low sequence specificity. Nevertheless, for different *in vitro* S/MAR–matrix binding systems a dissociation equilibrium constant (K_d) in the nanomolar range has been calculated (Das *et al.* 1993). Such high affinity interactions are thought to arise from cooperative binding of matrix proteins to multiple binding motifs along the S/MAR DNA. Accordingly, in various cases a correlation between binding strength and S/MAR fragment length has been noted (Käs *et al.* 1989; Amati *et al.* 1990; Mielke *et al.* 1990; Breyne *et al.* 1992; Romig *et al.* 1994).

In agreement with the involvement of characteristic structural features of S/MARs in their interaction with the nuclear matrix, different S/MARs have been shown to compete for binding to the same sites, albeit often with different efficiencies (Izaurralde *et al.* 1988; Ludérus *et al.* 1992; Das *et al.* 1993). Furthermore, S/MAR–matrix interaction appears to be evolutionarily conserved between yeast, animals, and plants, although S/MARs from different origins may display different binding affinities (Cockerill and Garrard 1986a,b; Izaurralde *et al.* 1988; Amati and Gasser 1990; Amati *et al.* 1990; Breyne *et al.* 1992; McKnight *et al.* 1992; Allen *et al.* 1993; Dietz *et al.* 1994; reviewed in Bode *et al.* 1995). These observations indicate that structural features of not only S/MARs themselves, but also of their binding sites on the nuclear matrix, are conserved across species barriers. In addition, they suggests an important biological role for S/MAR–matrix interactions.

If S/MARs are the prime elements that constitutively anchor chromatin loops to the nuclear matrix, one would expect to find S/MAR sequences distributed along the genome with an average spacing of roughly 85 kb, the estimated average size of chromatin loops in animal cells (Benyajati and Worcel 1976; Vogelstein *et al.* 1980; Jackson *et al.* 1990). To address this issue, matrix-bound sequences have been mapped along large, continuous regions of the *Drosophila* genome and were found at intervals that varied roughly between 10 kb and 100 kb (Mirkovitch *et al.* 1986; Surdej *et al.* 1990). Furthermore, based on a limited number of S/MAR mapping studies at different genomic loci of *Drosophila* and mammalian cells, an average spacing of 30 kb between S/MARs has been calculated (Garrard 1990). Recently, a conceptually novel procedure to locate DNA loop anchorage sites has been developed, which depends on chromosomal DNA loop excision by topoisomerase II-mediated DNA cleavage at matrix-attached regions (Gromova *et al.* 1995). Using this procedure, loop anchoring sites in a 500 kb region of the *Drosophila* X chromosome were found to be spaced by 20 to 90 kb (Iarovaia *et al.* 1996). Almost all of these sites overlapped with S/MARs that had previously been identified on the basis of selective matrix binding, but conversely, not every S/MAR constituted an anchoring site as defined by this procedure. Taken together and considering the experimental limitations, the available data could be compatible with an important chromatin anchoring role for S/MARs. However, the data neither exclude that there are S/MARs without a constitutive anchoring function, nor that there are additional types of attachment sequences. In this respect it should be mentioned that at least one non-S/MAR type of matrix attachment is constituted by actively transcribed DNA (Ciejek *et al.* 1983; Gerdes *et al.* 1994; Jackson *et al.* 1996). If S/MARs partition the entire genome in loops with an average length of roughly 85 kb, one would anticipate that at least 80 000 S/MAR binding sites are present per human nuclear matrix. So far, it has only been possible to estimate the number of S/MAR binding sites at the matrix from *in vitro* S/MAR titration experiments. Such binding studies have indicated that per nuclear matrix between 3000 and 23 000 binding sites are accessible for exogenously added S/MAR sequences (Cockerill and Garrard 1986a; Bode and Maass 1988; Ludérus *et al.* 1992; Das *et al.* 1993; Kay and Bode 1994; Romig *et al.* 1994). Izaurralde *et al.* (1988) have shown that in this type of assay genomic S/MARs are displaced from the matrix by the exogenously added S/MARs. It is possible that the determined numbers of binding sites are an underestimate due to incomplete displacement of endogenous S/MARs. Alternatively, there may be genomic regions that are not organized in S/MAR-based loops.

S/MAR binding proteins

In higher eukaryotes a number of proteins that bind S/MARs selectively *in vitro* have been identified. As expected from the characteristics of S/MAR–matrix interactions most of these proteins recognize structural features of S/MARs, rather than sequence motifs. However, there are also proteins that have more precise sequence requirements. An example of the latter type of S/MAR binding proteins is the attachment region binding protein (ARBP), which recognizes a sequence motif consisting of the core sequence 5'-GGTGT-3' flanked by AT-rich regions (Buhrmester *et al.* 1995). The AT-rich flanks are bound by ARBP in the minor groove and the two central guanines of the core motif are recognized through contacts in the major groove. ARBP is an evolutionarily conserved protein and a constituent of the nuclear matrix (Von Kries *et al.* 1991).

The proteins SATB1 (Dickinson *et al.* 1992) and nucleolin (Dickinson and Kohwi-Shigematsu 1995) belong to a group of proteins that specifically recognize base unpairing regions of S/MARs. Both proteins bind through contacts in the minor groove of the DNA. SATB1 binds to so-called ATC regions, AT-rich regions in which one strand consists of mixed As, Ts and Cs, but excluding Gs. The protein is the first example of a tissue-specific S/MAR binding protein. It is predominantly expressed in the thymus, where it is a component of the nuclear matrix. The overall S/MAR binding domain and the actual DNA contact sites of SATB1 have recently been identified (Nakagomi *et al.* 1994; Wang *et al.* 1995). In contrast to SATB1, which exclusively binds double-stranded S/MAR DNA, nucleolin also binds single-stranded S/MAR DNA, with a strong preference for the T-rich strand. In addition, single-stranded mammalian telomeric DNA and its cognate RNA version are bound with high affinity (Dickinson and Kohwi-Shigematsu 1995). Nucleolin is present both as a soluble protein and tightly bound to the nuclear matrix. Recently, a B cell-specific S/MAR binding protein with a recognition site similar to that of SATB1 has been identified. This protein, termed Bright, binds as a tetramer in the minor groove of a restricted ATC sequence (AT-rich region with a G/C strand bias) (Herrscher *et al.* 1995). This particular sequence has been found in the S/MARs of the immunoglobulin heavy chain (IgH) genes and Bright has been implicated as a transacting factor in the expression of these genes. A fourth protein that recognizes S/MAR sequences with a high unwinding propensity is p114, which is found only in human breast carcinomas (Yanagisawa *et al.* 1996).

For another group of S/MAR binding proteins, DNA fragment length and multiplicity of binding motifs are crucial as their binding strongly relies on cooperative interactions. This category of proteins includes histone H1, topoisomerase II, lamins, and hnRNP U. These are all abundant nuclear proteins which preferentially, but not exclusively, bind S/MAR DNA. Laemmli and colleagues have shown that histone H1 (Izaurralde *et al.* 1988) and topoisomerase II (Adachi *et al.* 1989) bind cooperatively to S/MARs *in vitro*, forming sedimentable protein–DNA complexes. A-tracts are the major determinants of these selective interactions. Histone H1 plays a crucial role in the compaction of chromatin through its association with internucleosomal linker DNA (see Chapter 3). The preferential interaction of histone H1 with S/MARs is thought to be instrumental in this function, since it nucleates further assembly of histone H1 onto flanking non-S/MAR DNA (Izaurralde *et al.* 1988). Only a minor fraction of the total amount of histone H1 co-fractionates with the nuclear matrix (Käs *et al.* 1989). Topoisomerase II, in contrast, is a major component of the interphase nuclear matrix (Berrios *et al.* 1985). The enzyme cleaves S/MAR DNA both *in vitro* (Adachi *et al.* 1989) and *in vivo* (Käs and Laemmli 1992), although different cleavage sequences are recognized. Compatible with the cooperative nature of its interaction with S/MARs and suggestive for a structural role of this protein in chromatin organization, topoisomerase II shows self-polymerization *in vitro*. In the absence of nucleic acid, formation of complexes of the size of tetramers and larger has been observed (Vassetzky *et al.* 1994). Also, A- and B-type lamins preferentially bind S/MAR fragments above a critical length (Ludérus *et al.* 1992, 1994). Lamins are the structural proteins of the nuclear lamina, which constitutes the peripheral structure of the nuclear matrix. Binding of S/MARs to *in vitro*-formed polymers, consisting of rat liver lamins A, B, and C, is partially inhibited by distamycin and can be competed by single-stranded random sequence DNA. This suggests that the minor groove of double-stranded S/MAR sequences as well as base-unpairing regions are involved. Before its identification as hnRNP U (Fackelmayer *et al.* 1994; Von Kries *et al.* 1994), this protein was detected as an S/MAR binding protein by three independent groups, who respectively named it p120 (Von Kries *et al.* 1991), SAF-A (Romig *et al.* 1992), and SP120 (Tsutsui *et al.* 1993). HnRNP U is part of macromolecular hnRNP complexes and a constituent of the nuclear matrix (Mattern *et al.* 1996). Although the protein preferentially binds double-stranded S/MAR DNA, it also shows considerable affinity for single-stranded DNA and RNA (Fackelmayer *et al.* 1994; Von Kries *et al.* 1994). The cooperative binding of hnRNP U to S/MAR DNA results in the formation of large aggregates. Under the electron microscope these appear as regular, unbranched protein filaments from which a halo of DNA strands radiates (Romig *et al.* 1992; Fackelmayer *et al.* 1994).

Considering the inhibiting effects of distamycin on most S/MAR–protein interactions, an interesting class of proteins are those that bind DNA in a distamycin-like fashion. These proteins recognize the narrow minor groove of oligo(dA)/oligo(dT) sequences. In contrast to the above-discussed S/MAR binding proteins, such as histone H1 and topoisomerase II, they do not depend on cooperative interactions and also bind relatively short runs of dA/dT base pairs with high affinity. One such distamycin analogue protein is the high mobility group protein HMGI/Y. This protein binds preferentially to A-tracts containing six or more dA/dT base pairs (Solomon *et al.* 1986). This selective interaction is mediated by minor groove contacts of three consensus regions called AT hooks (Reeves and Nissen 1990). HMGI/Y is abundant in nuclei of rapidly dividing cells and has been implicated in various nuclear functions. Investigations by Laemmli and colleagues have indicated that HMGI/Y and other distamycin analogue proteins may have an important role in the regulation of regional chromatin condensation by antagonizing S/MAR binding of histone H1 (Käs *et al.* 1993; Zhao *et al.* 1993). The protein CHD1, which has a binding preference for DNA segments containing tracts of 12 or more dA/dT base pairs, can also be considered as a distamycin analogue protein (Stokes and Perry 1995). Several CHD1 binding sites were found in the S/MAR flanking the intronic Ig κ gene enhancer. CHD1 fractionates as a constituent of bulk chromatin and is not a component of the nuclear matrix. In addition to a DNA binding domain, the protein possesses both a chromatin organization modifier (chromo) domain and a helicase/ATPase domain, pointing to a role in defining chromatin architecture.

As a tool to investigate the function of A-tracts in S/MARs, multi-AT hook (MATH) proteins have been synthesized (Strick and Laemmli 1995). *In vitro* these synthetic proteins bind with high affinity to the clustered A-tracts in S/MARs. When added to mitotic extracts containing sperm nuclei or somatic nuclei MATH proteins were found to block the shape determination and shape maintenance of chromosomes, but not their condensation *per se*. Although the specificity of MATH protein binding *in vivo* remains to be determined, these observations may point to a novel architectural role for S/MARs during chromosome assembly.

In conclusion, different types of S/MAR binding proteins can be distinguished based on their S/MAR DNA binding properties, nuclear abundance, tissue specificity, and association with the nuclear matrix. It is tempting to speculate that according to these properties different S/MAR–protein interactions have different functions. Some S/MAR binding proteins have properties suitable to fulfil a structural chromatin-anchoring role, while others are more likely to be regulators of chromatin structure, and again other S/MAR binding proteins could function similar to transcription factors. It should however be realized that most available information has been obtained *in vitro* and that final conclusions concerning the biological relevance of the various S/MAR–protein interactions have to await their demonstration in the living cell.

The role of S/MARs in gene expression

The organization of the eukaryotic genome in discrete loop domains is believed to be important for the regulation of gene expression. Transcriptional activity and associated alterations in chromatin structure are supposed to be regulated per individual chromatin domain. Domain boundaries are thus expected to harbour insulator elements that can block the action of *cis*-regulatory elements and the propagation of chromatin alterations from one domain to the other (see Chapter 5).

S/MARs as boundaries of chromatin structure

Transcriptional activation is known to involve both long-range and local structural alterations of the chromatin fibre. Long-range alterations include the formation of chromatin regions with generally elevated or reduced DNAase I sensitivity. These alterations, which are thought to result in the 'opening up' or 'closing down', respectively, of the chromatin fibre, usually extend far beyond the actual gene boundaries and include cognate regulatory elements.

S/MAR elements have been found to co-localize with the boundaries of such DNAase I-sensitive domains. The best documented examples are the S/MARs flanking the chicken lysozyme gene domain (Phi-Van and Strätling 1988; see Chapter 7), and the human apolipoprotein B gene domain (Levy-Wilson and Fortier 1989). These S/MARs could thus exemplify a class of genomic elements that constitute boundaries for long-range chromatin alterations associated with changes in transcriptional activity. Although this has not been investigated, it is possible that a similar co-habitation with the limits of DNAase I-sensitive domains is the case for a number of other S/MARs that reside in non-transcribed regions at the flanks of genes or gene clusters (Gasser and Laemmli 1986; Jarman and Higgs 1988; Hall *et al.* 1991; Breyne *et al.* 1992; Van der Geest *et al.* 1994). For yet another group of S/MARs, intronic location seems incompatible with such a boundary function (Cockerill *et al.* 1987; Käs and Chasin 1987; Jarman and Higgs 1988; Romig *et al.* 1994).

S/MARs as regulators of transcription

In contrast to the expression of most endogenous genes, which is rather uniform from cell to cell, expression of stably transfected transgenes can vary greatly depending on the genomic position of integration (see Chapters 7 and 8). These variations, known as position effects, can be explained by the absence of functional insulator elements on the transfected constructs, as a consequence of which the transgenes come under the influence of local genomic regulatory elements. To investigate whether S/MARs can insulate genes within a given chromatin domain from *cis*-regulatory influences of neighbouring domains, the effect of S/MARs flanking stably transfected genes has been studied. In this experimental set up, two criteria for insulator elements have been defined. First, when placed at the flanks of a transfected (enhancer)–promoter gene unit, insulators are expected to shield the gene from influences of chromosomal surroundings, and, hence, should confer position-independent expression. This in turn is expected to yield expression levels that are proportional to transgene copy number. Second, when placed between an enhancer and a promoter gene unit, bona fide insulator elements should block the influence of the enhancer on the promoter. To assure that the elements work on the chromatin level, both effects should only be seen for stably integrated transgene constructs and not for transiently transfected constructs.

Dampening of position effects

Such transformation experiments have shown that some S/MARs can confer position-independent expression of flanking transgenes (for a recent review see Bode *et al.* 1995). Interestingly, these appear to be S/MARs that in their genomic context reside at the flanks of genes and/or border DNAase I-sensitive chromatin domains. The first indication that the chicken lysozyme S/MARs, also known as

A-elements, could have insulator properties came from the observation that the complete gene domain for chicken lysozyme, including its S/MARs, is expressed in a position-independent fashion in transgenic mice (Bonifer *et al.* 1990; see also Chapter 7). Later it was shown that these S/MARs are able to reduce the variation in expression of a tissue-specific mouse gene in transgenic mice (McKnight *et al.* 1992) and of a GUS reporter gene in transgenic plants (Mlynárová *et al.*, 1994, 1995). Also, in stably transfected tissue culture cells the chicken lysozyme S/MARs have been shown to dampen the variation in expression of transgenes driven by homologous (Stief *et al.* 1989) and heterologous (Phi-Van *et al.* 1990) promoters. Stief *et al.* (1989) noted that an upstream lysozyme gene enhancer was required besides the S/MARs to establish high level and position-independent expression. Dampening of position effects has also been reported for several plant S/MARs in transgenic plants (Breyne *et al.* 1992; Van der Geest *et al.* 1994), as well as for the human apolipoprotein B S/MARs in tissue culture cells (Kalos and Fournier 1995). In transgenic mice, however, the apolipoprotein B S/MARs did not clearly isolate flanking transgenes from position effects (Brooks *et al.* 1994). An S/MAR element in the 3' untranslated region of a human neurofilament light (hNLF)/lacZ construct was recently found to be required for position-independent and tissue-specific expression of this construct in transgenic mice (Charron *et al.* 1995). Interestingly, this effect could be attributed to the unwinding region of the S/MAR, supporting the notion (reviewed in Bode *et al.* 1995) that unwinding elements are important features for S/MAR function.

Various S/MARs in their genomic context occur in close physical association with tissue-specific enhancer elements (Cockerill and Garrard 1986a, 1987; Gasser and Laemmli 1986; Jarman and Higgs 1988). The functional significance of this particular configuration is enhanced by the fact that for some S/MARs a synergistic collaboration with enhancer elements has been observed. For instance, the human globin 5' S/MAR by itself has little effect on transcription of a reporter gene in stably transformed cells, but when linked in *cis* with the polyoma virus enhancer their synergistic action results in enhanced expression, which is position independent and proportional to gene copy number (Yu *et al.* 1994). A synergistic effect has also been reported for the enhancer-linked intronic S/MARs of the murine Ig μ heavy chain locus. Although in stably transfected cell lines the intronic Ig μ enhancer is sufficient for proper μ gene expression and the flanking S/MARs are dispensable, in transgenic mice these S/MARs do have an essential function (Forrester *et al.* 1994, and references therein). Here, they collaborate synergistically with the flanking tissue-specific enhancer to ensure normal levels of position-independent and tissue-specific expression of a rearranged μ gene. The different effects of these Ig μ S/MARs in germline transformations versus stable cell transfections may be related to a difference in DNA methylation state. DNA that is integrated at the one-cell stage of mouse embryogenesis is methylated before the gene is expressed and demethylated at later developmental stages. In line with this hypothesis, transfection of *in vitro* methylated Ig κ gene constructs into cultured lymphoid B cells has shown that expression and demethylation of the κ gene depends on both the intronic enhancer and flanking S/MAR (Lichtenstein *et al.* 1994). However, in contrast to the findings for the Ig μ S/MARs, no dramatic effect of the Ig κ S/MAR could be detected in transgenic mice (Xu *et al.* 1989). A protein that has been implicated in the interplay between the Ig μ S/MARs and enhancer is the S/MAR binding protein Bright (Herrscher *et al.* 1995). This B cell-specific protein recognizes a specific subsequence within these S/MARs and in the context of the active wild-type Ig μ enhancer transactivates gene expression. Other proteins that have been shown to bind these particular S/MARs, and could thus contribute to the regulation of μ heavy chain transcription, are the S/MAR binding matrix protein MAR-BP1 (Zong and Scheuermann 1995), the thymus-specific S/MAR binding protein SATB1 (Dickinson *et al.* 1992), and the NF μ NR enhancer repressor, which is only present in non-B cells. The mechanism of action of enhancer-linked S/MAR elements may bear some resemblance to that of the 'facilitator elements' that bilaterally flank a T cell-specific enhancer domain in the human adenosine deaminase gene (Aronow *et al.* 1995). Together with the enhancer domain these 'facilitators' are considered to constitute a locus control region (LCR), generating gene copy-proportional and integration site-independent reporter gene expression in transgenic thymocytes.

Blocking of enhancer action

So far, results concerning the enhancer blocking potential of S/MARs are inconsistent. Bode *et al.* (1995) have recently reported that an S/MAR from the human interferon-β domain efficiently blocks the stimulatory effect of a CMV enhancer on transcription of a luciferase reporter gene in stably transformed cells. Enhancer blocking has also been observed in stably transformed cells for the constitutive 5' hypersensitive site 5 of the human β-globin locus, which contains an S/MAR (Jarman and Higgs 1988; Li and Stamatoyannopoulos 1994; see Chapter 8). However, it remains to be determined whether this blocking effect is actually due to S/MAR activity or to other sequence information on the transfected construct, considering that the analogous constitutive 5' hyper-sensitive site 4 of the chicken β-globin locus exhibits a similar enhancer blocking effect but does not contain an S/MAR (Chung *et al.* 1993). Negative enhancer blocking results have been obtained for two *Drosophila* S/MARs. Kellum and Schedl (1992) have shown that the *Drosophila* hsp70 S/MAR cannot block the action of the yp1 enhancer on the hsp70 promoter in transgenic flies, whereas in parallel assays several specialized chromatin structure (scs) insulator elements did show an efficient blocking effect. Likewise, in stably transfected human and mouse cells both the *Drosophila* hsp70 S/MAR and the histone repeat S/MAR were proven ineffective in enhancer blocking assays (Poljak *et al.* 1994). To gain insight into the molecular background of the apparent differences in enhancer blocking potential between different S/MARs it will be important to subject other natural and synthetic S/MAR sequences to this type of assay.

Quantitative effects on transcription level

An important conception that has emerged from these transfection experiments is that S/MARs generally increase the average expression level of flanking transgenes (Blasquez *et al.* 1989; Stief *et al.* 1989; Xu *et al.* 1989; Mielke *et al.* 1990; Phi-Van *et al.* 1990; Klehr *et al.* 1991; Allen *et al.* 1993; Brooks *et al.* 1994; Dietz *et al.* 1994; Forrester *et al.* 1994; Poljak *et al.* 1994; Kalos and Fournier 1995). The orientation of the S/MARs does not seem to be important for this effect (Klehr *et al.* 1991), similar to what has been found for the action of enhancers. However, it should be stressed that S/MARs do not act as classical enhancers since their effects on transcription are only observed for transgene constructs that have been integrated into the genome and not for transiently transfected constructs.

Several models to explain the stimulatory effect of S/MARs on transcription have been put forward. Laemmli and colleagues assign an important role to S/MARs in the unfolding of the chromatin fibre during transcriptional activation. Their 'chromatin switch model' proposes that S/MARs, by virtue of their A-tract richness, function as nucleation centres for the displacement of histone H1 from flanking chromatin regions by HMGI/Y and other distamycin analogue proteins (reviewed in Laemmli *et al.* 1992). In agreement with this view, formation of an extended DNAase I-sensitive domain around the rearranged Ig μ gene in transgenic mice has been shown to depend on the intronic Ig μ S/MARs (Forrester *et al.* 1994). Bode and collaborators have proposed that S/MAR elements may boost transcription by stabilizing the chromosomal topology which arises as a consequence of transcription (reviewed in Bode *et al.* 1995). This model incorporates the observation that the chicken lysozyme S/MAR elements are able to potentiate the stimulatory effect of butyrate on transgene transcription (Klehr *et al.* 1992; Schlake *et al.* 1994). Butyrate is an agent that through hyperacetylation of core histones generates unconstrained negatively supercoiled DNA, which in turn has been shown to stimulate various eukaryotic promoters. During transcription a negatively supercoiled domain is formed upstream of the transcription complex and a positively supercoiled domain downstream of it. This topology, known as 'twin domain structure', could be stabilized by S/MARs. By virtue of their tight binding to the nuclear matrix, S/MARs may inhibit the transmission of superhelicity from one domain to the other, and due to their tendency to separate strands and their ability to bind topoisomerase II, S/MARs could relax torsional stress that builds up during transcription. In support of this model, the stimulatory effect of S/MARs on transcription has been shown to require their unwinding and binding to the nuclear matrix (Mielke *et al.* 1990; Bode *et al.* 1992).

Evolutionary conservation

An essential observation is that, like the interaction of S/MARs with the nuclear matrix, their effects on gene expression also appear to be evolutionarily conserved. Various S/MARs have been shown to reduce the variation in expression level or to enhance transcription of stably transfected transgenes driven by heterologous promoters and/or in cells of a different species (Phi-Van *et al.* 1990; Klehr *et al.* 1991; McKnight *et al.* 1992; Allen *et al.* 1993; Brooks *et al.* 1994; Dietz *et al.* 1994; Mlynárová *et al.* 1994; Poljak *et al.* 1994; Kalos and Fournier 1995). The evolutionary conservation of S/MAR structure and function suggests that S/MARs are involved in a fundamental aspect of gene expression common to all eukaryotes.

Future directions

To gain further insight into the role of S/MARs in chromatin organization and gene expression it will be essential to identify causal relationships between sequence motifs and chromatin structure of S/MARs, S/MAR–protein inter-actions, and effects of S/MARs on gene expression. In view of the apparent diversity of S/MAR sequences, S/MAR binding proteins, genomic locations of S/MARs, and their effects on transgene expression, research may focus on specific S/MARs that reside within genetically well-defined genomic regions. Although adequate techniques need to be developed to assess the relevance of anticipated S/MAR–protein interactions *in vivo*, major insights into the relation-ship between specific S/MAR substructures and functions may be gained from a combination of *in vitro* DNA–protein binding experiments and detailed stable transformation studies. With respect to the latter type of experiment, special care should be taken to control for the increasing number of complicating phenomena that have been recognized in recent years. These problems include the prefer-ential integration of transgene constructs into transcriptionally poised chromatin (Blasquez *et al.* 1989; Dietz *et al.* 1994), the co-integration of co-transfected reporter and selector genes at the same chromosomal site (Kalos and Fournier 1995, and references therein), the potential presence of other matrix binding sequences on the construct besides S/MARs, and the tendency of transgenes to integrate in long tandem arrays, which in both plant (Allen *et al.* 1993) and animal cells (Dorer and Henikoff 1994; Kalos and Fournier 1995) are often transcriptionally inactive (for recent review see Bode *et al.* 1995).

Summary

S/MARs are specific AT-rich sequences. Their frequent distribution along the genome and their high binding affinity for the nuclear matrix make them plausible anchors of chromatin loops. *In vitro* binding studies have identified various candidate S/MAR binding proteins, most of which recognize character-istic structural features of S/MARs. Stable transformation studies have indicated that S/MARs are involved in gene expression. S/MARs enhance basal promoter functions in an orientation-independent manner, and this effect is only seen for transfected constructs that have been integrated into the genome and not for transiently transfected constructs. Underlying working mechanisms have not yet been firmly established. The finding that the structure of S/MARs, their interaction with the nuclear matrix, as well as their effects on gene expression are evolutionarily conserved suggests that S/MARs are involved in a very basic aspect of chromatin organization common to all eukaryotes. Current experi-mental evidence suggests that different classes of S/MARs with different, though probably related, functions may exist. Some S/MARs appear to have character-istics compatible with a function as boundaries of chromatin domains: they coincide with the borders of DNAase I-sensitive chromatin regions and are able to reduce position effects in stable transformants. Other S/MARs have been shown to collaborate synergistically with flanking tissue-specific enhancers to ensure position-independent, tissue-specific expression of cognate genes during development. Future efforts to elucidate S/MAR structure–function relationships are expected to give a deeper understanding of the role of S/MARs in chromatin organization and gene expression.

References

Adachi, Y., Käs, E., and Laemmli, U.K. (1989). Preferential, cooperative binding of DNA topoisomerase II to scaffold-associated regions. *EMBO Journal*, **8**, 3997–4006.

Allen, G.C., Hall Jr., G.E., Childs, L.C., Weissinger, A.K., Spiker, S., and Thompson, W.F. (1993). Scaffold attachment regions increase reporter gene expression in stably transformed plant cells. *Plant Cell*, **5**, 603–13.

Amati, B.B. and Gasser, S.M. (1990). Drosophila scaffold-attached regions bind nuclear scaffolds and can function as ARS elements in both budding and fission yeasts. *Molecular and Cellular Biology*, **10**, 5442–54.

Amati, B.B., Pick, L., Laroche, T., and Gasser, S.M. (1990). Nuclear scaffold attachment stimulates, but is not essential for ARS activity in *Saccharomyces cerevisiae*: analysis of the *Drosophila ftz* SAR. *EMBO Journal*, **9**, 4007–16.

Aronow, B.J., Ebert, C.A., Valerius, M.T., Potter, S.S., Wiginton, D.A., Witte, D.P., *et al.* (1995). Dissecting a Locus Control Region: facilitation of enhancer function by extended enhancer-flanking sequences. *Molecular and Cellular Biology*, **15**, 1123–35.

Benyajati, C. and Worcel, A. (1976). Isolation, characterization, and structure of the folded interphase genome of *Drosophila melanogaster. Cell*, **9**, 393–407.

Berrios, M., Osherhoff, N., and Fisher, P.A. (1985). *In situ* localization of DNA topoisomerase II, a major polypeptide component of the Drosophila nuclear matrix fraction. *Proceedings of the National Academy of Sciences of the USA*, **82**, 4142–6.

Blasquez, V.C., Xu, M., Moses, S.C., and Garrard, W.T. (1989). Immunoglobulin κ gene expression after stable integration. *Journal of Biological Chemistry*, **264**, 21183–9.

Bode, J. and Maass, K. (1988). Chromatin domain surrounding the human interferon-β gene as defined by scaffold-attached regions. *Biochemistry*, **27**, 4706–11.

Bode, J., Kohwi, Y., Dickinson, L., Joh, T., Klehr, D., Mielke, C., *et al.* (1992). Biological significance of unwinding capability of nuclear matrix-associating DNAs. *Science*, **255**, 195–7.

Bode, J., Schlake, T., Ríos-Ramírez, M., Mielke, C., Stengert, M., Kay, V., *et al.* (1995). Scaffold/matrix-attached regions: structural properties creating transcriptionally active loci. *International Review of Cytology*, **162A**, 389–454.

Bonifer, C., Vidal, M., Grosveld, F., and Sippel, A.E. (1990). Tissue-specific and position independent expression of the complete gene domain for chicken lysozyme in transgenic mice. *EMBO Journal*, **9**, 2843–8.

Boulikas, T. (1995). Chromatin domains and prediction of MAR sequences. *International Review of Cytology*, **162A**, 279–388.

Breyne, P., Van Montagu, M., Depicker, A., and Gheysen, G. (1992). Characterization of a plant scaffold attachment region in a DNA fragment that normalizes transgene expression in tobacco. *Plant Cell*, **4**, 463–71.

Brooks, A.R., Nagy, B.P., Taylor, S., Simonet, W.S., Taylor, J.M., and Levy-Wilson, B. (1994). Sequences containing the second-intron enhancer are essential for transcription of the human apolipoprotein B gene in the livers of transgenic mice. *Molecular and Cellular Biology*, **14**, 2243–56.

Buhrmester, H., von Kries, J. P., and Strätling, W. H. (1995). Nuclear matrix protein ARBP recognizes a novel DNA sequence motif with high affinity. *Biochemistry*, **34**, 4108–17.

Charron, G., Julien, J.-P., and Bibor-Hardy, V. (1995). Neuron specificity of the neurofilament light promoter in transgenic mice requires the presence of DNA unwinding elements. *Journal of Biological Chemistry*, **270**, 25739–45.

Chung, J.H., Whiteley, M., and Felsenfeld, G. (1993). A 5' element of the chicken β-globin domain serves as an insulator in human erythroid cells and protects against position effect in *Drosophila*. *Cell*, **74**, 505–14.

Ciejek, E.M., Tsai, M.-J., and O'Malley, B.W. (1983). Actively transcribed genes are associated with the nuclear matrix. *Nature*, **306**, 607–9.

Cockerill, P.N. and Garrard, W.T. (1986a). Chromosomal loop anchorage of the kappa immunoglobulin gene occurs next to the enhancer in a region containing topoisomerase II sites. *Cell*, **44**, 273–82.

Cockerill, P.N. and Garrard, W.T. (1986b). Chromosomal loop anchorage sites appear to be evolutionary conserved. *FEBS Letters*, **204**, 5–7.

Cockerill, P.N., Yuen, M.-H., and Garrard, W.T. (1987). The enhancer of the immunoglobulin heavy chain locus is flanked by presumptive chromosomal loop anchorage elements. *Journal of Biological Chemistry*, **262**, 5394–7.

Das, A.T., Ludérus, M.E.E., and Lamers, W.H. (1993). Identification and analysis of a matrix-attachment region 5' of the rat glutamate-dehydrogenase-encoding gene. *European Journal of Biochemistry*, **215**, 777–85.

Dickinson, L.A. and Kohwi-Shigematsu, T. (1995). Nucleolin is a matrix attachment region DNA-binding protein that specifically recognizes a region with high base-unpairing potential. *Molecular and Cellular Biology*, **15**, 456–65.

Dickinson, L.A., Joh, T., Kohwi, Y., and Kohwi-Shigematsu, T. (1992). A tissue-specific MAR/SAR DNA-binding protein with unusual binding site recognition. *Cell*, **70**, 631–45.

Dietz, A., Kay, V., Schlake, T., Landsmann, J., and Bode, J. (1994). A plant scaffold attached region detected close to a T-DNA integration site is active in mammalian cells. *Nucleic Acids Research*, **22**, 2744–51.

Dorer, D.R. and Henikoff, S. (1994). Expansions of transgene repeats cause hetero-chromatin formation and gene silencing in *Drosophila*. *Cell*, **77**, 993–1002.

Fackelmayer, F.O., Dahm, K., Renz, A., Ramsperger, U., and Richter, A. (1994). Nucleic-acid-binding properties of hnRNP-U/SAF-A, a nuclear-matrix protein which binds DNA and RNA *in vivo* and *in vitro*. *European Journal of Biochemistry*, **221**, 749–57.

Forrester, W.C., van Genderen, C., Jenuwein, T., and Grosschedl, R. (1994). Dependence of enhancer-mediated transcription of the immunoglobulin: gene on nuclear matrix attachment regions. *Science*, **265**, 1221–5.

Garrard, W.T. (1990). Chromosomal loop organization in eukaryotic genomes. In *Nucleic acids and molecular biology*, Vol. 4 (ed. F. Eckstein and D.M.J. Lilley), pp 163–75. Springer-Verlag, Berlin.

Gasser, S.M. and Laemmli, U.K. (1986). Cohabitation of scaffold binding regions with upstream/enhancer elements of three developmentally regulated genes of *D. melanogaster*. *Cell*, **46**, 521–30.

Gerdes, M.G., Carter, K.C., Moen, Jr., P.T., and Lawrence, J.B. (1994). Dynamic changes in the higher-level chromatin organization of specific sequences revealed by *in situ* hybridization to nuclear halos. *Journal of Cell Biology*, **126**, 289–304.

Gromova, I.I., Thompsen, B., and Razin, S.V. (1995). Different topoisomerase II antitumour drugs direct similar specific long-range fragmentation of an amplified c-myc gene locus in living cells and in high salt-extracted nuclei. *Proceedings of the National Academy of Sciences of the USA*, **92**, 102–6.

Hall, G., Allen, G.C., Loer, D.S., Thompson, W.F., and Spiker, S. (1991). Nuclear scaffolds and scaffold-attachment regions in higher plants. *Proceedings of the National Academy of Sciences of the USA*, **88**, 9320–4.

Herrscher, R.F., Kaplan, M.H., Lelsz, D.L., Das, C., Scheuermann, R., and Tucker, P.W. (1995). The immunoglobulin heavy-chain matrix-associating regions are bound by Bright: a B-cell-specific trans-activator that descibes a new DNA-binding protein family. *Genes and Development*, **9**, 3067–82.

Iarovaia, O., Hancock, R., Lagarkova, M., Miassod, R., and Razin, S.V. (1996). Mapping of genomic DNA loop organization in a 500-kilobase region of the *Drosophila* X chromosome by the topoisomerase II-mediated DNA loop excision protocol. *Molecular and Cellular Biology*, **16**, 302–8.

Izaurralde, E., Käs, E., and Laemmli, U.K. (1988). Highly preferential nucleation of histone H1 assembly on scaffold-associated regions. *Journal of Molecular Biology*, **210**, 573–85.

Jackson, D.A., McCready, S.J., and Cook, P.R. (1984). Replication and transcription depend on attachment of DNA to the nuclear cage. *Journal of Cell Science Supplement*, **1**, 59–79.

Jackson, D.A., Dickinson, P., and Cook, P.R. (1990). The size of chromatin loops in HeLa cells. *EMBO Journal*, **9**, 567–71.

Jackson, D.A., Bartlett, J., and Cook, P.R. (1996). Sequences attaching loops of nuclear and mitochondrial DNA to underlying structures in human cells: the role of transcription units. *Nucleic Acids Research*, **24**, 1212–19.

Jarman, A.P. and Higgs, D.R. (1988). Nuclear scaffold attachment sites in the human globin gene complexes. *EMBO Journal*, **7**, 3337–44.

Kalos, M. and Fournier, R.E.K. (1995). Position-independent transgene expression mediated by boundary elements from the apolipoprotein B chromatin domain. *Molecular and Cellular Biology*, **15**, 198–207.

Käs, E. and Chasin, L.A. (1987). Anchorage of the chinese hamster dihydrofolate reductase gene to the nuclear scaffold occurs in an intragenic region. *Journal of Molecular Biology*, **198**, 677–92.

Käs, E. and Laemmli, U.K. (1992). *In vivo* topoisomerase II cleavage of the *Drosophila* histone and satellite III repeats: DNA sequence and structural characteristics. *EMBO Journal*, **11**, 705–16.

Käs, E., Izaurralde, E., and Laemmli, U.K. (1989). Specific inhibition of DNA binding to nuclear scaffolds and histone H1 by distamycin. *Journal of Molecular Biology*, **210**, 587–99.

Käs, E., Poljak, L., Adachi, Y., and Laemmli, U.K. (1993). A model for chromatin opening: stimulation of topoisomerase II and restriction enzyme cleavage of chromatin by distamycin. *EMBO Journal*, **12**, 115–26.

Kay, V. and Bode, J. (1994). Binding specificity of a nuclear scaffold: supercoiled, single-stranded, and scaffold-attached-region DNA. *Biochemistry*, **33**, 367–74.

Kellum, R. and Schedl, P. (1992). A group of scs elements function as domain boundaries in an enhancer-blocking assay. *Molecular and Cellular Biology*, **12**, 2424–31.

Klehr, D., Maass, K., and Bode, J. (1991). Scaffold-attached regions from the human interferon β domain can be used to enhance the stable expression of genes under the control of various promoters. *Biochemistry*, **30**, 1264–70.

Klehr, D., Schlake, T., Maass, K., and Bode, J. (1992). Scaffold-attached regions (SAR elements) mediate transcriptional effects due to butyrate. *Biochemistry*, **31**, 3222–9.

Laemmli, U.K., Käs, E., Poljak, L., and Adachi, Y. (1992). Scaffold-associated regions: cis-acting determinants of chromatin structural loops and functional domains. *Current Opinion in Genetics and Development*, **2**, 275–85.

Levy-Wilson, B. and Fortier, C. (1989). The limits of the DNase I-sensitive domain of the human apolipoprotein B gene coincide with the locations of chromosomal anchorage loops and define the 5' and 3' boundaries of the gene. *Journal of Biological Chemistry*, **264**, 21196–204.

Li, Q. and Stamatoyannopoulos, G. (1994). Hypersensitive site 5 of the human β locus control region functions as a chromatin insulator. *Blood*, **5**, 1399–401.

Lichtenstein, M., Keini, G., Cedar, H., and Bergman, Y. (1994). B cell-specific demethylation: a novel role for the intronic κ chain enhancer sequence. *Cell*, **76**, 913–23.

Ludérus, M.E.E., de Graaf, A., Mattia, E., den Blaauwen, J.L., Grande, M.A., de Jong, L., *et al.* (1992). Binding of matrix attachment regions to lamin B.. *Cell*, **70**, 949–59.

Ludérus, M.E.E., den Blaauwen, J.L., de Smit, O.J.B., Compton, D.A., and van Driel, R. (1994). Binding of matrix attachment regions to lamin polymers involves single-stranded regions and the minor groove. *Molecular and Cellular Biology*, **14**, 6297–305.

Mattern, K.A., Humbel, B.M., Muijsers, A.O., de Jong, L., and van Driel, R. (1996). HnRNP proteins and B23 are the major proteins of the internal nuclear matrix of HeLa S3 cells. *Journal of Cellular Biochemistry*, **62**, 275–89.

McKnight, R.A., Shamay, A., Sankaran, L., Wall, R.J., and Hennighausen, L. (1992). Matrix-attachment regions can impart position-independent regulation of a tissue-specific gene in transgenic mice. *Proceedings of the National Academy of Sciences of the USA*, **89**, 6943–7.

Mielke, C., Kohwi, Y., Kohwi-Shigematsu, T., and Bode, J. (1990). Hierarchical binding of DNA fragments derived from scaffold-attached regions: correlation of properties *in vitro* and function *in vivo*. *Biochemistry*, **29**, 7475–85.

Mirkovitch, J., Mirault, M.-E., and Laemmli, U.K. (1984). Organization of the higher order chromatin loop: specific DNA attachment sites on nuclear scaffold. *Cell*, **39**, 223–32.

Mirkovitch, J., Spierer, P., and Laemmli, U.K. (1986). Genes and loops in 320,000 base-pairs of the *Drosophila melanogaster* chromosome. *Journal of Molecular Biology*, **190**, 255–8.

Mlynárová, L., Loonen, A., Heldens, J., Jansen, R. C., Keizer, P., Stiekema, W. J., *et al.* (1994). Reduced position effect in mature transgenic plants conferred by the chicken lysozyme matrix-associated region. *Plant Cell*, **6**, 417–26.

Mlynárová, L., Jansen, R.C., Conner, A.J., Stiekema, W.J., and Nap, J.-P. (1995). The MAR-mediated reduction in position effect can be uncoupled from copy number-dependent expression in transgenic plants. *Plant Cell*, **7**, 599–609.

Nakagomi, K., Kohwi, Y., Dickinson, L.A., and Kohwi-Shigematsu, T. (1994). A novel DNA-binding motif in the nuclear matrix attachment DNA-binding protein SATB1. *Molecular and Cellular Biology*, **14**, 1852–60.

Phi-Van, L. and Strätling, W. H. (1988). The matrix attachment regions of the chicken lysozyme gene co-map with the boundaries of the chromatin domain. *EMBO Journal*, **7**, 655–64.

Phi-Van, L., von Kries, J.P., Ostertag, W., and Strätling, W.H. (1990). The chicken lysozyme 5' matrix attachment region increases transcription from a heterologous promoter in heterologous cells and dampens position effects on the expression of transfected genes. *Molecular and Cellular Biology*, **10**, 2302–7.

Poljak, L., Seum, C., Mattioni, T., and Laemmli, U.K. (1994). SARs stimulate but do not confer position independent gene expression. *Nucleic Acids Research*, **22**, 4386–94.

Reeves, R. and Nissen, M.S. (1990). The A·T-DNA-binding domain of mammalian high mobility group I chromosomal proteins. *Journal of Biological Chemistry*, **265**, 8573–82.

Romig, H., Fackelmayer, F.O., Renz, A., Ramsperger, U., and Richter, A. (1992). Characterization of SAF-A, a novel nuclear DNA binding protein from HeLa cells with high affinity for nuclear matrix/scaffold attachment DNA elements. *EMBO Journal*, **11**, 3431–40.

Romig, H., Ruff, J., Fackelmayer, F.O., Patil, M.S., and Richter, A. (1994). Characterization of two intronic nuclear-matrix-attachment regions in the human DNA topoisomerase I gene. *European Journal of Biochemistry*, **221**, 411–19.

Schlake, T., Klehr-Wirth, D., Yoshida, M., Beppu, T., and Bode, J. (1994). Gene expression within a chromatin domain: the role of core histone hyperacetylation. *Biochemistry*, **33**, 4197–206.

Solomon, M.J., Strauss, F., and Varshavsky, A. (1986). A mammalian high mobility group protein recognizes any stretch of six A·T base pairs in duplex DNA. *Proceedings of the National Academy of Sciences of the USA*, **83**, 1276–80.

Stief, A., Winter, D.M., Strätling, W.H., and Sippel, A.E. (1989). A nuclear DNA attachment element mediates elevated and position-independent gene activity. *Nature*, **341**, 343–5.

Stokes, D.G. and Perry, R.P. (1995). DNA-binding and chromatin localization properties of CDH1. *Molecular and Cellular Biology*, **15**, 2745–53.

Strick, R. and Laemmli, U.K. (1995). SARs are cis DNA elements of chromosome dynamics: synthesis of a repressor protein. *Cell*, **83**, 1137–48.

Surdej, P., Got, C., Rosset, R., and Miassod, R. (1990). Supragenic loop organization: mapping in *Drosophila* embryos, of scaffold-associated regions on a 800 kilobase DNA continuum cloned from the 14B-15B first chromosome region. *Nucleic Acids Research*, **18**, 3713–22.

Tsutsui, K., Tsutsui, K., Okada, S., Watarai, S., Seki, S., Yasuda, T., *et al.* (1993). Identification and characterization of a nuclear scaffold protein that binds the matrix attachment region DNA. *Journal of Biological Chemistry*, **268**, 12886–94.

Van der Geest, A.H.M., Hall, Jr., G.E., Spiker, S., and Hall, T.C. (1994). The β-phaseolin gene is flanked by matrix attachment regions. *Plant Journal*, **6**, 413–23.

Vassetsky, Y.S., Dang, Q., Benedetti, P., and Gasser, S.M. (1994). Topoisomerase II forms multimers *in vitro*: effects of metals, β-glycerophosphate, and phosphorylation of its C-terminal domain. *Molecular and Cellular Biology*, **14**, 6962–74.

Vogelstein, B., Pardoll, D., and Coffey, D.S. (1980). Supercoiled loops and eukaryotic DNA replication. *Cell*, **22**, 79–85.

Von Kries, J.P., Buhrmester, H., and Strätling, W.H. (1991). A matrix/scaffold attachment region binding protein: identification, purification, and mode of binding. *Cell*, **64**, 123–35.

Von Kries, J.P., Buck, F., and Strätling, W.H. (1994). Chicken MAR binding protein p120 is identical to human heterogeneous nuclear ribonucleoprotein (hnRNP) U. *Nucleic Acids Research*, **22**, 1215–20.

Wang, B., Dickinson, L.A., Koivunen, E., and Kohwi-Shigematsu, T. (1995). A novel matrix attachment region DNA binding motif identified using a random phage peptide library. *Journal of Biological Chemistry*, **270**, 23239–42.

Xu, M., Hammer, R.E., Blasquez, V.C., Jones, S.L., and Garrard, W.T. (1989). Immunoglobulin _ gene expression after stable integration. *Journal of Biological Chemistry*, **35**, 21190–5.

Yanagisawa, J., Ando, J., Nakayama, J., Kohwi, Y., and Kohwi-Shigematsu, T. (1996). A matrix attachment region (MAR)-binding activity due to a p114 kilodalton protein is found only in human breast carcinomas and not in normal and benign breast disease tissue. *Cancer Research*, **56**, 457–62.

Yu, J., Bock, J.H., Slightom, J.L., and Villeponteau, B. (1994). A 5' β-globin matrix-attachment region and the polyoma enhancer together confer position-independent transcription. *Gene*, **139**, 139–45.

Zhao, K., Käs, E., Gonzalez, E., and Laemmli, U.K. (1993). SAR-dependent mobilization of histone H1 by HMG-I/Y *in vitro*: HMG-I/Y is enriched in H1-depleted chromatin. *EMBO Journal*, **12**, 3237–47.

Zong, R.-T. and Scheuermann, R.H. (1995). Mutually exclusive interaction of a novel matrix attachment region binding protein and the NF-μNR enhancer repressor. *Journal of Biological Chemistry*, **270**, 24010–18.

7

The chicken lysozyme chromatin domain

Constanze Bonifer, Nicole Faust, Matthias C. Huber, Harald
Saueressig, and Albrecht E. Sippel

Introduction

The investigation of the molecular mechanisms of differential gene expression
will ultimately lead to a detailed understanding of the basis of differentiation
decisions within a developing organism. For many years research focused on
identifying and characterizing individual *cis*-regulatory elements responsible for
tissue-specific gene expression. This approach led to the discovery and characterization of the molecular anatomy of enhancers, silencers, and promoters. The
minimal requirements for tissue-specific gene expression were determined,
mostly by transient transfection assays, and in many cases were assigned to the
immediate promoter region of a gene. However, with increasing stringency of
assay systems and the need to obtain reproducibly high and stable gene
expression (for instance in gene therapy approaches), it became clear that many
gene constructs shown to work well in a transient transfection assay do not show
any activity when assayed stably integrated in a chromosomal environment
(Wilson *et al.* 1990). Moreover, differences were observed when the same
constructs were analysed stably transfected into cultured cells and transgenic
organisms (Zimmerman *et al.* 1990). It was therefore feasible to assume that
dynamic changes in chromatin structure of a gene locus play an essential role in
the regulation of its transcriptional activation in development. For these reasons
we set out to answer the following general questions.

1. What structural changes in chromatin take place during differentiation-
   dependent activation of an entire gene locus?

2. Which *cis*-regulatory elements are responsible for this activation and how do
   they interact during cellular differentiation?

3. What are the prerequisites for correct tissue-specific regulation of a gene
   locus in a transgenic organism?

4. How does the introduction of mutations, such as deletions, affect expression
   pattern and chromatin structure?

The model we used to answer these questions is the chicken lysozyme locus.

Chromatin structure of the chicken lysozyme locus undergoes dramatic changes during activation of the gene locus during development

The chicken lysozyme gene is expressed in two different tissues: the mature oviduct and cells of the myeloid lineage (Schütz *et al.* 1978; Hauser *et al.* 1981). However, the mode of regulation in the two tissues is remarkably different. In the oviduct gene expression is strongly stimulated by steroid hormones, whereas in myeloid cells the gene is steroid unresponsive and is upregulated during differentiation of multipotent myeloid progenitor cells to mature granulocytes and macrophages (Sippel *et al.* 1988; Huber *et al.* 1995). Analysis of the chromatin structure of the gene locus in oviduct and in myeloid cells (Fritton *et al.* 1983, 1984; Sippel *et al.* 1988) revealed an overlapping, but non-identical set of DNAase I-hypersensitive chromatin sites (DHSs) in the two tissues. It is well established that in most cases the presence of a tissue-specific DHS in chromatin is an indication for the activity of a *cis*-regulatory element. The complex DHS pattern in chicken lysozyme chromatin pointed to a surprising number of *cis*-regulatory elements, all involved in the regulation of the gene locus in different cell types. Transfection analysis revealed the presence of two enhancers (Es; –6.1 kb and –2.7 kb), a hormone-responsive element (HRE; –1.9 kb), a silencer (S; –2.4 kb) and a complex promoter (P) (Fig. 7.1) (Hecht *et al.* 1988; Altschmied *et al.* 1989; Baniahmad *et al.* 1990; Bonifer *et al.* 1991; Grewal *et al.* 1992). Recently, an additional enhancer at –3.9 kb was identified, which is marked by the presence of a DHS at –3.9 kb in myeloid cells (Saueressig 1994; M.C. Huber and C. Bonifer, unpublished).

The formation of DHSs is not the only structural change in chromatin that is observed along with transcriptional activation of the lysozyme gene locus. Another phenomenon, of which the molecular basis is still poorly understood, is the increase in general DNAase I sensitivity over an area of more than 20 kb around the transcribed region (Jantzen *et al.* 1986). The transition regions from high to average DNAase I sensitivity are the same in oviduct and in macrophage chromatin (Sippel *et al.* 1993), indicating that defined structural features might determine the size of the general DNAase I-sensitive domain. It could indeed be shown that the borders of the DNAase I-sensitive domain coincide with sequences having a high affinity for the nuclear matrix *in vitro*, called scaffold/matrix-associated regions or S/MARs (Phi-Van and Strätling 1988; see Chapter 6).

The recent development of retrovirally transformed chicken cell lines, resembling multipotent myeloid progenitor cells (Graf *et al.* 1992), which can be induced to differentiate into non-myeloid and more mature myeloid cells, gave us the unique opportunity to follow the development of an active chromatin configuration during myeloid differentiation. We were also able to determine the developmental onset of activity of the various *cis*-regulatory elements (Huber *et al.* 1995). The results of these experiments are summarized in Fig. 7.1. It turned

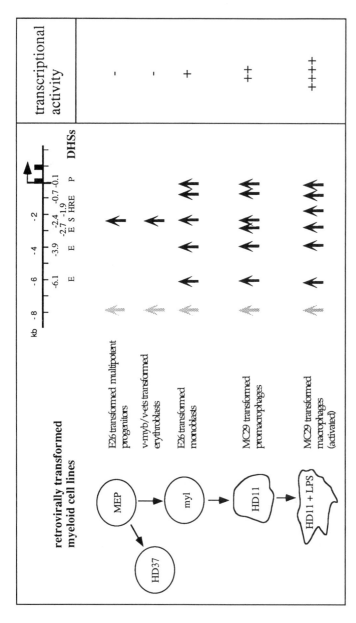

Fig. 7.1 Chromatin structure of the chicken lysozyme locus is reorganized during myeloid differentiation. The upper panel indicates the overall structure of the 5' regulatory region of the lysozyme locus with the transcriptional start site (indicated by a horizontal arrow) and the exons (black bars). The positions of the DNAase-hypersensitive sites (DHSs) in kb are indicated by numbers below the map, the location and the nature of *cis*-regulatory elements are indicated further below. E, enhancer; S, silencer; P, promoter; HRE, hormone-responsive element. The left panel depicts the various differentiation stages of the myeloid lineage, as represented by retrovirally transformed chicken cell lines. The middle panel indicates the pattern of DHSs in these cell lines. DHSs are symbolized by black arrows; the grey arrows indicate the position of the constitutive DHS at −7.9 kb, which is not relevant for this analysis. The right panel demonstrates the mRNA level of the lysozyme gene in the various cell lines.

out that the lysozyme gene is transcriptionally inactive in multipotent progenitor (MEP) cell lines and exhibits a chromatin configuration characteristic for non-expressing cells, as resembled by the erythroid cell line HD37. Only the DHSs at the −2.4 kb silencer and at −7.9 kb are visible. At the myeloblast stage, represented by HD50 myl cells, DHSs appear at the −6.1 kb enhancer, at the −3.9 kb enhancer, and at the promoter. A later stage in macrophage differentiation, the promacrophage stage, is represented by the MC29 transformed HD11 cell line, which can be further differentiated to macrophage-like cells after stimulation with bacterial lipopolysaccharide (LPS). Here, a switch in chromatin structure in the region between −2.7 kb and −1.9 kb upstream of the transcriptional start site is observed. The DHS at −2.4 kb disappears and a new DHS at −2.7 kb is formed. Upon LPS stimulation an additional DHS at −1.9 kb is formed. Along with the gradual reorganization of the 5' flanking chromatin, the transcriptional activity of the gene is upregulated from a very low level in myeloblast-like cells to an almost 100-fold higher activity in LPS-stimulated macrophage-like cells.

The complete chicken lysozyme locus is the regulatory unit of transcription: all *cis*-regulatory elements are required for tissue-specific and copy number-dependent expression

To obtain detailed information about the role of the various *cis*-regulatory elements in the developmentally controlled activation of the lysozyme locus we employed the most stringent assay system, i.e. the transgenic mouse. All previous experiments had indicated that the chicken lysozyme gene is located in a structurally confined chromatin domain harbouring all necessary *cis*-regulatory elements. In order to be able to draw relevant conclusions we used the following strategy. We knew that many gene constructs do not function correctly when stably integrated into the genome. Depending on the chromosomal location their expression levels vary, a phenomenon called genomic position effect (Wilson *et al.* 1990; see Chapters 8 and 13). A second difficulty with gene constructs not fulfilling the requirements for correct gene expression is ectopic expression, i.e. transcriptional activation of a gene in tissues where it is normally not expressed (Bonifer *et al.* 1994). Thus, chromosomal position effects would obstruct any attempt to correlate structure and expression of a locus. From other systems, particularly from the β-globin locus (Grosveld *et al.* 1987; Strouboulis *et al.* 1992), we were aware that the correct regulation of a gene locus in transgenic mice requires much larger chromosomal regions than anticipated from transient transfection analysis. Hence, instead of analysing distinct individual *cis*-regulatory elements or combinations thereof, we first determined which type of construct was expressed specifically in the right cell type (i.e. in macrophages) and was not influenced by chromosomal position effects. The complete, structurally defined lysozyme locus, including upstream and downstream S/MARs, fulfilled this criterion (Bonifer *et al.* 1990). It was expressed in the correct cell type (macrophages) and at the same level per gene copy in different independently derived mouse lines.

Subsequently, we modified the locus systematically to determine the contribution of the individual *cis*-regulatory elements. The fact that the lysozyme locus is small (approximately 21 kb) made it possible to analyse the role of each element in cell differentiation-dependent transcriptional activation in the context of the entire gene locus, and without having to severely disturb the spatial organization of the locus (Bonifer *et al.* 1994). Figure 7.2 summarizes the results of these experiments. We could show that constructs with a deletion of one enhancer are not expressed in a copy number-dependent fashion and are thus subject to chromosomal position effects (Fig. 7.2(b)). This finding has been confirmed in another experimental system. Totipotent mouse embryonic stem (ES) cells were transfected with the wild-type construct (i.e. the complete lysozyme locus) or deletion mutants lacking the distal or the medial enhancer regions (Fig. 7.2(c)). Using an *in vitro* differentiation system supporting macrophage growth and development, we subsequently induced several independently derived cell clones to differentiate to macrophages and examined cell-specific and copy number-dependent lysozyme gene expression (Faust *et al.* 1994). Undifferentiated ES cells do not express the lysozyme transgene. Differentiation to macrophages leads to correct activation of the wild-type lysozyme locus. With the mutants the same result as in transgenic mice was observed: integration site dependence of expression is lost as soon as one of the regulatory elements is deleted.

Deletion of the domain borders, harbouring matrix attachment regions (S/MARs), does not abolish copy number-dependent expression of the lysozyme gene. However, if in addition to the domain borders one of the enhancers is also deleted, a substantially higher incidence of ectopic expression is observed compared to the constructs containing the domain borders; this is in addition to abrogation of position independency. Taken together, our results show that only the presence of the complete set of *cis*-regulatory elements and their cooperative interaction guarantees position-independent expression of the lysozyme gene in transgenic mice.

The structural basis of genomic position effects: lessons from transgenic mice

Chromosomal position effects are the major obstacles in experiments trying to gain information about the influence of *cis*-regulatory changes on gene expression and chromatin structure in a chromosomal environment. The influence of the chromosomal environment on chromatin structure of a randomly integrated gene is most pronounced with transgenes that are subject to position effect variegation (PEV) (see Chapter 13). Here, transgenes are silenced by the spreading of juxtaposed telomeric or centromeric heterochromatin. Each cell exhibits a

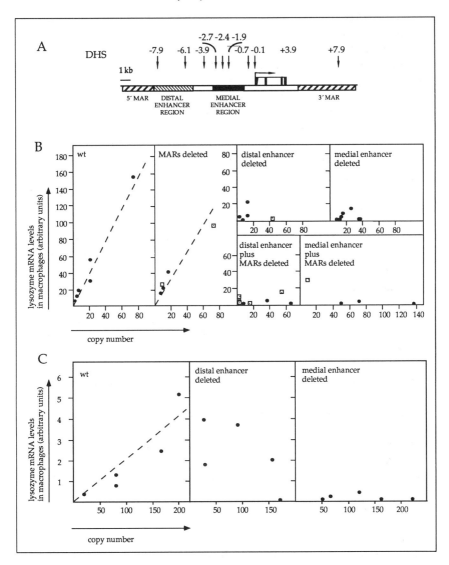

Fig. 7.2 Expression of the chicken lysozyme locus in macrophages derived from transgenic mice and from mouse embryonic stem cells. (a) Map of the chicken lysozyme locus indicating the positions of DNAase-hypersensitive sites (DHSs) and *cis*-regulatory regions. MAR, matrix attachment regions. (b) Expression of the wild-type (wt) lysozyme locus and of deletion mutants in macrophages of transgenic mice. Filled circles: transgenes showing no ectopic expression; open squares: transgenes showing ectopic expression. (c) Expression of the wild-type (wt) lysozyme locus and deletion mutants in macrophages derived from mouse embryonic stem cells. mRNA levels were measured by S1 protection analysis; transgene copy numbers were determined by Southern blotting. Expression levels are plotted as a function of transgene copy number.

different level of expression, indicating that heterochromatin has spread over variable distances, differing from cell to cell. However, distance from the heterochromatic chromosomal region and promoter strength influence the overall level of silencing. This indicates that activating and silencing forces are in equilibrium. It is reasonable to assume that a very similar situation exists at the random chromosomal positions harbouring a transgene. By analysing the local chromatin structure of transgenes that are lacking one of the essential *cis*-regulatory elements and that are expressed in a position-dependent manner, we could show that the formation of DHSs correlates with the expression level per gene copy (Huber *et al.* 1994). In other words, in a chromosomal position unfavourable for gene expression the reorganization of local chromatin structure that normally leads to gene activation, as monitored by the formation or disappearance of DHSs, is suppressed. Transgenes harbouring the complete set of *cis*-regulatory elements do not show this behaviour; they are expressed in a copy number-dependent manner and all DHSs are formed at the right position and with the correct relative frequency. Measuring lysozyme expression on the single cell level by *in situ* hybridization demonstrated that expression levels are uniform within a cell population (Huber *et al.* 1994). Hence, lysozyme transgenes containing incomplete loci are not subject to PEV. All mouse lines carry clusters of transgene copies, raising the possibility that the variable efficiency of DHS formation is the result of a variable number of loci in a multicopy transgene cluster, which form DHSs (Huber *et al.* 1994). To estimate what fraction of the integrated transgene copies are able to form DHSs, we analysed the chromatin structure of lysozyme transgenes by micrococcus nuclease (MNase) digestion (Huber *et al.* 1996). MNase cleaves preferentially in nucleosome linker regions (Noll 1974), thus enabling the mapping of specifically located (phased) nucleo-somes by indirect endlabelling. By comparing nucleosome phasing patterns in cells derived from individual mouse cell lines that carried the same construct, but expressed the gene at different levels, we could analyse whether all transgene copies within a cluster of integrated transgenes adopt the same chromatin structure, or whether MNase patterns characteristic for active and for inactive loci are superimposed. Our analysis shows that, unlike what is found for wild-type transgene loci (complete locus), position-dependent expressed transgenes exhibit a mixed nucleosomal pattern (Huber *et al.* 1996). This indicates the presence of active and inactive transgenes within the same multicopy cluster.

We assume that the factors bound to a complete set of *cis*-acting elements form a large complex whose structure and composition is shaped by evolution and which leaves no room for influences from outside factors (Fig. 7.3(a)). Once damaged by removing one essential component, or possibly even by spatial changes, this complex becomes susceptible to positional influences (Fig. 7.3(b), (c)). This can lead to a stimulatory influence of enhancers located outside the transgene locus on transgene expression, resulting in ectopic expression (Fig. 7.3(b)), or in macrophages to a disturbance in developmentally controlled chromatin reorganization (Fig. 7.3(c)). In the latter case, depending on the

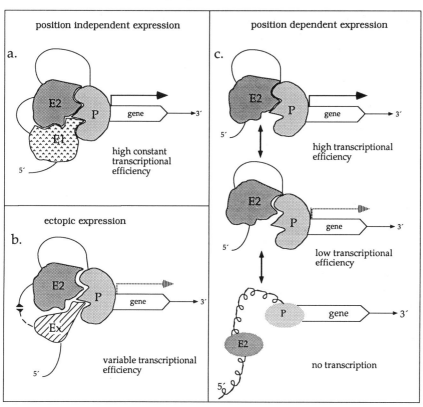

Fig. 7.3 Chromosomal position effects and their explanation. This figure symbolizes the strength of interaction of different enhancer complexes on a transgene (E1, E2) or on an enhancer located outside the transgene locus (Ex) with a promoter (P). (a) Here a single locus in a multicopy transgene cluster is shown in which all transgene copies are transcriptionally active, independent of the genomic position of integration. (b) In the case of ectopic expression, a *cis*-regulatory element located outside the transgene locus activates expression at ectopic sites. (c) In the case of position-dependent expression with constructs lacking one *cis*-regulatory element, the interaction between upstream regulatory elements and the promoter becomes unstable and only a subset of transgene copies are in the transcriptionally active state (upper panel); the others are inactive (lowest panel). The formal possibility also exists that transgenes are expressed at a different transcriptional frequency compared to fully active transgenes (middle panel). The proportion of active genes versus inactive copies (c) is determined by the chromosomal position. Variable transcription efficiencies are symbolized by the grey level of the horizontal arrow as indicated in the figure (black represents a high level of transcription).

chromosomal environment and the nature of the introduced mutation, a variable efficiency in overcoming silencing forces is observed, leading to a heterogeneous chromatin structure in a multicopy cluster of transgenes.

Towards the molecular basis of the cooperative activation of the chicken lysozyme locus

Our chromatin studies indicate that the activation of the lysozyme locus occurs in several steps, whereby only the full activity of all stimulatory elements leads to maximal transcriptional activity of the gene. Each *cis*-regulatory element has its unique developmental activation stage, indicating that transitory differentiation stages might exist in which the gene is transcribed at a lower frequency compared to the fully activated stage. An indication for this idea is our finding that at the myeloblast stage the transcriptional level is low, but all cells seem to exhibit hypersensitivity at the –6.1 kb enhancer, the –3.9 kb stimulatory element, and the promoter (Huber *et al.* 1995, 1996). In addition, we performed MNase cleavage pattern analysis of the entire 5' regulatory region of the lysozyme locus. We could demonstrate the presence of phased nucleosomes over the upstream as well as over medial enhancer region. In each case one nucleosome is relocated during cell differentiation (Huber *et al.* 1996). While at the –6.1 kb enhancer all cells show nucleosome reorganization from the myeloblast stage onwards, the situation at the –2.7 kb/–2.4 kb region is different. Here, nucleosome reorganization leading to activation of the enhancer is only complete in mature macrophages. Myeloblasts and unstimulated HD11 cells seem to represent a mixed cell population exhibiting either one or the other chromatin structure. After LPS stimulation all cells reach a homogeneous differentiation state where the DHS at the –2.4 kb silencer element has disappeared and transcription frequency is high.

Several transcription factors regulating chicken lysozyme gene expression have been identified. The –6.1 kb enhancer is composed of at least six protein binding sites, two of which are bound by members of the CCAAT/enhancer binding protein (C/EBP) family (Grewal *et al.* 1992; Goethe and Phi-Van 1994). One of the elements is similar to an AP-1 binding site, a sequence matching the consensus sequence for Myb binding is present, and another element has been identified as the binding site for members of the ubiquitously expressed nuclear factor 1 (NFI) transcription factor family (Borgmeyer *et al.* 1984). The newly identified –3.9 kb enhancer also contains an NFI binding site (Nowock and Sippel 1982). Sequence analysis has revealed the presence of potential binding sites for C/EBP and Myb. Interestingly, the only NFI binding sites of the chicken lysozyme 5' region are thus present in the two *cis*-elements that are activated early, i.e. at the myeloblast stage. It is tempting to speculate that NFI is involved in this early cell stage-specific activation. However, high level expression of the gene at later stages of cell differentiation is driven by other proteins, most likely by members of the C/EBP family. Binding sites for C/EBP have also been identified in the promoter region (Altschmied *et al.* 1989; Natsuka *et al.* 1992) and at the –2.7 kb enhancer (Altschmied *et al.* 1989; N. Faust, C. Bonifer and A.E. Sippel, unpublished). The latter also carries a binding site for an Ets-like protein (most likely Pu.1). However, C/EBPs seem to be the main mediators of stimulation of enhancer and promoter activity by LPS (Altschmied *et al.* 1989; Goethe and Phi-Van 1994; N. Faust, C. Bonifer and A.E. Sippel, unpublished). C/EBPs play a major role in chicken lysozyme gene activation, a notion that is substantiated by results from experiments in which ectopic expression of C/EBP along with Myb was sufficient to activate the gene in embryonic fibroblasts, where the gene is normally transcriptionally inactive (Ness *et al.* 1993). C/EBPs are the only proteins that can bind to most *cis*-regulatory elements. Since all *cis*-regulatory elements have to cooperate to achieve correct, high level chicken lysozyme gene expression, the idea is intriguing that C/EBPs may be the main mediators of cooperativity between the various *cis*-regulatory elements of the lysozyme locus.

Future directions

Our analysis of DNAase I and MNase cleavage patterns, employing the different chicken myeloid cell lines, has revealed a surprising number of drastic as well as many subtle changes in chromatin structure occurring during activation of the lysozyme locus, each of which is developmentally regulated (Huber *et al.* 1995, 1996). In the transcriptionally inactive state chromatin around the –6.1 kb and the –2.7 kb enhancer is organized in an array of phased nucleosomes, far exceeding the *cis*-element borders. Upon activation each enhancer element reorganizes nucleosomes in a different fashion. Our results suggest that this extended nucleosome phasing pattern, and as a consequence also the presence of sequences flanking the regulatory elements, is of importance for correct developmentally controlled gene activation.

Supporting evidence for this idea comes from experiments demonstrating a requirement for enhancer flanking sequences in transgenic mice carrying various constructs of the human ADA gene (Aronow *et al.* 1995). These sequences are not DNAase I hypersensitive but are absolutely essential for correct regulation of transgene expression. It is possible that in addition to tissue-specific transcription factors, accessory DNA binding proteins are involved in global control of chromatin architecture. A careful analysis of transgene chromatin structure might therefore reveal another layer of gene expression control mechanisms hitherto neglected but nevertheless crucial for correct gene regulation. Another question related to the elucidation of the molecular mechanism of cooperative action of *cis*-regulatory elements concerns the extent to which they can reorganize chromatin in the absence of transcription. Last but not least, transcription factor–nucleosome interactions have to be examined at high resolution in order to reveal molecular details of the concerted action of lysozyme *cis*-regulatory elements. The results of these experiments will shed new light on basic principles governing tissue-specific activation of eukaryotic gene loci, and provide the means for a molecular explanation of cell differentiation.

Summary

The chicken lysozyme locus is transcriptionally activated during macrophage differentiation. A stepwise reorganization of chromatin structure is observed. High level and position-independent expression of the gene in transgenic mice requires the concerted action of all *cis*-regulatory elements in the locus. As soon as one *cis*-regulatory region is deleted, position independency of expression is lost. Genomic position effects observed in transgenic mice carrying deletion mutants are caused by an impediment of chromatin reorganization. Mutant transgenes organized in multicopy clusters show a heterogeneous chromatin structure with variable ratios of inactive and active gene loci. Each *cis*-regulatory element has its unique activation stage during cell differentiation, whereby maximal transcriptional activity of the gene is only observed when all *cis*-elements are active. C/EBPs, which bind to most *cis*-regulatory elements of the lysozyme locus analysed so far, are most likely the main mediators of cooperative locus activation.

References

Altschmied, J., Müller, M., Baniahmad, S., Steiner, S., and Renkawitz, R. (1989). Cooperative interaction of chicken lysozyme enhancer sub-domains partially overlapping with a steroid receptor binding site. *Nucleic Acids Research*, **17**, 4975–91.

Aronow, B.J., Ebert, C.A., Valerius, T.M., Potter, S.S., Wiginton, D.A., Witte, D.P., *et al.* (1995). Dissection of a locus control region: facilitation of enhancer function by extended enhancer-flanking sequences. *Molecular and Cellular Biology*, **15**, 1123–35.

Baniahmad, A., Müller, M., Steiner, C., and Renkawitz, R. (1990). Modular structure of chicken lysozyme silencer: involvement of an unusual thyroid receptor binding site. *Cell*, **61**, 729–40.

Bonifer, C., Vidal, M., Grosveld, F., and Sippel, A.E. (1990). Tissue specific and position independent expression of the complete gene domain for chicken lysozyme in transgenic mice. *EMBO Journal*, **9**, 2843–8.

Bonifer, C., Hecht, A., Saueressig, H., Winter, D.M., and Sippel, A.E. (1991). Dynamic chromatin: the regulatory domain organisation of eukaryotic gene loci. *Journal of Cellular Biochemistry*, **47**, 99–108.

Bonifer, C., Yannoutsos, N., Krüger, G., Grosveld, F., and Sippel, A.E. (1994). Dissection of the locus control function located on the chicken lysozyme gene domain in transgenic mice. *Nucleic Acids Resarch*, **22**, 4202–10.

Borgmeyer, U., Nowock, J., and Sippel, A.E. (1984). The TGGCA-binding protein: a eucaryotic nuclear protein recognizing a symmetrical sequence in double stranded linear DNA. *Nucleic Acids Research*, **12**, 4295–311.

Faust, N., Bonifer, C., Wiles, M.V., and Sippel, A.E. (1994). An *in vitro* differentiation system for the examination of transgene activation in mouse macrophages. *DNA and Cell Biology*, **13**, 901–7.

Fritton, H.P., Sippel, A.E., and Igo-Kemenes, T. (1983). Nuclease-hypersensitive sites in the chromatin domain of the chicken lysozyme gene. *Nucleic Acids Research*, **11**, 3467–85.

Fritton, H.P., Igo-Kemenes, T., Nowock, J., Strech-Jurk, U., Theisen, M., and Sippel, A.E. (1984). Alternative sets of DNase I-hypersensitive sites characterize the various functional states of the chicken lysozyme gene. *Nature*, **311**, 163–5.

Goethe, R. and Phi-Van, L. (1994). The far upstream chicken lysozyme enhancer at –6.1 kilobase, by interacting with NF-M, mediates lipopolysaccharide-induced expression of the chicken lysozyme gene in chicken myelomonocytic cells. *Journal of Biological Chemistry*, **269**, 31302–9.

Graf, T., McNagny, K., Brady, G., and Frampton, J. (1992). Chicken 'erythroid' cells transformed by the gag-myb-ets-encoding E26 leukemia virus are multipotent. *Cell*, **70**, 201–13.

Grewal, T., Theisen, M., Borgmeyer, U., Grussenmeyer, T., Rupp, R.A.W., Stief, A., *et al.* (1992). The –6.1-kilobase chicken lysozyme enhancer is a multifactorial complex containing several cell-type-specific elements. *Molecular and Cellular Biology*, **12**, 2339–50.

Grosveld, F., Blom van Assendelft, G., Greaves, D.R., and Kollias, G. (1987). Position-independent high level expression of the human β-globin gene in transgenic mice. *Cell*, **51**, 975–85

Hauser, H., Graf, T., Greiser-Wilke, I., Lindenmeier, W., Grez, M., Land, H., *et al.* (1981). Structure of lysozyme gene and expression in the oviduct and macrophages. *Hematology and Blood Transfusion*, **26**, 175–8.

Hecht, A., Berkenstam, A., Stršmstedt, P.-E., Gustafsson, J.-A., and Sippel, A.E. (1988). A progesterone responsive element maps to the far upstream steroid dependent DNase I hypersensitive site of the chicken lysozyme gene. *EMBO Journal*, **7**, 2063–73.

Huber, M.C., Bosch, F., Sippel, A.E., and Bonifer, C. (1994). Chromosomal position effects in chicken lysozyme gene transgenic mice are correlated with suppression of DNAse I hypersensitive site formation. *Nucleic Acids Research*, **22**, 4195–201.

Huber, M.C., Graf, T., Sippel, A.E., and Bonifer, C. (1995). Dynamic changes in the chromatin of the chicken lysozyme gene domain during differentiation of multipotent progenitors to macrophages. *DNA and Cell Biology*, **14**, 397–402.

Huber, M.C., Krüger, C., and Bonifer, C. (1996). Genome position effects lead to an inefficient reorganization of nucleosomes in the 5' regulatory region of the chicken lysozyme locus in transgenic mice. *Nucleic Acids Research*, **24**, 1443–53.

Jantzen, K., Fritton, H.P., and Igo-Kemenes, T. (1986). The DNase I sensitive domain of the chicken lysozyme gene spans 24 kb. *Nucleic Acids Research*, **14**, 6085–99.

Natsuka, S., Akira, S., Nishio, Y., Hashimoto, S., Sugita, T., Isshiki, H., *et al.* (1992). Macrophage differentiation-specific expression of NF-IL6, a transcription factor for Interleukin-6. *Blood*, **79**, 460–6.

Ness, S.A., Kowenz-Leutz, E., Casini, T., Graf, T., and Leutz, A. (1993). Myb and NF-M: combinatorial activators of myeloid genes in heterologous cell types. *Genes and Development*, **7**, 749–59.

Noll, M. (1974). Subunit structure of chromatin. *Nature*, **251**, 249–51.

Nowock, J. and Sippel, A.E. (1982). Specific protein–DNA interaction at four sites flanking the chicken lysozyme gene. *Cell*, **30**, 607–15.

Phi-Van, L. and Strätling, W.H. (1988). The matrix attachment regions of chicken lysozyme gene co-map with the boundaries of the chromatin domain. *EMBO Journal*, **7**, 655–64.

Saueressig, H. (1994). Der Mechanismus der Lokuskontrollfunktion des Hühnerlysozymgens: Charakterisierung in stabil transfizierten Zellinien. PhD thesis, University of Freiburg.

Schütz, G., Nguyen-Huu, M.C., Giesecke, K., Hynes, N.E., Groner, B., Wurtz, T., *et al.* (1978). Hormonal control of egg-white protein messenger RNA synthesis in the chicken oviduct. *Cold Spring Harbor Symposia on Quantitative Biology*, **42**, 617–24.

Sippel, A.E., Theisen, M., Borgmeyer, U., Strech-Jurk, U., Rupp, R.A.W., Püschel, A.W., *et al.* (1988). Regulatory function and molecular structure of DNaseI-hypersensitive elements in the chromatin domain of a gene. In *The architecture of eukaryotic genes* (ed. G. Kahl), pp. 355–69. Verlagsgesellschaft Chemie,Weinheim.

Sippel, A.E., Schäfer, G., Faust, N., Saueressig, H., Hecht, A., and Bonifer, C. (1993). Chromatin domains constitute regulatory units for the control of eukaryotic genes. *Cold Spring Harbor Symposia on Quantitative Biology*, **58**, 37–44.

Strouboulis, J., Dillon, N. and Grosveld, F. (1992). Developmental regulation of a complete 70kb human β-globin locus in transgenic mice. *Genes and Development*, **6**, 1857–64.

Wilson, C., Bellen, H.J., and Gehring, W.J. (1990). Position effects on eucaryotic gene expression. *Annual Review of Cell Biology*, **6**, 679–714.

Zimmerman, K., Legony, E., Stewart, V., Depinho, R., and Alt, F.W. (1990). Differential regulation of the N-myc gene in transfected cells and transgenic mice. *Molecular and Cellular Biology*, **256**, 48–178.

8

Locus control regions

Frank Grosveld and Peter Fraser

Introduction

The nucleosome is the basic repeat structure of chromatin and individual nucleosomes are joined by a short region of linker DNA. Histone H1 associates with the linker DNA at the border of the nucleosome. In the presence of H1 the nucleosomes are compacted to form the 30 nm fibre visible in the electron microscope (Thoma *et al.* 1977). Non-histone proteins are thought to organize the 30 nm fibre into higher order structures. When depleted of histones, metaphase chromosomes show loops of DNA attached to a central protein scaffold (Laemmli *et al.* 1977). Chromatin, therefore, appears to be organized into discrete domains where each domain is a 30–100 kb loop attached at their base to a nuclear scaffold (Gasser and Laemmli 1987). One of the major proteins attached to the scaffold is topoisomerase II (Gasser and Laemmli 1987), which suggests that torsional stress induced during transcription or replication may be released at the base of the loops.

The eukaryotic genome can roughly be divided further into two cytologically distinguishable states: euchromatin and heterochromatin (e.g. Eissenberg *et al.* 1995). Euchromatin regions are decondensed in interphase. They replicate early and contain mostly single copy sequences and genes. In contrast, hetero-chromatin regions remain condensed throughout the cell cycle, usually replicate late, and contain a high proportion of middle and highly repetitive sequences (John 1988). The normal state of chromatin, when histone H1 is bound, appears to be a compacted structure that is repressive to gene transcription (Bogenhagen *et al.* 1982; Schlissel and Brown 1984). However, chromatin is not a passive structural scaffold (Pillus and Grunstein 1995) and the questions, therefore, are how the state of chromatin modulates gene transcription, replication, and repair, and how the necessary machinery to carry out these processes gets access to the DNA template.

It is known that actively transcribed chromatin is relatively low in histone H1 and more accessible to endonucleolytic and chemical cleavage than silent chromatin (Groudine and Weintraub 1976), while regulatory elements are associated with DNAase-hypersensitive sites (Wu *et al.* 1979). These phenomena have been interpreted to mean that active chromatin is decondensed and that regular nucleosome structure is disrupted by the binding of transcription factors. Several chemical changes to the primary components, i.e. DNA and histones, have been documented. For example, expressed DNA is associated with hypomethylation of the DNA (Bird 1992) and hyperacetylation of histones (Durrin *et al.* 1991; Jeppesen and Turner 1993). Hypomethylation changes the ability of certain transcription factors to bind to DNA and results in a failure of general methylated-DNA binding proteins to bind to DNA (Bird 1992). The acetylation of histone H3 and H4 tails also influences transcription factor access to nucleosomal DNA (e.g. Durrin *et al.* 1991; Jeppesen and Turner 1993; see Chapter 3), and may therefore play an important role in modulating the local structure of chromatin. Interestingly, methylation-free islands (CpG islands) have a relatively low histone H1 content and contain more acetylated histone H3 and H4 than the bulk of the chromatin (Tazi and Bird 1990). However, it is not known whether methylation or acetylation play a primary role in causing chromatin activation, or whether they are the result of activation. Quite recently a number of cofactors have been described that may assist sequence-specific transcription factors in remodelling chromatin and that counteract the repressive effects of nucleosomal histones on transcription. Two of these complexes, SWI/SNF and NURF (Kornberg and Lorch 1995; Tsukiyama and Wu 1995; see Chapter 4), have been identified and require ATP for remodelling. It would not be surprising if acetylases were associated with such complexes.

It is attractive to speculate that a small set of key transcription factors exist, which in a given situation may interact with factors like SWI/SNF or NURF and thereby gain access to the closed repressive chromatin by inducing a transition to the transcriptionally active chromatin state. However, this is probably a too simplistic view. The process of inducing changes in chromatin structure that results in an activated state most likely requires the cooperation of a large number of protein factors. Such combinations of factors are expected to occur in transcription-regulating genomic elements that are capable of activating chromatin, i.e. locus control regions (LCRs). At present, there is not sufficient detailed knowledge to describe these elements in terms of combinations of protein factors and of the biochemical activity of each of the factors, to explain chromatin activation. Nevertheless, our knowledge about the basic properties of such regulatory elements, i.e. LCRs, is rapidly increasing and will be summarized below.

Locus control regions

Enhancers are *cis*-acting DNA elements that stimulate transcription when positioned either upstream or downstream of the site of transcription initiation. Enhancers were originally identified by transient transfection experiments with the SV40 enhancer (Banerji *et al.* 1981), a situation in which the normal chromatin composition is only partially reconstituted. Cellular enhancer elements have been identified for many genes and they consistently appear to be complex structures made up of a number of transcription factor binding motifs. Often there appears to be functional redundancy for these sites, e.g. the IgH enhancer contains a tandem array of protein binding motifs and no single factor appears to be crucial for enhancer activity (Lenardo *et al.* 1987). Complex enhancers consist of elements that have different cell-type specificities, and genes with complex expression patterns appear to contain several enhancers with different tissue or developmental specificity (Ondek *et al.* 1987; Gerster *et al.* 1987). Almost all enhancers retain their activity when introduced stably into chromatin, as is the case in stable transfection assays. However, it should be realized that in such experiments the enhancer (and the transgene) are coupled to a selectable marker and only cells that express the marker gene will survive selection. Hence, only the cells that integrated the transgene (and the selector gene) in an active chromatin environment are analysed. A different result is obtained when no selection procedures are used, such as after microinjection of a construct into a fertilized egg to obtain transgenic mice. This often results in variable and non-physiological levels of expression or complete absence of gene activity, when different mice containing the same enhancer construct are compared. This variation has been named 'position effect' and is thought to be due to positive or negative effects exerted by the local chromatin structure at the site of integration of the transgene. Such position effects are frequently observed and pose a significant problem in the study of gene regulation (see below).

Studies on the human β-globin gene locus led to the discovery of the LCR, which is able to overcome position effects if integrated into transgene constructs (Grosveld *et al.* 1987). The globin LCR is defined by its ability to confer site of integration-independent and physiological levels of erythroid-specific expression to a gene linked in *cis*, which is directly proportional to the transgene copy number in transgenic mice. The ability of the globin LCR to override position effects and confer transgene copy number-dependent expression suggests that it can open silent chromatin and establish and maintain a transcriptionally competent chromatin domain in an erythroid cell environment (Grosveld *et al.* 1987; Felsenfeld 1992; Dillon and Grosveld 1994). A growing number of gene loci have been described that appear to meet the criteria of LCR function, including ensuring a fully reproducible activity of single or low copy number transgenes. These include the T cell-specific CD2 (Greaves *et al.* 1989) and TCR loci (Diaz *et al.* 1994), the B cell-specific MHC class II Ea gene (Carson and Wiles 1993), the macrophage-specific lysozyme gene (Bonifer *et al.* 1990), the neuronal specific S100β gene (Friend *et al.* 1992), the liver-specific LAP gene (Talbot *et al.* 1994), and a number of others. Although initially thought to have an LCR, the human α-globin locus may represent a different type of chromatin domain with a telomeric location (Higgs *et al.* 1990; Sharpe *et al.* 1993).

The β-globin LCR happens to be present as one contiguous piece of DNA with five tissue-specific hypersensitive sites (small regions of 200–300 bp). In other cases different numbers of hypersensitive sites have been found and they need not be adjacent to each other. The hypersensitive regions bind transcription factors, and the suggestion is that this binding is responsible for an open

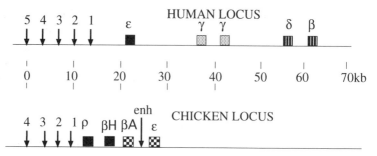

Fig. 8.1 Schematic representation of the human and chicken β-globin loci. The DNAase-hypersensitive sites constituting the LCRs are indicated by vertical arrows. The 3' hypersensitive site in the chicken locus has been indicated as enhancer (enh).

chromatin configuration by altering the local nucleosome structure. As a result DNA becomes more accessible to exogenously added nucleases. It should be noted that not all LCRs have been defined and analysed by the same functional criteria, and it is therefore not surprising that some of the identified LCRs turn out to be incomplete. The most common problems in defining an LCR are due to the fact that these genomic elements can be spread over considerable distances (more than 20 kb) as a collection of regulatory elements, and that a complete LCR need not be located in one unique position (upstream or downstream) relative to the genes. The simplest example of this fact is seen in a comparison of the human and the chicken β-globin loci. In the human locus all the elements of the LCR are clustered on the 5' side of the locus, whereas in the chicken the LCR has been split: one of the essential elements is located between the β- and the ε-genes at the 3' side of the locus (Fig. 8.1). It is therefore understandable that the human β-globin LCR was the first to be defined, whereas it is much more difficult for complicated multigene loci, such as the Hox gene clusters where the elements are probably interspersed among the genes. Although this may give problems in terms of definition, it probably makes no difference in terms of mechanism of LCR action (see below) and the role of individual elements that together constitute an LCR.

To date only the human, mouse and chicken β-globin LCRs have been characterized thoroughly. Each of these LCRs consists of a number of tissue-specific DNAase-hypersensitive sites. Each hypersensitive site has a central core region of 200–300 bp, which has a high density of transcription factor binding sites. The presence of the LCR is required for the expression of each of the genes in the locus and also determines the timing of replication in erythroid cells. When absent, replication switches from early to late (Forrester *et al.* 1990). If the individual hypersensitive sites are tested for their ability to stimulate transcription in transient transfection assays, only one of the sites (HS2) has a significant enhancer activity (Tuan *et al.* 1989; Hug *et al.* 1992; Hardison *et al.* 1993). The other sites are only active when integrated into the genome, i.e. only in the context of a normal chromatin configuration (Collis *et al.* 1990). However, it should be noted that the transfection experiments involve the selection for a drug resistance marker in *cis*, and hence that such experiments select for integration into an active area of the host cell genome. When each of the hypersensitive regions individually are coupled to a β-globin gene and introduced into mice, each of the hypersensitive sites (with the exception of HS5) maintains position-independent expression, if integrated in multiple copies into the host genome (Fraser *et al.* 1990; Ellis *et al.* 1996, and references therein). When tested as a single copy integrant the sites no longer function as an LCR, although HS3 appears to be the most consistent in inducing activity of the transgene (Ellis *et al.* 1996). When combinations of DNAase-hypersensitive sites are tested (Collis *et al.* 1990; Fraser *et al.* 1990), it appears that the activity of the individual sites is additive, with HS2 and HS3 being the most active. These results, in combination with the fact that the presence of the LCR does not prevent aberrant expression due to position effects in non-erythroid cell types, suggest that the combination of HS1, HS2, HS3, and HS4 is responsible for site-of-integration-independent expression through a dominant activation effect in erythroid cells, rather than insulation from the surrounding chromatin.

Insulation

An interesting case is that of the remaining DNAase-hypersensitive site (HS5 in human and mouse, HS4 in chicken). HS5 in the human locus was originally described as a non-tissue-specific site (Tuan *et al.* 1985), but later studies of transgenic mice and human tissues showed that the site is tissue specific (Zafarana *et al.* 1995). One group reported on the ability of HS5 to cooperate with the polyama virus enhancer to give position-independent expression in transfection experiments (Yu *et al.* 1994), suggesting that it may have a property different from the other hypersensitive sites. Such a role was suggested by experiments with chicken HS4 (Chung *et al.* 1993), which marks a boundary of the region of overall DNAase sensitivity and a local transition in the level of histone acetylation in the chicken β-globin domain (Hebbes *et al.* 1994). When this site (and also the human HS5) was tested in an enhancer blocking assay, it resulted in a lowering of expression of the selectable marker that was flanked by these sites. When the site was subsequently tested with the *white* gene in *Drosophila*, it showed insulating properties of the *white* gene (Chung *et al.* 1993), similar to that observed with the *Drosophila* scs and scs' border elements (Kellum and Schedl 1992; see also Chapter 5). However, when the human HS5 element and the *Drosophila* scs elements were tested in an enhancer blocking or an insulation assay they showed no activity (Zafarana *et al.* 1995). Moreover, extensive nuclease sensitivity assays have shown no change in DNAase sensitivity for at least 100 kb 5' of the human β-globin LCR (Pruzina 1996).

At present it is unclear why different results were obtained with the chicken and the human hypersensitive elements (HS4 and HS5, respectively) that are located at the extreme 5' end of the β-globin loci. It may indicate that the chicken and human 5' elements are not functionally equivalent. Alternatively, it could be due to differences in the assay systems used for each study: one used drug selection in tissue culture cells while the other examined β-globin expression in transgenic animals. Another factor that may be important is the strength of the properties of the activating in the tester constructs: in one case only part of the LCR was used, while in the other the entire LCR was used. Another difference is that the human HS5 was not only in an artificial construct, but also in its natural position relative to the LCR. That the properties of the enhancer might be relevant is suggested by experiments on scs and scs' (J. Vazquez and P. Schedl, personnel communication) and on su(Hw) (Roseman *et al.* 1993), which indicate that the efficacy of enhancer blocking depends upon features of the enhancer.

Hence, caution should be exercised in the interpretation of these results, particularly when differences between species have not been rigorously examined and excluded. For example, the chicken A elements are located near the edge of the chromatin domain of the lysozyme gene and mark the limits of a domain of overall DNAase I sensitivity in the chicken genome. In tissue culture cells the A elements are found to block enhancer action and are required for position-independent gene expression, albeit in an exponential rather than linear fashion (Stief *et al.* 1989). However, a different result was obtained when the lysozyme A elements were tested in transgenic mice (Bonifer *et al.* 1994; see also Chapter 7). In this case the A elements did not appear to be required for a high level of integration site-independent expression in the appropriate cells. In contrast, integration site-independent expression was conferred by the combined action of several enhancers, which together constitute a locus control element. The incidence of ectopic expression increased when the construct contained a deletion in the lysozyme LCR or lacked the A elements (see below).

The mechanism of LCR action

Two models can be postulated to explain how an LCR achieves activation of an entire locus. In both models transcription factors are assumed to bind to the LCR region, thereby preventing the formation of or altering the repressive chromatin structure by removing nucleosomes from the DNA and forming nuclease-sensitive areas. This may occur as an all-or-none phenomenon in which DNAase-hypersensitive sites are formed stochastically (Boyes and Felsenfeld 1996). One model predicts that this will lead to a general change in chromatin structure over large distances, resulting in a change in overall DNAase sensitivity of the complete locus. Such opening up would allow transcription factors to bind to promoter and enhancer sequences and result in active transcription of genes in the locus. In the alternative model, after the change in chromatin structure the LCR operates through formation of a direct chromatin–chromatin interaction between the LCR and individual genes via looping over large distances (at least

50 kb in the case of β-globin), resulting in gene activation. This mechanism is analogous to the looping model that has been suggested for enhancers (Müller *et al.* 1989; Bickel and Pirotta 1990, and references therein).

In the looping model there are two possibilities: either each individual hypersensitive site of the LCR interacts with different genes (split LCR model; Engel 1993), or at any moment in time all of the hypersensitive sites interact simultaneously with one and the same gene (holo LCR model; Ellis *et al.* 1996). Interestingly, part of the developmental regulation of the locus is achieved through competition between the genes for the LCR, with a competitive advantage for the genes proximal to the LCR (Hanscombe *et al.* 1991). This property, which was first described in transfection experiments (DeVilliers *et al.* 1982; Wasylyk *et al.* 1983), provides an excellent model system to distinguish between the two different mechanisms described above.

Competition

Polar competition, which is dependent on relative position of a gene with respect to the LCR, would be explained if the LCR (or a process starting at the LCR) has an intrinsic preference for interaction with the more proximally located gene. A search along the chromosome would result in a preferential interaction with the first gene it encounters. As a result, the overall activity of a gene will be determined by the frequency with which it is activated by the LCR that scans the genome in a polar way.

A second possibility is the formation of defined chromatin structures that bring gene and LCR together. Such a mechanism would probably require specific and developmentally regulated functions for the spacer regions in the locus. A comparison of the phenotypes of various deletions in the locus (for review see Poncz *et al.* 1989), together with the transgenic mouse data that test the effect of different spacer sequences and changes in gene spacing distance, argue against such a role of intergene sequences. The alternative and much simpler possibility is that the key parameter is the relative frequency of contact between genes and LCR in a looping model (Hanscombe *et al.* 1991). When each of two genes has a significant capacity to form a stable interaction with the LCR, a difference in their frequency of contact with the LCR will dramatically affect the competition between them. This frequency of contact will, according to the model, depend on the relative distance of the genes to the LCR (Hanscombe *et al.* 1991). During the fetal stage the interaction of the LCR with the γ gene is likely to be more frequent and stronger than that with the β gene, due to their difference in distance to the LCR and to the action of stage-specific factors. The combination of these two parameters (frequency and strength of interaction) would allow the γ genes to largely compete out β-globin gene expression. In the adult stage, although the β gene would now have the stronger interaction with the LCR, its lower frequency of contact (due to its longer distance to the LCR) would make it difficult for the β gene to compete out a γ gene if that gene retained a significant capability to

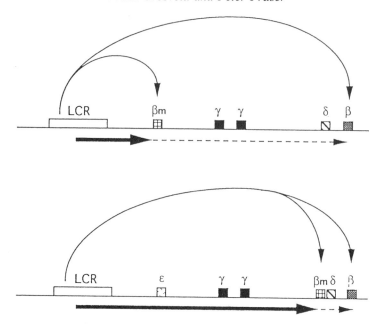

Fig. 8.2   Schematic representation of two β-globin loci, that were introduced in transgenic mice. The top locus contains a marked β-globin gene (βm) replacing the ε-globin gene. The bottom locus contains the same βm but now positioned just 5' to the δ-globin gene. The solid arrows indicate either a direct contact between the LCR and genes via looping or a linear chromatin spreading effect (scanning). The dashed arrows indicate the same processes but taking place at a much lower level.

form such interactions. It is possible that the β gene could still compete out the γ gene by a strong interaction of the β-globin gene with the LCR, but it seems unlikely that such an interaction would have evolved specifically to silence γ expression, since single point mutations in the γ promoter are sufficient for high γ-globin expression. The more likely mechanism is promoter-mediated silencing of the early genes (γ genes) by stage-specific factors. The transgenic mouse data support that mechanism (Raich *et al.* 1990; Dillon and Grosveld 1991).

In order to distinguish between the different possibilities, a transgenic mouse experiment was set up in which a marked β-globin gene was placed at different positions in the complete locus (Fig. 8.2; Dillon *et al.* 1995; N. Dillon, T. Trimborn, J. Strouboulis, P. Fraser, and F. Grosveld, submitted). Results show that there is indeed strong competition, dependent on the relative distance to the LCR. When the marked gene is in the proximal position, it strongly inhibits the expression of all more distal genes at all developmental stages. This is in agreement with experiments in which selectable marker genes are integrated into the mouse globin locus (Kim *et al.* 1992; Shehee *et al.* 1993). In contrast, when the marked β-globin gene is in a more distal position, it only competes with the distal endogenous β-gene, but much less effectively if compared to a construct

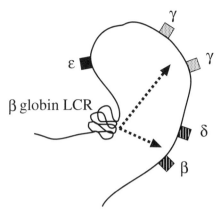

β globin LCR

Fig. 8.3 A model of the LCR monogenic interaction with the genes via a stochastic looping mechanism. The LCR is indicated as a squiggle to indicate that the hypersensitive regions of the LCR could act together or form a holocomplex to establish an interaction of the LCR with one of the genes.

where it is positioned proximal to the LCR. These results confirm that competition does play a role in regulation in the complete globin locus, and provide a strong argument for a looping model with stochastic interactions between the genes and the LCR (Dillon *et al.* 1995; N. Dillon, T. Trimborn, J. Strouboulis, P. Fraser, and F. Grosveld, submitted).

Dynamics of chromatin interactions

The LCR holocomplex model described above explains why the activity of the DNAase-hypersensitive sites in the LCR is additive and why each of the genes requires all of the hypersensitive sites for full expression. Importantly, it explains the fact that there is balanced competition for LCR function between the genes. A key requirement in this model is that the LCR either is limited to activating only a single gene at a time or, alternatively, that it splits its function (Engel 1993) and activates multiple genes simultaneously (Bresnick and Felsenfeld 1994; Furukawa *et al.* 1994). This question was addressed by *in situ* detection of short-lived globin precursor RNAs in erythroid tissues, derived from a transgenic mouse line carrying a single copy of the complete human β-globin locus. The results show that the LCR activates only one gene at a time, indicating that the LCR–gene interaction is monogenic (Fig. 8.3; Wijgerde *et al.* 1995). These interactions are dynamic, with transcription-stimulating interactions forming, breaking, and reforming in a type of flip-flop mechanism. Interestingly, these experiments provided an estimate of the average duration of the LCR–gene interaction: of the order of 5–50 min.

Although these experiments used a completely different approach compared to the competition experiment described above, they again are most readily explained by a looping mechanism in which the LCR stably complexes one single gene at a time and switches back and forth between genes in a flip-flop type of mechanism. These results also suggest that the LCR is complexed with a gene almost all of the time. Of course these experiments do not tell whether some kind of chromatin opening process, in a linear fashion along the chromatin fibre, is operative during the very first time the locus is activated.

Position effects

Sometimes chromosomal rearrangements result in a juxtaposition of hetero-chromatic and euchromatic regions, which can cause a change in expression of genes located in the regions of the breakpoint (Wilson *et al.* 1990). One type of position effect, known as position effect variegation (PEV; see also Chapter 13), occurs through localization of a gene into a heterochromatic environment, resulting in mosaic expression of the gene. This phenomenon, which was first observed in *Drosophila* and yeast, is clonal and is heritable in daughter cells (see Karpen 1994). PEV also occurs in mammals and has been shown to affect expression of the β-globin locus (see Milot *et al.* 1996 and references therein). Variegation can be altered by the product of a number of genes that are thought to change chromatin packaging. The spreading of such factors along the chromatin fibre into neighbouring areas is assumed to modulate the ability of transcription factors to bind their target DNA sequences.

Chromatin-mediated silencing in competition with activation, as observed in PEV, can often be distinguished from other position effects (see Milot *et al.* 1996). The effect of PEV, measured at the level of a single cell, is an all-or-none phenomenon, i.e. the relevant gene in a particular cell is either completely 'on' or completely 'off'. In contrast, if measured in a population of cells, instead of at the single cell level, PEV appears as an overall decrease of expression and is indistinguishable from position effects caused by juxtaposition of a transgene to the regulatory region of another gene.

Position effects have been frequently observed in transformation systems and transgenic animals (Wilson *et al.* 1990). Since LCRs are defined as elements that allow a gene to be expressed independently of its site of integration in the genome (Grosveld *et al.* 1987), it is perhaps surprising that there have been reports about position effects or severe impairment of expression in transgenic mice where LCR sequences were used in the construct. However, all of these constructs either contained deletions in the LCR (Bonifer *et al.* 1994; Robertson *et al.* 1995; Bungert *et al.* 1995; Festenstein *et al.* 1996; Milot *et al.* 1996), or carried additional regulatory elements, forming unnatural combinations with the LCR (Elliott *et al.* 1995). Four of the cases reported PEV (Robertson *et al.* 1995; Elliott *et al.* 1995; Festenstein *et al.* 1996; Milot *et al.* 1996). Although it was not pointed out in that publication, Robertson *et al.* (1995) used an incomplete β-globin LCR (Philipsen *et al.* 1990; Pruzina *et al.* 1991). Elliott *et al.* (1995) used the CD2 LCR in combination with the Ig enhancer. The use of the CD2

LCR alone does not result in PEV (Lang *et al.* 1988). Of the remaining cases that reported position effects of transgene constructs that contain an LCR, Bonifer *et al.* (1994) showed that position effects only become important when deletions in the LCR are made, while Bungert *et al.* (1995) reported only a severe impairment of expression of the transgene. Although no firm conclusions can be made, the suggestion from all these results is that interference with the LCR in the form of deletions or additions makes it sensitive to position effects.

An analysis of the effect of deletions in the LCR was carried out in transgenic mice, and gene expression was measured at the single cell level and correlated to the position of integration in the host genome (Festenstein *et al.* 1996). In the case of the CD2 LCR, deletion of the most 3' hypersensitive site (HS3 downstream of the gene) yielded a number of mice with classical, clonally inheritable PEV (Festenstein *et al.* 1996). In the case of the β-globin LCR, the effect of deletion of each of the individual DNAase-hypersensitive sites was tested. Again, position effects were found at a high frequency (Milot *et al.* 1996). The most interesting observation of the experiments with the two LCRs was the correlation between the observed position effects and integration of deletion constructs into heterochromatic regions of the host genome. Interestingly, in the case of the β-globin LCR deletions two different types of position effects were observed. Some mice showed PEV, as was the case with the CD2 LCR deletion. Perhaps more interestingly, other mice showed a novel type of position effect. All of the erythroid cells appeared to have an active transgene locus, but the genes were transcribed only part of the time, rather than continuously. It was clear that this effect is not stochastic. It is likely that it is related to the cell cycle, although effects of the state of terminal differentiation and cessation of proliferation of the cells have not been excluded.

These results have important implications for the proposed mechanisms of LCR-induced activation of gene expression. As discussed above, the different DNAase-hypersensitive regions of the LCR may act together as a single functional unit, which interacts with a single gene in the locus at any given moment in time. Initiation of transcription of a competent gene in the 'open' chromatin domain of the β-globin locus appears to originate from complex formation with the LCR with the promoter of that gene. Maintenance of transcriptional activity of that gene requires continued association between LCR and gene (see also Wijgerde *et al.* 1995). The results of experiments using incomplete LCRs show that deletions that decrease the stability of LCR–gene interaction shorten the duration of such associations, thereby reducing the transcriptional output from a given gene by decreasing the time interval the gene is activated by the LCR. This effect is most obvious in heterochromatic regions of the genome, where the LCR spends a significant amount of time uncoupled from any globin gene, resulting in lower levels of expression. Hence, we propose that the LCR determines the level of gene expression by determining the frequency and the duration of transcription periods, rather than only controlling the rate of transcription initiation.

Future developments

Obviously, there is still a large gap between understanding the chromatin activation process summarized in the introduction of this chapter and our present knowledge about LCRs. It will be very important to study the individual protein factors that bind to LCR DNA and to analyse their interaction with proteins that are associated with chromatin or are involved in the modification of chromatin structure. It will necessarily require reductionist and *in vitro* approaches to study the function of individual proteins and protein interactions, before the complicated overall picture can be understood. It will be necessary to gain insight into mechanisms of very high levels of complexity and to develop methodologies that allow analysis of gene expression at very high levels of spatial and temporal resolution, e.g. methods that allow visualization and direct measurement of interactions between different regions of the genome as a function of time in living cells.

Summary

In summary, we can draw a number of conclusions about the properties of LCRs. They are combinations of regulatory sequences that, if integrated in a transgene construct, provide reproducible expression of a gene, independent of the site of its integration into the host genome. LCRs are able to overcome hetero-chromatin-induced silencing, but only when the complete LCR sequence is present. It is not yet understood at the molecular level why an LCR is capable of activating chromatin, but it is clear that the level of transcription is controlled by ensuring that the locus is active in all of the relevant cells at all times, rather than directly controlling the rate of transcription. In other words, an LCR controls the frequency of transcriptionally active periods. When multiple genes are regulated by a single LCR, individual genes compete for the LCR and this competition appears not only to be dependent on regulatory regions directly flanking the genes, but also on their relative distance to the LCR. The most plausible mechanism that underlies these conclusions is that the LCR and the genes make direct contact via DNA looping and that continued interaction is required to allow loading of the basic transcriptional machinery on the DNA.

References

Banerji, J., Rusconi, S., and Schaffner, W. (1981). Expression of a β-globin gene is enhanced by remote SV40 sequences. *Cell*, **33**, 729–40.

Bickel, S. and Pirotta, V. (1990). Self-association of the *Drosophila* zeste protein is responsible for transvection effects. *EMBO Journal*, **9**, 2959–67.

Bird, A. (1992). The essentials of methylation. *Cell*, **70**, 5–8.

Bogenhagen, D., Wormington, W., and Brown, D. (1982). Stable transcription complexes of *Xenopus* 5S RNA genes: a means to maintain the differentiated state. *Cell*, **28**, 413–21.

Bonifer, C., Vidal, M., Grosveld, F., and Sippel, A. (1990). Tissue specific and position independent expression of the complete gene domain for chicken lysozyme in transgenic mice. *EMBO Journal*, **9**, 2843–8.

Bonifer, C., Yannoutsos, N., Kruger, G., Grosveld, F., and Sippel, A.E. (1994). Dissection of the locus control function located on the chicken lysozyme gene domain in transgenic mice. *Nucleic Acids Research*, **22**, 4202–10.

Boyes, J. and Felsenfeld, G. (1996). Tissue-specific factors additively increase the probability of the all-or-none formation of a hypersensitive site. *EMBO Journal*, **15**, 2496–507.

Bresnick, E.H. and Felsenfeld, G. (1994). Dual promoter activity by the human β globin locus control region. *Proceedings of the National Academy of Sciences of the USA*, **91**, 1314–17.

Bungert, J., Dave, U., Lim, K.C., Lieuw, K., Shavit, J., Liu, Q., *et al.* (1995). Synergistic regulation of human beta-globin gene switching by locus control region elements HS3 and HS4. *Genes and Development*, **9**, 3083–96.

Carson, S. and Wiles, M. (1993). Far upstream regions of Class II MHC Ea are necessary for position independent copy dependent expression of Ea transgenesis. *Nucleic Acids Research*, **21**, 2065–72.

Chung, J., Whiteley, M., and Felsenfeld, G. (1993). A 5' element of the chicken β globin domain serves as an insulator in human erythroid cells and protects against position effects in *Drosophila*. *Cell*, **74**, 505–14.

Collis, P., Antoniou, M., and Grosveld, F. (1990). Definition of the minimal requirements within the human β globin gene and the dominant control region for high level expression. *EMBO Journal*, **9**, 233–40.

DeVilliers, J., Olson, C., Banerji, J., and Schaffner, W. (1982). Analysis of the transcriptional enhancer effect. *Cold Spring Harbor Symposia on Quantitative Biology*, **47**, 911–9.

Diaz, P., Cado, P., Winoto, A. (1994). A locus control region in the T cell receptor α/δ locus. *Immunity*, **1**, 207–17.

Dillon, N. and Grosveld, F. (1991). Human γ-globin genes silenced independently of other genes in the β-globin locus. *Nature*, **350**, 252–4.

Dillon, N. and Grosveld, F. (1994) Chromatin domain as potential units of eukaryotic gene function. *Current Opionion in Genetics and Development*, **4**, 260–4.

Dillon, N., Strouboulis, J., and Grosveld, F. (1995). The regulation of human β globin gene expression: polarity of transcriptional competition in the human β-globin locus. In *Proceedings of the ninth conference on hemoglobin switching* (ed. G. Stamatoyanno-poulos), pp. 23–8. Intercept, Andover.

Durrin, L., Mann, P., Kayne, P., and Grunstein, M. (1991). Yeast histone H4 N-terminal sequence is required for promoter activation *in vivo*. *Cell*, **65**, 1023–31.

Eissenberg, J.C., Elgin, S.C.R,. and Paro, R. (1995). In *Chromatin structure and gene expression* (ed. S.C.R. Elgin), pp. 147–71. Oxford University Press, Oxford.

Elliott, J.I., Festenstein, R., Tolaini, M., and Kioussis, D. (1995) Random activation of a transgene under the control of a hybrid hCD2 locus control region/Iq enhancer regulatory element. *EMBO Journal*, **14**, 575–84.

Ellis, J., Tan-Un, K.C., Harper, A., Michalovich, D., Fraser, P., Yannoutsos, N., *et al.* (1996). A dominant chromatin opening activity in 5' hypersensitive site 3 of the human β-globin locus control region. *EMBO Journal*, **15**, 562–8.

Engel, L. (1993). Developmental regulation of human β globin gene transcription: switch of loyalties? *Trends in Genetics*, **9**, 304–9.

Felsenfeld, G. (1992). Chromatin as an essential part of the transcriptional mechanism. *Nature*, **335**, 219–24.

Festenstein, R., Tolaini, M., Corbella, P., Mamalaki, C., Parrington, J., Fox, M., *et al.* (1996). Locus control region function and heterochromatin-induced position effect variegation. *Science*, **271**, 1123–5.

Forrester, W.C., Epner, E., Driscoll, M.C., Enver, T., Brice, M., Papayannopoulou, T., *et al.* (1990). A deletion of the human β-globin locus activation region causes a major alteration in chromatin structure and replication across the entire β-globin locus. *Genes and Development*, **4**, 1637–49.

Fraser, P., Hurst, J., Collis, P., and Grosveld, F. (1990). DNase I hypersensitive sites 1, 2 and 3 of the human β-globin dominant control region direct position-independent expression. *Nucleic Acids Research*, **18**, 3503–8.

Friend, W., Clapoff, S., Landry, C., Becher, L., O'Hanlon, D., Allore, R., *et al.* (1992). Cell specific expression of high levels of human S100 beta in transgenic mouse brains is dependent on gene dosage. *Journal of Neuroscience*, **12**, 4337–46.

Furukawa, T., Zitnik, G., Leppig, K., Papayannopoulou, T., and Stamatoyannopoulos, G. (1994). Coexpression of gamma and beta globin mRNA in cells containing a single human beta globin locus: results from studies using single-cell reverse transcription polymerase chain reaction. *Blood*, **83**, 1412–19.

Gasser, S. and Laemmli, U. (1987) A glimpse of chromosomal order. *Trends in Genetics*, **3**, 16–22.

Gerster, T., Matthias, P., Thali, M., Jiricny, J., and Schaffner, W. (1987). Cell type specificity elements of the immunoglobulin heavy chain enhancer. *EMBO Journal*, **6**, 1323–30.

Greaves, D., Wilson, F., Lang, G., and Kioussis, D. (1989). Human CD2 3'-flanking sequences confer high level, T cell specific, position independent expression in transgenic mice. *Cell*, **56**, 979–86.

Grosveld, F., Blom van Assendelft, G., Greaves, D.R. and Kollias, G. (1987). Position-independent, high-level expression of the human β-globin gene in transgenic mice. *Cell*, **51**, 975–85.

Groudine, M. and Weintraub, H. (1976). Chromosomal subunits in active genes have an altered conformation. *Science* **193**, 846–56.

Hanscombe, O., Whyatt, D., Fraser, P., Yannoutsos, N., Greaves, D., Dillon, N., *et al.* (1991). Importance of globin gene order for correct developmental expression. *Genes and Development*, **5**, 1387–94.

Hardison, R., Xu, J., Jackson, J., Mansberger, J., Selifonova, O., Grotch, B., *et al.* (1993). Comparitive analysis of the locus control region of the rabbit β globin gene cluster. *Nucleic Acids Research*, **21**, 1265–72.

Hebbes, T., Clayton, Thorne, A., and Crane-Robinson, C. (1994). Core histone acytelation correlates with generelized DNase sensitivity in the chicken β globin chromosomal domain. *EMBO Journal*, **13**, 1823–30.

Higgs, D.R., Wood, W.G., Jarman, A.P., Sharpe, J., Lida, J., Pretorius, I-M., and Ayyub, H. (1990). A major positive regulatory region located far upstream of the human α-globin gene locus. *Genes and Development*, **4**, 1588–601.

Hug, B., Moon, A., and Ley, T. (1992). Structure and function of the murine βglobin locus control region 5'HS3. *Nucleic Acids Research*, **21**, 5771–8.

Jeppesen, P. and Turner, B. (1993). The inactive X chromosome in female mammals is distinguished by a lack of histone H4 acytelation, a cytogenetic marker for gene expression. *Cell*, **74**, 281–9.

John, B. (1988). In *Heterochromatin: molecular and structural aspects* (ed. R.S. Verma), p. 1. Cambridge University Press, Cambridge.

Karpen, G.H. (1994). Position effect variegation and the new biology of heterochromatin. *Current Opinion in Genetics and Development*, **4**, 281–91.

Kellum, R. and Schedl, P. (1992). A group of scs elements function as domain boundaries in an enhancer blocking assay. *Molecular and Cellular Biology*, **12**, 2424–31.

Kim, C.G., Epner, E., Forrester, W.C., and Groudine, M. (1992). Inactivation of the human β-globin gene by targeted insertion into the β-globin locus control region. *Genes and Development*, **6**, 928–38.

Kornberg, R. and Lorch, Y. (1995). Interplay between chromatin structure and transcription. *Current Opinion in Cell Biology*, **7**, 371–5.

Laemmli, U., Cheng, S., Adolph,K., Paulseon, J., Brown, J., and Baumbach, W. (1977). Metaphase chromosome structure: the role of non-histone proteins. *Cold Spring Harbor Symposia on Quantitative Biology*, **42**, 351–60.

Lang, G., Wotton, D., Owen, M.J.M., Sewell, W.A., Brown, M.H., Mason, D.Y., et al. (1988). The structure of the human CD2 gene and its expression in transgenic mice. *EMBO Journal*, **7**, 1675–82.

Lenardo, M., Pierce, J., and Baltimore, D. (1987). Protein binding sites in immunoglobulin gene enhancers determine transcriptional activity and inducibility. *Science*, **236**, 1573–7.

Milot, E., Strouboulis, J., Trimborn, T., Wijgerde, M., de Boer, E., Langeveld, A., et al. (1996). Heterochromatin effects on the frequency and duration of LCR-meditated gene transcription. *Cell*, **87**, 105–14.

Müller, S.H., Sogo, J.M., and Schaffner, W. (1989). An enhancer stimulates transcription in trans when attached to the promoter via a protein bridge. *Cell*, **58**, 767–77.

Ondek, B., Shepard, A., and Herr, W. (1987). Discrete elements within the SV40 enhancer region display different cell-specific enhancer activities. *EMBO Journal*, **6**, 1017–25.

Philipsen, S., Talbot, D., Fraser, P., and Grosveld, F. (1990). The β-globin dominant control region: hypersensitive site 2. *EMBO Journal*, **9**, 2159–67.

Pillus, L. and Grunstein, M. (1995). Chromatin structure and epigenetic regulation in yeast. In *Chromatin structure and gene expression* (ed. S.C.R. Elgin), pp. 123–46. Oxford University Press, Oxford.

Poncz, M., Henthorn, P., Stoeckert, C., and Surrey, S. (1989). Globin gene expression in hereditary persistence of fetal hemoglobin and delta-beta thalassaemia. In *Oxford surveys on eukaryotic genes* (ed. N. McLean), pp. 163–203. Oxford University Press, Oxford.

Pruzina, S. (1996). PhD Thesis, Open University, London.

Pruzina, S., Hanscombe, O., Whyatt, D., Grosveld, F., and Philipsen, S. (1991). Hypersensitive site 4 of the human β globin locus control region. *Nucleic Acids Research*, **19**, 1413–9.

Raich, N., Enver, T., Nakamoto, B., Josephson, B., Papayannopoulou, T., and Stamatoyannopoulos, G. (1990). Autonomous developmental control of human embryonic globin gene switching in transgenic mice. *Science*, **250**, 1147–9.

Robertson, G., Garrick, D., Wu, W., Kearns, M., Martin, D., and Whitelaw, E. (1995). Position-dependent variegation of globin transgene expression in mice. *Proceedings of the National Academy of Sciences of the USA*, **92**, 5371–5.

Roseman, R.R., Pirrotta, V., and Geyer, P.K. (1993). The su(Hw) protein insulates expression of the *Drosophila melanogaster white* gene from position effects. *EMBO Journal*, **12**, 435–42.

Schlissel, M. and Brown, D. (1984). The transcriptional regulation of Xenopus 5S RNA genes in chromatin: the roles of active stable transcription complexes and histone H1. *Cell*, **37**, 903–13.

Sharpe, J.A., Wells, D.J., Whitelaw, E., Vyas, P., Higgs, D.R. and Wood, W.G. (1993). Analysis of the human α-globin gene cluster in transgenic mice. *Proceedings of the National Academy of Sciences of the USA*, **90**, 11262–6.

Shehee, W.R., Oliver, P., and Smithies, O. (1993). Lethal thalassaemia after insertional disruption of the mouse major adult β-globin gene. *Proceedings of the National Academy of Sciences of the USA*, **90**, 3177–81.

Stief, A., Winter, D.M., Stratling, W.H., and Sippel, A.E. (1989). A nuclear DNA attachment element mediates elevated and position-independent gene activity. *Nature*, **341**, 343–5.

Talbot, P., Descombes, P., and Schibler, U. (1994). The 5' flanking region of the rat LAP gene can direct high-level, position independent, copy number dependent expression in multiple tissues in transgenic mice. *Nucleic Acids Research*, **22**, 756–66.

Tazi, J. and Bird, A. (1990). Alternative chromatin structure at CpG islands. *Cell*, **60**, 909–20.

Thoma, F., Koller, T. and Klug, A. (1977). Involvement of histone H1 in the organisation of the nucleosome and of the H1-dependent superstructures of chromatin. *Journal of Cell Biology*, **83**, 403–27.

Tsukiyama, T. and Wu, C. (1995). Purification and properties of an ATP-dependent nucleosomee remodeling factor. *Cell*, **83**, 1011–20.

Tuan, D., Solomon, W., Li, Q., and London, I. (1985). The 'β-like globin' gene domain in human erythroid cells. *Proceedings of the National Academy of Sciences of the USA*, **82**, 6384–8.

Tuan, D., Solomon, W., London, I., and Lee, D. (1989). An erythroid specific developmental stage independent enhancer far upstream of the human β-like globin genes. *Proceedings of the National Academy of Sciences of the USA*, **86**, 2554–8.

Wasylyk, B., Wasylyk, C., Augerean, P., and Chambon, P. (1983). The SV40 72 BP repeat preferentially potentiates transcription starting from proximal natural or substitute promoter elements. *Cell*, **32**, 503–14.

Wijgerde, M., Grosveld, F., and Fraser, P. (1995). Transcription complex stability and chromatin dynamics *in vivo*. *Nature*, **377**, 209–13.

Wilson, C., Hugo, G., Bellen, H., and Gehring, W. (1990). Position effects on eukaryotic gene expression. *Annual Review of Cell Biology*, **6**, 679–714.

Wu, C., Wong, Y., and Elgin, S. (1979). The chromatin structure of specific genes: disruption of chromatin structure during gene activity. *Cell*, **16**, 807–14.

Yu, J., Bock, J.H., Slightom, J.L., and Villeponteau, B. (1994). A 5' beta-globin matrix-attachment region and the polyoma enhancer together confer position-independent transcription. *Gene*, **139**, 139–45.

Zafarana, G., Raguz, S., Pruzina, S., Grosveld, F., and Meijer, D. (1995). The regulation of human β-globin gene expression: the analysis of hypersensitive site 5 (HS5) in the LCR. *Proceedings of the ninth conference on hemoglobin switching* (ed. G. Stamatoyannopoulos), pp. 39–44. Intercept, Andover.

9

Interphase chromosome positions and structure during silencing, transcription and replication

Laura Manuelidis

Introduction

Before 1975 there was little appreciation of the high degree of chromosomal folding within interphase nuclei. Classical cytologists were limited to descriptions of different chromatin staining patterns in cells of different lineage. Condensed chromatin, which was stained darkly by basic dyes, was known to include the inactivated X chromosome in female cells (Barr body). Such dense or heterochromatic regions of the nucleus were all assumed to be genetically silent. The other major compartment for DNA was called euchromatin. This 'good' chromatin was thought to be 'open' and possibly completely unravelled, a necessary structural prelude for RNA transcription and subsequent synthesis of proteins essential for life.

Biochemical fractionation of chromatin from disrupted nuclei and measurements of accessibility of chromatin to various molecules such as DNAases yielded insight into basic histone–DNA interactions. At the same time sophisticated Fourier transformation and reconstitution experiments elucidated the structural details of DNA–histone complexes as they form nucleosome fibres. However, nucleosome threads were beyond the limits of conventional light microscopy, having diameters or widths of about 10 nm (Manuelidis and Chen 1990; see Chapters 1–3). How these threads are folded into visible interphase chromosome fibres, or placed within large euchromatic and heterochromatic regions of the nucleus, remains a fundamental mystery.

Several paradoxes suggest that the term euchromatin may be too broad or insufficiently precise. For example, only a small portion of the mammalian genome (less than 5 per cent) is directly involved in transcription for protein production, yet in some cells, such as large neurones, the nucleus is almost entirely euchromatic. Therefore, at least some transcriptionally inactive regions of DNA appear to be in a structurally similar state as active genes or exons. Second, metaphase chromosomes contain many bands that are relatively heterochromatic or euchromatic, based on their staining properties as well as their molecular signatures (Manuelidis 1990). Each of these visible intrachromosomal metaphase bands contains $\geq 300\,000$ bp of DNA (Manuelidis and Chen 1990), yet these bands cannot be correlated in any simple way with the large (1–3 μm) cohesive regions of euchromatin and heterochromatin in each nucleus. Because each chromosome lies within a defined nuclear space or territory (Manuelidis 1985b; Cremer *et al.* 1988), its internal heterochromatic bands cannot be spatially related to the few large heterochromatic compartments in the nucleus. Indeed, when nuclear structure is preserved, euchromatin does not diffusely extend from each chromosome, but remains in close proximity to its parent.

The purpose of this chapter is to suggest that only very small structural changes are needed for transcriptional activity. More specifically, the following points will be made. First, both gene activity and silencing can involve highly local and subtle structural modulations, i.e. transcription can be associated with very small and possibly rapid transitions in structure that are not readily visible. In this process, the higher order folding of the chromosome remains intact. Second, the dominance or proportion of specific types of non-coding ('junk') DNA can define either the silencing or the specific recruitment of large intrachromosomal bands in the functioning nucleus. Third, chromosome arms with an overwhelming preponderance of condensed heterochromatin can have a high degree of transcriptional activity. Surprisingly, many of these highly folded regions actively produce RNA transcripts. Fourth, in cultured cells there is a remarkable congression of these heterochromatic chromatids with each other and with the nuclear membrane. This suggests that recognition of like chromosome domains is achieved not only at a molecular level, but also at a global level of nuclear organization.

The above observations imply that our concepts of heterochromatin and euchromatin need major revision. They also beg for a more exact understanding of the types of chromosome unfolding that are required for transcription and replication. In this context I examine the predictions of a chromosome model for its fidelity to new observations. The structural changes that can be seen in chromosomes caught in the act of replication are compared with those seen during transcription. Transcription appears to be far more conservative of structure when judged within the confines of current techniques. In the following overview I cover some of the molecular and structural approaches currently used in our laboratory, and present a very limited sample of unpublished data concerning the above points.

Specific methods: advantages and limitations

The structure and nuclear location of different chromosome domains could not be addressed adequately before non-isotopic methods for labelling nucleic acids were introduced. There were only a few identified antibodies against nuclear proteins that could decorate specific chromosome regions (e.g. centromeres), as well as general fluorescent stains for DNA and RNA. The resolving power of non-isotopic detection of specific sequences in the nucleus was already apparent by 1982 (Manuelidis *et al.* 1982). Additional methods for tagging DNA and RNA led to many applications, but the most pertinent for delineating interphase chromosome structure are three-dimensional (3D) preservation, and the use of various tags to delineate different functional states such as DNA replication or RNA transcription. Replication can be monitored by incorporation of BrdU during cellular DNA synthesis. At the same time, a specific chromosome region can be specifically labelled with another tag such as biotin or digoxigenin (Manuelidis and Borden 1988). Thus one can focus on events in a defined chromosome domain with known attributes. Similarly, RNA transcription can be simultaneously monitored with antisense probes for RNA, while the active source DNA is labelled with sense probes. However, because DNA is highly folded in the intact nucleus, one will not resolve molecularly adjacent motifs, or visualize very subtle structural changes at the most basic nucleosomal level. Indeed, only with swelling or disruption of native structure is it possible to resolve DNA loci separated by 50–200 kb (e.g. Yokota *et al.* 1995; Lawrence *et al.* 1990), a span greater than most exons. Furthermore, even when nuclear structure is preserved by isotonic aldehyde fixation, the necessary melting of DNA to provide access and reannealing of a labelled probe will distort structure. Fortunately, this change is minimal or not apparent by inspection, even at the ultrastructural level (Manuelidis 1984, 1991; Borden and Manuelidis 1988). Thus the first limitation for all sequence-specific DNA detection is its inability to describe very subtle structural alterations at the nucleosome or folded nucleosomal (solenoid) level. On the other hand, major unravelling of the larger compact interphase chromosome fibres can be evaluated, as shown later.

A second limitation is that fixation is required to capture the cell at an instant in time. A dynamic view is gained only by following a given domain over a relevant period of time. This can easily be done in a developmental setting to show large changes in nuclear organization, as was done in initial studies of neuronal maturation (Manuelidis 1985a). Nonetheless, there may be extremely rapid and local changes in substructure that cannot be evaluated by this approach. Alternative methods for the study of specific sequences in living cells are needed to address this issue.

Various enzyme-linked detectors can be used to examine hybridized domains at the ultrastructural level. We have found that peroxidase rather than larger gold ligands are most useful for this purpose, because they penetrate more consistently into highly folded and contracted heterochromatic regions (e.g. see Borden and Manuelidis 1988). Although electron microscopic (EM) examination provides a very good indication that the conditions we use for fixation and hybridization are non-disruptive, serial EM sections are usually required to delineate each labelled 3D chromosome domain. This approach is very labour intensive and is subject to alignment errors. High voltage EM of whole nuclei is also often inadequate, because more condensed chromosomes as well as other nuclear bodies can obscure the region of interest. Furthermore, the preparation of whole mount nuclei removes the fundamental relationships and orientations to surrounding cells and to culture substrates. An example of these types of oriented relationships that would have been lost with nuclear isolation is illustrated in Plate 2 (e–g).

At the present time the best trade-off in terms of resolution and 3D visualization is confocal microscopy. More than one probe can be simultaneously evaluated and this is especially useful when the second fluorescent reporter is used to define a functional state. When the optical sectioning capacities of the confocal microscope are enhanced by deconvolution, and structures are studied in 3D rotations after reconstruction, a significant number of details and relationships can be brought forth. On the other hand, non-specific background fluorescence can be very high in tissues such as mammalian brain and this currently limits dual fluorescent detection to simpler samples such as cultured cells. Nevertheless, a number of experimental tissue culture models can be used for insertion of inducible genes, with controlled exposure to appropriate chemicals or treatments to affect cell physiology. We have used this approach to evaluate changes in an inducible and large transcriptional domain created by recombinant techniques, as briefly discussed below.

Confocal analysis

Figure 9.1 shows an overview of the process we use. For *in situ* hybridization we take great care to avoid swelling, protease treatment, and any drying of cells that can distort 3D structure. Only short detergent treatment after paraformaldehyde fixation is required to allow penetration of 100–400 nt long tagged sequences (Manuelidis and Ward 1984). We use a Leica confocal microscope because it has a very precise and reproducible piezoelectric stepper for collection of sequential optical sections. This machine can also rapidly display x–z as well as x–y sections for orientation of domains in the nucleus (see Plate 2). The correct z-axis proportions are periodically checked using 0.5 μm fluorescent beads. To ensure accurate overlays of each fluorescent signal, two fluorescent channels are simultaneously evaluated (green and red, typically FITC and either rhodamine or Texas red reporters). The whole nucleus is then resectioned with UV excitation for Dapi fluorescence of the whole nucleus (Fig. 9.1, step 3) using the same z-step origin and increments (0.15 or 0.25 μm steps). Our video sit camera (Cohu) for Dapi images has been mounted with a variable zoom lens and adjustment screws for reasonably precise video alignment (within two pixels) at any magnification. It can also be used to rapidly verify fluorochrome alignments of the red and green confocal channels by epifluorescence. A minimum of four frames of 512×512 pixels are collected for every z-axis step in each of the three fluorescent channels. The largest magnification to accommodate the structure of interest is used for each group of stacks, the laser is run at the lowest power setting possible to minimize photobleaching, and the pinhole is set to a relatively small size for optimal resolution. The video memory TCD board in our system

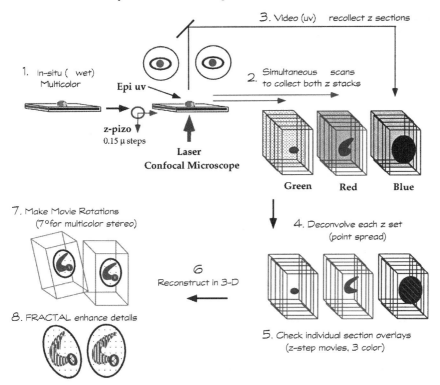

Fig. 9.1 Steps in the collection and analysis of confocal sections. For details see text.

allows us to collect as many as 40 slices for each of the three channels and is also later used to play three colour movies (Fig. 9.1, steps 5 and 7).

To obtain the most detailed information a point spread deconvolution is done on each serial image in the three channel stacks. This further removes out-of-focus fluorescence, but it is important that the brightness, contrast, and degree of deconvolution do not remove information (Fig. 9.1, step 4). Rapid deconvolutions are made possible using a 486 PC with an array processor (PL800, EighteenEight Labs) and deconvolution software (Vaytek Inc.). The resulting stacks are then displayed as individual three-colour sections to verify overlapping signals. The stacks, or relevant portions of the stacks, are also displayed as stereo pair rotations using maximum projection to reconstruct the 3D image for each stack. These primary data are far more informative and exact than interactive reconstructions used previously (Manuelidis and Borden 1988; Borden and Manuelidis 1988). Each rotation is also viewed as a movie or stereo movie with two or three colour channel overlays (Fig. 9.1, step 7). Fractal enlargement is sometimes used on regions of interest (software from Images Inc.). In principle, fractal analysis will find statistically connected spatial patterns. Details such as loops made of solenoid-like fibres within, or extending only minimally from the basic interphase fibre, can be improved with this routine (see Fig. 9.3). For reference a solenoid fibre is about 30 nm thick and consists of a wound nucleosome fibre (see Fig. 9.2). Finally, no sharpening or other filters are used except for removal of random single pixel camera noise (Adobe Photoshop software).

Shared sequence signatures

The alphabet of coding DNA specifies gene products, but is insufficient for defining the orderly recruitment of genes on different chromosomes. Transcribed sequences are punctuated by non-coding sequences, and it is likely that paragraphs and chapters in the genome are recognized at least in part through repeated non-coding sequences. Different repeat motifs predominate in euchromatic and heterochromatic chromosome bands, and each repeat type is interspersed with particular classes of coding motifs. In human cells for example, tissue-specific genes that are developmentally regulated are generally found in the more heterochromatic and late replicating bands with abundant LINE sequences. In contrast, shorter interspersed Alu repeats and GC-rich sequences strongly favour the more euchromatic bands (Manuelidis and Ward 1984; Chen and Manuelidis 1989). Additional less numerous interspersed repeats can define a more limited subset of bands on only a few chromosomes. These types of observations led to the suggestion that different repeat motifs or 'sequence signatures' may be key elements in the regulation of genetic activity (Manuelidis 1990). In principle, as few as 10 unique repeats can combinatorially specify over 3×10^6 distinct or partially related domains on different chromosomes. Thus their strategic positioning could be part of a global indexing system. This system may provide the necessary recognition for multiple sets of genes recruited at a given time, as for example during development or stress.

Some of the following experiments show that it is possible to create different types of chromosome domains using recombinant approaches for inserting repeated DNA motifs. These domains are correctly recognized by the cell, and remarkably are organized in particular ways in the interphase nucleus. I will discuss the systematic examination of very heterochromatic to more euchromatic domains with reference to shared sequence signatures, replication time, and spatial positioning within the interphase nucleus. At the same time I will address the visible features of structural change that accompany a motif in G_1 and during transcription and replication. To understand or evaluate these transitions it is necessary to refer to the pertinent features of a folded chromosome model previously detailed (Manuelidis and Chen 1990; Manuelidis 1990).

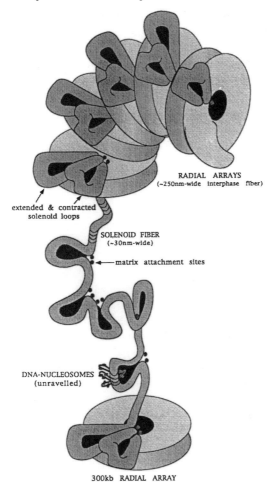

RADIAL ARRAYS
(~250nm-wide interphase fiber)

extended & contracted
solenoid loops

SOLENOID FIBER
(~30nm-wide)

matrix attachment sites

DNA-NUCLEOSOMES
(unravelled)

300kb RADIAL ARRAY

Fig. 9.2 Simplified model of part of an interphase and puffed haploid chromatid (top) giving rise to more unravelled solenoid fibres and completely unravelled nucleosomes (middle of diagram). Likely sites where matrix proteins may bind or tether DNA are also depicted at a few sites (filled circles). Alternatives addressed here include a puffing model where extended solenoid loops of about 120 kb permit access to proteins needed for transcription, but maintain the integrity of compact radial arrays. In contrast, greater unravelling disrupts the basic chromatid structure. Only a short portion of the curving intact chromatid is shown (about 250 nm in width when all its solenoid loops are contracted). In super heterochromatin the chromatid can coil upon itself (Manuelidis 1990). Such tight coils yield ~750 μm wide metaphase chromatids. For scale models and DNA compaction calculations see Manuelidis and Chen (1990).

How extended is an interphase chromosome?

In principle, the extent of chromosome unfolding needed for transcription or replication can be extreme. Figure 9.2 shows an idealized and simplified diagram of the basic haploid interphase chromosome fibre or chromatid. We previously proposed that this fibre is formed by loops of smaller solenoid fibres that can extend out or 'puff' during transcription, as depicted in Fig. 9.2. Each extended solenoid loop contains about 120 kb of DNA and such puffed loops (top) would only slightly increase the diameter of the interphase fibre, to ~350 nm. For comparison, this fibre is about 250 nm wide in its most contracted and transcriptionally silent configuration during prophase. Each horizontal segment or turn is arbitrarily defined as a radial array and can accommodate about 300 Mb of DNA (Manuelidis and Chen 1990). Thus only a simple relaxation or puffing of the radial array would be needed to synthesize RNA from any surface of the ~30 nm solenoids (see interior black spaces that are in continuity with the surface uncoloured space). In contrast, a more conventional view is that solenoid fibres are very unravelled or completely open during transcription and replication, as depicted in the central portion of Fig. 9.2. In this case solenoid fibres would not be organized in coherent radial arrays, and thus DNA should be found at a distance from the interphase chromatid. Given the current methods, it is possible to find whether transcriptionally active or replicating DNA is very closely associated with the large and reasonably compact chromatids, or alternatively if it is unravelled. In the latter case the chromatid would be either interrupted by less structured solenoid and nucleosome threads, or these threads would extend well away from the chromosome periphery. The transcriptional studies below support a modulated small relaxation or puffing model for transcription, rather than major unravelling. Furthermore, even in the most extended fibres that are decorated with domain-specific probes, loops of small 30–40 nm wide fibres are visible (see Fig. 9.3, right arrow). Whether these solenoid loops are wound exactly as proposed (Manuelidis and Chen 1990) cannot be ascertained at the confocal level of resolution (*vide supra*).

At the other extreme are silent regions. These maintain a metaphase or contracted folding pattern. In this case the interphase chromatid has an additional helical turn (Manuelidis and Chen 1990; Manuelidis 1990). Some regions of the chromosome, such as those containing long silent tandem repeats of DNA, can be tightly coiled in interphase. Previous studies have shown that 7–11 Mb of silent DNA is organized in the predicted coils in interphase (Manuelidis 1991). Very tight and contracted coils have the final well-described width of each sister chromatid in metaphase (about 750 nm), a time at which genetic silencing is complete. Replication may represent the other extreme of permissible unfolding. Our model proposes that after replication each chromatid will be nested with its sister to yield a cohesive single thicker fibre in which each of the chromatids cannot be morphologically distinguished. The pulse labelling studies I have done with short exposures to BrdU are consistent with this view. However, during replication, which progresses very rapidly, some labelled regions are clearly unravelled. The reasons for greater, albeit transient, disruption of chromosome structure during replication compared to transcription are probably best understood from a biological perspective, as later discussed.

Long tandem repeats define super heterochromatin

Experiments with high copy number tandem repeats, such as major satellite DNAs on different chromosomes, have shown that these chromosome regions can adhere to each other in the interphase nucleus. The degree of association is extreme. Centromeric regions with little or possibly no coding DNA, as well as subcentromeric domains with ≥ 9 Mb of multiple short tandem DNA repeats, are compartmentalized together in the nucleus. Indeed, these repeats can be coalesced to such a degree that individual chromosome regions within these dense bodies are not recognizable as separate structural elements (Manuelidis 1984). Only by labelling repeats on a single chromosome is it possible to define the limits of each individual element in the large heterochromatic body. Interestingly, the 3D positioning of these highly heterochromatic bodies in the nucleus (hereafter called 'super heterochromatin') can be tissue and cell lineage specific. Different patterns of congregation are characteristic and unique for different neuronal and glial populations (Manuelidis 1984; Manuelidis and Borden 1988; Borden and Manuelidis 1988). Moreover, the pattern of super heterochromatin in each neuronal subtype is maintained in evolution regardless of the details of the alphabet or sequence of the tandem repeat (Manuelidis and Borden 1988).

To further define the necessary requirements for super heterochromatin we studied transgenic mice with an inserted repeat of about 11 Mb in length (Manuelidis 1991). This DNA was confined to a single chromosome locus that was distant from the centromere. A β-globin exon that was devoid of interspersed repeats was the sole element. Although each unit of this tandem repeat was relatively GC rich and longer than many major satellite DNAs, we hypothesized that this construct would make its chromosome locus behave as other centromeric loci with natural tandem repeats. Thus, this new domain should maintain a very heterochromatic or condensed state in G_1 cells. It should also be positioned together with other natural satellites in large bodies of super heterochromatin in the nucleus. These predictions were fulfilled. This enormous transgene was completely without effect on mouse development. Silencing appeared to be solidified by the physical segregation of this transgene into coalesced bodies with other centromeric satellite DNAs in the nucleus. Furthermore, the transgene was organized in a very cell type-specific manner as determined by assessing nuclear positions in different types of neurones (Manuelidis 1991). Electron microscopy also confirmed that these inserts formed more coiled fibres, as predicted by the model.

Although the experiments have not yet been done, I would suspect that interspersion of only a few copies of a promoter–exon sequence into these long tandem arrays would be insufficient to override the dominant parameters of silencing created by the long tandem repeats. In contrast, insertion of many tandem repeats of DNA containing promoter–exon–enhancer sequences can override the natural transcriptional characteristics of the local domain into which they insert. Moreover, such genetically competent inserts can be specifically activated to produce RNA given the proper stimulus (*vide infra*). This is the contrary form of sequence dominance.

Interspersed repeats that define heterochromatic chromosome arms

It is not yet possible to insert multiple interspersed repeats in defined regions of the genome to test the importance of these repeats in chromosome recognition and function. However, nature has provided us with a remarkable example of almost completely 'heterochromatic' chromosome arms for study. In the Syrian hamster a defined set of chromosome arms show a 'constitutive' heterochromatic Giemsa staining pattern. I use the term β-heterochromatin here to distinguish these regions from large arrays of silent and truly constitutive super hetero-chromatin. All the β-heterochromatic arms of the Syrian hamster contain many interspersed copies of an endogenous retroviral intracisternal A particle (IAP) sequence, while other chromosome arms have few, if any, of these interspersed repeats (Taruscio and Manuelidis 1991). This feature makes it possible to completely delineate these arms without using dextran sulphate, a chemical necessary for adequate detection of single copy genes that disrupts and distorts interphase chromosome structure. IAP-labelled probes decorate the entire length of 10 different chromosome p-arms, as well as the entire q-arm of the X chromosome. In a comparative study of repeats it was pertinent to evaluate the structure and overall organization of these chromatids in interphase nuclei. For example, would they be organized together as like domains of super hetero-chromatin, or would they instead be widely dispersed in the nucleus? Double label studies of cells in G_1 and in S phase were also used to test the accuracy of the nested chromosome replication model. Additionally, because we had found that IAP sequences are actively transcribed both in the brain and in cultured cells, it was relevant to see if IAP RNA transcripts were produced only by a few unravelled domains, or if they arose from numerous sites along the compactly organized chromatids.

There was a remarkable congression of these β-heterochromatic IAP-rich chromosome arms in G_1 nuclei of cultured cells. These findings emphasize that 3D interphase collections of like chromosome domains are not restricted to super heterochromatic and transcriptionally inactive motifs. Thus, similar chromosome domains, defined by repeated coding DNA, can be organized together in the nucleus, possibly for functional purposes. The majority of the IAP-rich chromatids associated together at the nuclear edge, although some were in more interior portions of nucleus. Some of the arms so closely apposed the nuclear envelope that they could be used to define the peripheral border. In most cells the majority of these arms were at the top of the nucleus (pointing to the feeding medium) and only a few were at the bottom nearest coverslip attachment sites. Plate 2(g) for example shows this typical pattern where the IAP hybridization sites are depicted in red. In this reconstruction, representing about 2 μm of

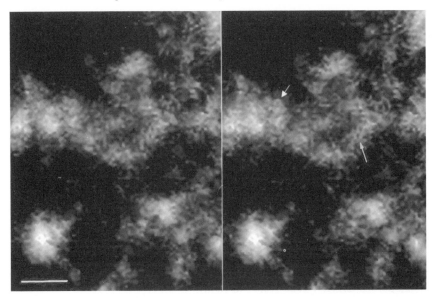

Fig. 9.3   Stereo pair of endogenous retroviral intracisternal A particle (IAP)-labelled chromatids in the nucleus after point spread deconvolution of serial sections and fractal magnification to visualize fine details. Cells were treated with Aza C for 4 h to increase IAP transcripts. The most extended labelled chromatids are shown, with a more compact chromatid at lower left for comparison. Note small loops (small circular features at right arrow), interpreted as solenoid loops of the chromatid. Left arrow points to a C-shaped interphase fibre, again containing small integral solenoid-like loops. Bar is 1 μm. Stereo for cross-eye viewing (turn upside down for viewing with stereo glasses).

through focus sections, the individual labelled chromatids are not well resolved from each other, especially because they have not been deconvolved as in Fig. 9.3. However, detailed structural studies showed more coiled 0.8–1 μm wide fibres (like super heterochromatin), as well as more relaxed interphase chromosome fibres that are about 0.4 μm wide in G_1 nuclei (Fig. 9.3). Note the numerous small loops or circles of smaller solenoid-width fibres within and at the surface of the relaxed interphase fibres (Fig. 9.3, right arrow). For reference, at the lower left of Fig. 9.3 a very compact and probably supercoiled IAP region is seen. None of the IAP solenoid loops extended far from their origin in G_1 cells. Interestingly, the degree of chromatid coiling was not readily related to the 3D position of the IAP chromatid in the nucleus.

Studies of cells pulsed with BrdU for only 15 min and then arrested by fixation were used to assess changes in the structure of these chromatids during replication. The general replication pattern of BrdU incorporation in early S-phase nuclei is characteristically diffuse, whereas when BrdU is incorporated later in replication, large and more compact bodies are typically labelled (e.g. see Bravo and Macdonald-Bravo 1987; Nakayasu and Berezney 1989). In Syrian hamster cells all the β-heterochromatic arms, with the exception of the active X chromosome, replicate late in S phase (Taruscio and Manuelidis 1991). In late S-phase nuclei therefore most late replicating domains should be resolved as IAP chromatids, while in early S phase, unreplicated IAP-rich chromatids can be evaluated.

Plate 1(a) shows the structure of IAP-labelled chromatids in early S phase (in red, with no BrdU incorporation). The characteristic more diffuse pattern of early replication is seen in the rest of the nucleus (in green). When both BrdU (green) and IAP probe detectors co-localize the signal is yellow. Note many IAP chromatids are slender (0.3–0.4 μm wide) and some form semicircular motifs and more condensed thicker coils (arrows). Unlike super heterochromatin, many of the associating IAP chromatids maintain their individual structural identities. Although the surfaces of these chromatids are not completely smooth, there is not an extensive or diffuse halo of labelled surrounding fibres. In this nucleus the top portion with the nuclear membrane was omitted for greater clarity. Nevertheless, two clusters of chromatids are still seen abutting the nuclear membrane (at bottom and at arrow). Interestingly, a collection of early replicating domains also abuts the membrane (green bodies at right).

In late replicating cells the IAP chromatids are wider and sister chromatids cannot be distinguished from each other. Such intertwined thicker yellow double chromatids are seen in Plate 1(b), most of which are clustered in groups at the nuclear membrane and in the centre of the nucleus. Some replicating IAP chromatids however showed a more detailed fibre substructure suggestive of unravelling (e.g. Plate 1(b), arrow). Some of these small fibres had solenoid fibre widths (about 0.05 μm). These structural motifs may represent domains caught during active replication, a very rapid process. Interestingly, not all IAP chromatids replicated at exactly the same time. Plate 1(b) shows some red-only IAP chromatids that have not yet replicated. Differences in the time of IAP chromatid replication were not readily related to their 3D position in the nucleus. Replication laggards were seen in both central and peripheral positions.

The above findings are consistent with the nested replication model, indicate that more than one DNA sequence signature (besides the IAP motif) may determine precise timing of replication, and suggest that active DNA synthesis involves a transient partially unravelled state. The extent of unravelling with replication was solidified in studies of smaller and simpler chromosome domains (*vide infra*).

Rather unexpectedly, IAP RNAs were transcribed from highly compact chromosome domains. RNA transcripts coincided and closely adhered to compact DNA chromatids. Furthermore, transcription originated from multiple sites on different chromatids. It was also not restricted to chromatids with a particular position, such as those in more interior regions of the nucleus. For these studies we used both unstimulated cells, as well as cells exposed to 5-aza-2'-deoxycytidine (Aza C) for 16 h. In the latter case the IAP RNA transcripts are increased 5- to 10-fold as determined by Northern blotting (data not shown). The

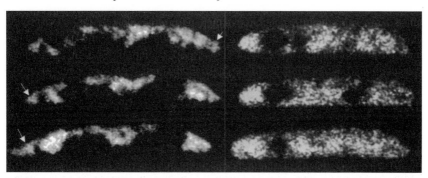

Fig. 9.4 Individual serial *x–z* sections (not deconvoluted) showing IAP RNA transcripts in non-denatured preparations at left and corresponding confocal sections stained with propidium iodide at right. Most of the RNA corresponds to preferred positions of the IAP chromatids at the top of the nucleus. Arrows point to regions of cytoplasmic RNA as determined by merging sections. Aza C treatment for 4 h.

IAP cDNA was inserted into an SP6/T7 plasmid to allow synthesis of labelled antisense RNA run-offs (for hybridization to RNA transcripts) as well as synthesis of labelled probes in a sense orientation (for labelling of chromatid DNA). The recombinant plasmid was cleaved with appropriate restriction enzymes to ensure each labelled probe represented either the 3' or 5' end of the IAP sequence, and labelling controls (e.g. switching biotin or digoxigenin for RNA and DNA detection) gave comparable results.

To evaluate possible artefacts with denaturation, RNA transcripts alone were first evaluated. With no denaturation of nuclear DNA, antisense probes labelled numerous RNA transcripts in the nucleus and many of these were focally clustered in chromatid-size domains. Plate 2(a) shows a nucleus with relatively few transcripts (yellow) in a preparation counterstained with propidium iodide (red) to reveal total nuclear DNA. Figure 9.4 shows serial *x–z* sections in a cell with more abundant IAP transcripts (left), where the nuclear DNA of the corresponding slice is shown on the right. Note the transcripts correspond to the typical IAP positions at the top of the nucleus. In these *x–z* sections RNA was also detected in the remaining cytoplasm (arrows). In contrast, control sense probes yielded only a few random background dots in non-denatured nuclei (data not shown). The close association of the IAP transcripts with IAP chromatids was further solidified by double label studies. Plate 2(b) and (c) show a single confocal slice with IAP DNA in green and RNA transcripts in red. Although the antisense probe can also label denatured DNA, some regions were only green or red (i.e. in non-contiguous structures). This confirmed that each label differentially decorated RNA and DNA. In Plate 2(e), a merged image of RNA and DNA from two sequential sections of this nucleus, note that RNA is being transcribed from a compact central IAP chromatid as well as from reasonably compact IAP chromatids at the nuclear periphery. Thus, these studies gave the first indication that reasonably compact chromatids with negligible unravelling are capable of transcription. Furthermore, transcription was not dependent on nuclear position and involved many different IAP chromatids. Therefore, the attributes needed for transcription at these multiple sites might be universally designated by the IAP sequence, or by other shared sequence motifs.

Several principles can be drawn from the above studies on repeated DNAs. First, the cell can recognize chromosomal repeats provided they are sufficiently long or abundant. Second, this recognition is not dependent on the exact nucleotide sequence, but rather on the overall sequence organization. Insertion of long tandem repeats of exons lacking promoters, enhancers, or interspersed Alu family repeats creates a new super heterochromatic domain. Such domains are compartmentalized into highly coalesced bodies in the nucleus, a strong structural and probably rigid form of silencing. Third, high copy interspersed repeats (e.g. more than 1000 copies IAP) can define a different type of structure and association. They can designate β-heterochromatin over distances as long as an entire chromatid arm. These interspersed repeats are involved in the congression of similarly constituted chromatids in the nucleus. However, β-heterochromatic chromatids retain their individual spatial identities. The less coalesced association of β-heterochromatic chromatids probably provides for a flexible and rapid induction of transcripts in response to environmental or developmental signals. Fourth, although replication times are not precisely synchronous, closely apposed heterochromatic chromatids can replicate with extraordinary synchrony, as shown in Plate 1(b). Interestingly, early replicating chromatids can also cluster together as shown in Plate 1(a) (large peripheral green body). Finally, attachment to the nuclear envelope is not specific for different types of chromatin.

Telomeric repeats

Thus far I have covered repeats that can modulate the position and replication time of very large (Mb) chromosome domains. To contrast the organization of smaller domains, we studied a unique set of repeats with limited copy numbers. Such studies were helpful in understanding some of the features needed to determine nuclear position and replication time. The less numerous tandem repeats found at the ends of chromosomes (telomeres) were chosen for this purpose. A canonical TTAGGG sequence is found at the end of mammalian metaphase chromosomes (Moyzis *et al.* 1988). Several studies have suggested that truncation of these telomere repeats leads to cell senescence, whereas additional telomeric copies are typically correlated with cell immortality (see Boeke 1990). Therefore, the biological consequences of these repeats could be very significant. However, this senescence–immortality concept may be more complex than originally indicated in studies where total telomere copies per cell were calculated. Individual chromosomes can have quite variable telomere repeat copies, as shown by pulsed field gel electrophoresis of large chromosome domains (Manuelidis 1994). In the same study, some immortal cell lines with a limited life span *in vitro* also had fewer telomere copies than normal cells. Additionally, it is often assumed that telomeres are universally positioned on the nuclear membrane, but this is clearly not true in mammalian, plant, and yeast cells (Manuelidis and Borden 1988; Rawlings *et al.* 1991; Vourch *et al.* 1993; Palladino *et al.* 1993; Manuelidis 1994).

Momentarily putting aside the different arrangements of interphase chromosomes in different organisms, as well as those specific for cells of different lineage, it is possible to ask if small telomeric repeats define a unique pattern of organization, i.e. one that is independent of the variant local sequences on each chromosome. Alternatively, is the behaviour of each telomere determined by attributes of its associated domain, as a submissive rather than dominant partner? This question has relevance for genetic silencing, best known through position effect variegation in *Drosophila* (see Chapter 13). It may also be relevant for spreading transcriptional activation or recruitment of Mb regions of DNA at the termini of chromosomes. In this case, for example, a more euchromatic determinant (such as many adjacent Alu repeats) should dominate the behaviour of its telomere locus.

If telomere repeats are submissive because of their relatively short length, they could be controlled by adjacent heterochromatic repeats. In Syrian hamster cells telomeres that inhabit terminal regions of β-heterochromatic chromatids should collect together with their respective chromatids at positions near to the nuclear membrane at the top of the cell. In contrast, telomeric repeats at the ends of more euchromatic chromatids might be more separated and/or at different nuclear positions. Furthermore, submissive telomeres should have replication times that are determined by their specific contextual domains, i.e. they should replicate in synchrony with their individual chromosome locales. Replication time is one of the best indicators of functional activity, with most housekeeping genes replicating early and tissue-specific genes replicating later in the cell cycle (Holmquist 1989). I am aware of only two studies on the replication time of telomeres, and both studies used hybridization to blots of synchronized cells for this determination. Each showed vastly different results, with exclusively late or very variable replication times of telomeres (McCarroll and Fangman 1988; ten Hagen *et al.* 1990). The structural studies below show that both replication time and spatial positioning of telomeres in the nucleus reflect their contextual DNA domains. Moreover, the degree of unravelling that can occur with replications was further clarified in the study of these small elements.

Canonical telomeric repeats, as well as polymerase chain reaction amplification products synthesized with 5'-IAP and consensus mammalian telomere primers, were generated as probes. In both instances characteristic small foci of hybridization were seen at metaphase chromosome termini (data not shown). Plate 2(e) shows a reconstruction from 10 serial 0.2 μm *x–z* sections at the centre of an early replicating nucleus (BrdU in green, telomeres in red). This nucleus has a characteristic dispersed early replication pattern. For comparison, a late replicating cell is shown in Plate 2(f), where the sections were collected starting close to the lateral edge of the nuclear membrane. Characteristically, the later replication pattern shows label in larger and more discrete regions of the nucleus. In both images the size of the telomeres is variable, with larger red domains indicating close apposition of telomeres from several chromosomes. This conclusion was further solidified by counting telomeres in interphase. The number of telomeres was equal to or less than one-third of the number of telomeres in metaphase spreads. Additionally, the size of most telomere domains in interphase was typically more than three-fold greater than found in metaphase chromosomes. Thus telomeres often associate with each other in mammalian interphase nuclei.

Several large and small telomere domains were not attached to the nuclear membrane. Even with very large telomere aggregates only a small portion of the domain was attached to or near the nuclear membrane (Plate 2(e) and (f)). This indicates that self recognition is more important for these termini than envelope association. Interestingly, many telomeres were oriented near the bottom of the nucleus (facing the coverslip) in both early and late replicating cells (Plate 2(e) and (f)). A similar nuclear position was also often seen in G_1 cells (data not shown). Thus telomeric domains can be positioned in the nucleus with a particular 3D orientation. This orientation, however, is not strictly polar or Rabl-like, in accord with previous studies of mammalian cells (e.g. Manuelidis and Borden 1988; Manuelidis 1994). In contrast, the larger telomere aggregates collected at the top of the nucleus in regions known to harbour β-heterochromatic chromatids. For example, Plate 2(f) shows two large telomere collections at the top of the nucleus. Thus the orientation of particular groups of telomeres in the nucleus appears to be specified by their local chromosome residence, with greater aggregation of telomeres on β-heterochromatin. IAP or other interspersed repeats are probably key signals for these associations and variant positions. Similarly, the numerous hamster B1 family of Alu repeats could have a role in the more limited congression of different chromatid termini at the bottom of the nucleus.

The replication time of telomeric regions was variable in these cells. Some telomeric clusters, especially those near the bottom of the cell, replicated early in S phase, while fewer of these replicated later in S (Plate 2(e) and (f), yellow versus red structures in each). These differences indicate that replication time is also determined by the characteristics of the adjacent DNA domain. For example, telomere clusters from β-heterochromatic chromatids at the top of the cell in Plate 2(f) (red and yellow) are late replicating (compare the position of these arms shown in red in Plate 2(g)), as are a few smaller and more dispersed telomeres at the bottom of the cell. The variable replication time of telomeres shown here is entirely in accord with blotting studies of telomeres in human cells (ten Hagen *et al.* 1990). In human cells telomeres lie beside both heterochromatic and euchromatic bands. Although the late-only replication pattern in yeast cells may seem discordant, yeast telomeres all localize in heterochromatic domains

(McCarroll and Fangman 1988). Thus the uniform late replication time in yeast may be a consequence of their surrounding DNA motifs.

Finally, small details of structure were seen within replicating but compact telomeres. Some double-labelled solenoid-like fibres showed a helical or winding pattern (Plate 2(e), arrow) in accord with the proposed model (see Fig. 9.2). Double-labelled threads with similar widths could also be unravelled. Indeed, in some cases the small telomeric domain was no longer recognizable as a cohesive structure (Plate 2(f), arrow). This image probably represents capture of a rapid transient unravelling within an actively replicating domain. It also demonstrates that the methods used are capable of detecting unravelled fibres. We have not yet observed this degree of unravelling with RNA transcription, even in single large recombinant regions, where transcription was experimentally induced (*vide infra*). Therefore, replication appears to be more disruptive of structure than transcription.

Insertion of tandem coding domains

In the systematic evaluation of different chromosome domains, it becomes essential to compare events in a large and well-defined euchromatic domain that is transcriptionally active. Because there were no available probes that decorated only a single euchromatic domain of more than 200 kb, we chose to create a large inducible coding domain in the above hamster cells (D. Hanlon and L. Manuelidis, unpublished data). To create domains that would be as easy to detect as some of the above repeats, we chose to concatomerize a single transcriptionally competent element to produce tandem arrays totalling more than 400 kb in length. These concatomers were transfected into cells. For rapid monitoring of experimental induction of transcripts we used the β-galacosidase reporter under the control of a heat shock promoter. The details of heat induction, RNA transcription kinetics, and subcloning of cells to select for large inserts positioned in a few selected β-heterochromatic or euchromatic loci will be reported elsewhere. However, these studies confirmed the maintenance of a cohesive chromatid structure with active transcription. Each large locus was similarly inducible, regardless of its local environment. Typically these loci formed round bodies in G_1. With heat-induced transcription these sites remained compact, but had slightly greater diameters than their silent uninduced counterparts. One of the few more uncoiled fibres induced to transcribe is shown in Plate 2(g) (heat shock domain in green and the IAP chromatids labelled in red). The decorated green fibre is consistent with a more relaxed or uncoiled interphase chromatid as previously modelled. Double labelling of heat shock RNA transcripts and construct DNA confirmed transcription from these relaxed as well as more coiled heat shock loci (data not shown). Thus, this more refined experiment confirmed the ability of compact folded chromatids to support robust transcription.

Interestingly, these recombinant heat shock loci were positioned in the interior of the nucleus and did not attach to the nuclear membrane. Furthermore, although some of the artificial inserts were over 1 Mb, they did not associate with each other. Each haploid chromatid bearing the insert was positioned in separated regions of the nucleus. Thus, these repeats were insufficient to override the segregated positioning of their parent chromatids. Studies on replication in these loci were consistent with the structural observations made on other loci, as described above. At least some of these heat-inducible loci were late replicating, as would be expected for an inducible exon. The complete lack of expression of the heat shock β-galactosidase reporter in uninduced cells, even when inserts were in euchromatic positions, shows that these large domains are correctly recognized by the cell. It remains to be seen if smaller constructs in euchromatic loci are similarly recognized, or are inappropriately recruited with other housekeeping genes, i.e. are subject to spreading activation.

Discussion

The above studies show that the characteristics of reiterated DNA are important determinants of chromosomal positioning and replication time. In the transgenic model system new super heterochromatin was created by insertion of long tandem repeats and was correctly assigned to congealed large bodies of heterochromatin in interphase nuclei (Manuelidis 1991). These bodies represent the classic form of heterochromatin that is associated with silencing. How the cell recognizes, silences, and compartmentalizes these domains is not clear. However, the structure of these large bodies suggests that specific proteins participate in this process. Such proteins must include more than centromere-specific proteins that function during mitosis, because newly created DNA inserts that were far from the centromere were similarly compartmentalized in the nucleus. Indeed, they coalesced with centromere-rich bodies in a cell type-specific pattern. Interestingly, the yeast SWI6 and SIR proteins are intimately associated with centromeres, silent mating-type loci, and telomeres (see Chapter 15), and it is most pertinent that telomeres form similarly cohesive hetero-chromatic bodies in yeast nuclei (Palladino *et al.* 1993; Ekwall *et al.* 1995). I assume that homologous proteins are involved in the organization of super heterochromatin in mammalian cells. The cell type-specific positioning of these bodies in the nucleus could also involve additional proteins or binding elements that are found only in mammalian cells.

This type of silencing may be viewed as the relegation of uninteresting junk sequences into a meaningless nuclear compartment. This view is predicated on the notion that function is restricted to exons that encode protein products. However, this may be an unnecessarily narrow concept of function. It is possible that transcriptionally silent sequences, such as very long tandem repeats, are actively used in interphase to organize specific groups of chromatids in mammalian nuclei (Manuelidis 1991). This could provide a functional advan-tage, especially in cementing particular patterns of expression in cells of different lineage. Highly refined cells, such as different types of neurones, may depend on these locking mechanisms to prevent their slippage into less exacting (or less cell-specific) patterns. I presume these alternate patterns would be detrimental for survival of the organism.

At the same time, the ability to organize a particular subset of chromosomes together in interphase may also provide a greater range or flexibility of expression. The variable positioning of groups of chromosomes in different mammalian cells contrasts with the simpler and more uniform Rabl organization (polar centromere orientation) of interphase chromosomes in *Drosophila* embryos (Ellison and Howard, 1981). However, transcriptionally silent regions are not the sole signals for chromosome segregation in mammals. The obvious congression of entire chromatids with β-heterochromatin-specific repeats exemplifies another form of chromosome-specific segregation in interphase. Congression of specific chromatid domains appears to be driven by their specific sequence identities rather than by some recognition element at the nuclear membrane. All classic forms of chromatin (euchromatin, β-heterochromatin and super heterochromatin) have been shown to focally adhere to the nuclear membrane, but most of the individual elements in these bodies have no direct attachment to the membrane. Thus the driving force of the nuclear envelope or its proteins is less substantial in this organization.

The effect of more numerous repeats that define super heterochromatin and β-heterochromatin was compared to less numerous repeats at defined loci. Telomere-repeat domains behaved as submissive elements. They appear to be incapable of overriding their local sequence environment and acquire both the position and replication time of their individual locales. Nevertheless, they can also form cohesive aggregates with each other and in this sense they mimic their longer tandem cousins. In yeast cells many telomeres also collect together, and the repression of gene expression that correlates with this spatial association can be destroyed in mutants with deficiencies in the SIR3 and SIR4 proteins (Palladino *et al.* 1993). Moreover, these protein deficiencies lead to a more scattered or diffuse organization of telomeres throughout the nucleus. This is further evidence that aggregation indicates silencing and is solidified by specific protein associations. It also is a strong indication that greater positional dispersion can be part of the process of transcriptional activation.

Because telomeres in normal yeast cells are uniformly late replicating and next to heterochromatin, it will be of interest to find if these telomeres acquire an early replication pattern after deletion of the SIR3 and SIR4 genes. Such a study would clarify the role of these proteins in the designation of replication time. It is also pertinent to find whether telomere repeats that are inserted within chromosome arms replicate in synchrony with their locale, and whether insertion of many telomere copies can override or alter the native replication pattern. Such determinations are most relevant for mobile or variable copy number intra-chromosomal repeats. Finally, the instability of telomeres may provide one mechanism for differential expression in cells of different lineage. Although shortening of telomeres has been associated with senescence, the inherent instability of telomeric silencing (Chien *et al.* 1993) could be part of a differentiation mechanism. Similarly, small intrachromosomal repeats that are unstable could likewise have modulatory effects on transcription, not only in differentiation but also in disease. It remains to be seen if accumulating short tandem repeats within exons, as in Huntington's disease (Gusella and MacDonald 1995), can lead to a spreading change in genetic activity.

Although Giemsa (G)-dark bands on mammalian chromosomes are often assumed to be transcriptionally silent (e.g. Palladino *et al.* 1993), transcriptional studies of very heterochromatic G-dark chromatids show this assumption is unwarranted. RNA emanates from numerous regions on most of these chromatids. Furthermore, the actively transcribing DNA is remarkably compact and unravelled chromatids are not apparent. This indicates that structural disruption of the basic interphase chromatid is not required for transcription. Study of more discrete recombinant loci with more than 400 kb of coding repeats arrayed in tandem further verified the maintenance of the basic interphase fibre during transcription. Detailed studies with deconvolution showed structures consistent with a limited puffing or extension of radial arrays, rather than chromatid unravelling to the solenoid or nucleosomal level. Neither of these studies in 3D preserved nuclei support the notion of pulled out long solenoid loops of more than 2 Mb *in vivo*, an interpretation based on mapping studies of hypotonically swollen nuclei (Yokota *et al.* 1995).

The visualization of more unravelled structures with replication shows that small solenoid-like fibres are detectable, as shown for example in the telomere domains. The reason for obvious unravelling during replication, but not during transcription, is a matter of speculation. A biological explanation, however, seems most appropriate. In a committed cell the conservation of structure during transcription would be advantageous for preserving individual and unique cell functions. In contrast, during replication the cell must be able to reformulate or reorganize its chromatin for proper functioning in new environments, or in new developmental roles. Thus, the greater flexibility during replication may provide a window of opportunity for more extensive remodelling with consequent changes in function. Moreover, multiple sites that replicate at the same time can be modified together. From the above studies such changes are likely to be at the molecular level and affect submicroscopic nucleosome structure, leaving larger chromatid folding relatively intact. Remodelling with acetylated histones (see Chapter 3) or phosphorylated chromosomal proteins for example, along with DNA modifications such as methylation (see Chapter 11), may be key players in this act. These types of changes are amply discussed by others in this volume.

Future developments

Several principles of interphase chromosome organization have become apparent in the last 15 years, yet many fundamental questions remain. The precise relationship between function and structure, and the molecular underpinnings of chromosome position effects need to be addressed. Most pertinent for such studies is the choice of meaningful biological systems that can be experimentally manipulated. Clearly, the molecular insertion of defined genetic domains in cultured cells, or in transgenic animals, will continue to be informative. Such experimental approaches are likely to clarify several key features of developmental regulation. Nevertheless, the largest challenge at present is to develop chromosome-specific probes that can be used in living cells. These would allow one to investigate responses to more transient and controlled physiological stimuli and to give a more comprehensive view of rapid structural and protein interactive changes.

Summary

Classic concepts of heterochromatin and euchromatin are inadequate for distinguishing the many functionally diverse regions of a chromosome. Three-dimensional confocal microscopy hybridization studies have illuminated the structure of different chromosomal segments in interphase nuclei. More specifically, genetically silent, transcriptionally competent, and actively replicating domains of large size were used to evaluate degrees of chromatid unfolding. Several conclusions can be made. First, although many metaphase chromosome regions stain with the characteristics of 'constitutive' heterochromatin, only some of these are completely silent in interphase. Silent large regions, such as those at the centromere, lose their individual identity as they are congealed together in the nucleus in visible dense bodies. This form of 'super heterochromatin' is driven by very long tandem repeats. It can be created *de novo* by insertion of long tandem repeats of promoterless open reading frames, even at non-centromeric sites. In contrast, very heterochromatic chromosome arms with many long interspersed repeats maintain their individual structural identity in interphase. Unlike super heterochromatin, these chromatids cannot be assigned to G-dark or ultrastructurally dense bodies in the nucleus. Nonetheless, these chromatids congregate with each other, most often at the upper nuclear periphery of cultured cells. Furthermore, abundant transcripts are produced along these β-heterochromatic arms. Remarkably, there was only a minimal change in their overall structure during transcription. Because these chromatids contain specific and abundant interspersed repeats, these repeats may signal a different type of congression that is used for the orderly recognition and recruitment of their cell-type specific genes.

Second, for comparison, more euchromatic large loci were created by recombinant means. Each of these megabase DNA domains contained multiple tandem repeats of a competent reporter gene under a heat shock promoter. These large domains did not associate with each other in interphase, regardless of their chromosome position in G-dark or G-light bands. However, they again maintained a coherent 3D structure with transcriptional activation. Only a small increase in width, consistent with puffing of the chromatid fibre, was observed and there was no evidence for long unravelled loops of solenoid/nucleosome fibres. Thus again, transcription was conservative of structure.

Third, less numerous, conserved tandem repeats at chromosome termini were also evaluated. These termini behaved in an analogous manner as their longer 'super heterochromatic' cousins in that they often adhered to each other in interphase. However, telomeres were submissive in their position and replication patterns. For example, those next to β-heterochromatin domains replicated late, while telomeres on more euchromatic arms replicated early in S phase.

Finally, several defined domains were caught in the act of replication. During this process the chromatid became more unravelled than during transcription. The above studies suggest that genetic activity is controlled through 3D positioning of chromosomes with specific sequence signatures. Transcriptional activation has only a minimal effect on the overall structure of the interphase chromatid. In contrast, the more dramatic unravelling of the chromatid during S phase can be part of the process of molecular remodelling that provides a way for the cell to flexibly adapt to new environmental stimuli. During replication the cell can modify molecular details of the chromosome to recruit alternate gene families for differentiation or in response to disease.

Acknowledgements

I thank Peter Borden and Liz Corbett for technical assistance. This work was supported by NIH grant CA15044.

References

Boeke, J.D. (1990). Reverse transcriptase, the end of the chromosome and the end of life. *Cell*, **61**, 193–5.

Borden, J. and Manuelidis, L. (1988). Movement of the X chromosome in epilepsy. *Science*, **242**, 1687–91.

Bravo, R. and Macdonald-Bravo, H. (1987). Existence of two populations of cyclin/ proliferating cell nuclear antigen during the cell cycle: association with DNA replication sites. *Journal of Cell Biology*, **105**, 1549–54.

Chen, T.L. and Manuelidis, L. (1989). SINEs and LINEs cluster in distinct DNA fragments of Giemsa band size. *Chromosoma*, **98**, 309–16.

Chien, C-T., Buck, S., Sternglanz, R., and Shore, D. (1993). Targeting of SIR1 protein establishes transcriptional silencing at HM loci and telomeres in yeast. *Cell*, **75**, 531–41.

Cremer, T., Lichter, P., Borden, J., Ward, D.C., and Manuelidis, L. (1988). Detection of chromosome aberrations in metaphase and interphase tumor cells by *in situ* hybridization using chromosome-specific library probes. *Human Genetics*, **80**, 235–46.

Ekwall, K., Javerzat, J.P., Lorentz, A., Schmidt, H., Cranston, G., and Allshire, R. (1995). The chromodomain protein Swi6: a key component at fission yeast centromeres. *Science*, **269**, 1429–31.

Ellison, J.R. and Howard, G.C. (1981). Nonrandom position of the AT rich DNA sequences in early embryos of *Drosophila virilis*. *Chromosoma*, **83**, 555–61.

Gusella, J.F. and MacDonald, M.E. (1995). Huntington's disease. *Seminars in Cell Biology*, **6**, 21–8.

Holmquist, G.P. (1989). Evolution of chromosome bands: molecular ecology of non-coding DNA. *Journal of Molecular Evolution*, **28**, 469–86.

Lawrence, J.B., Singer, R.H., and McNeil, J.A. (1990). Interphase and metaphase resolution of different distances within the human dystrophin gene. *Science*, **249**, 928–32.

Manuelidis, L. (1984). Different CNS cell types display distinct and non-random arrangements of satellite DNA sequences. *Proceedings of the National Academy of Sciences of the USA*, **181**, 3123–7.

Manuelidis, L. (1985a). Indications of centromere movement during interphase and differentiation. *Annals of the New York Academy of Sciences*, **450**, 205–21.

Manuelidis, L. (1985b). Individual interphase domains revealed by *in situ* hybridization. *Human Genetics*, **71**, 288–93.

Manuelidis, L. (1990). A view of interphase chromosomes. *Science*, **250**, 1533–40.

Manuelidis, L. (1991). Heterochromatic features of an 11 Mb transgene in brain cells. *Proceedings of the National Academy of Sciences of the USA*, **88**, 1049–53.

Manuelidis, L. (1994). Genomic stability and instability in different neuroepithelial tumors: a role for chromosome structure? *Journal of Neuro-Oncology*, **18**, 225–39.

Manuelidis, L. and Borden, J. (1988). Reproducible compartmentalization of individual chromosome domains in human CNS cells revealed by *in situ* hybridization and three-dimensional reconstruction. *Chromosoma*, **96**, 397–401.

Manuelidis, L. and Chen, T.L. (1990). A unified model of eukaryotic chromosomes. *Cytometry*, **18**, 8–25.

Manuelidis, L. and Ward, D.C. (1984). Chromosomal and nuclear distribution of the Hind III1.9kb repeat segment. *Chromosoma*, **91**, 28–38.

Manuelidis, L., Langer-Safer, P.R., and Ward, D.C. (1982). High resolution mapping of satellite DNA using biotin-labeled DNA probes. *Journal of Cell Biology*, **95**, 619–25.

McCarroll, R.M. and Fangman, W.L. (1988). Time of replication of yeast centromeres and telomeres. *Cell*, **54**, 505–13.

Moyzis, R.K., Buckingham, J.M., Cram, L.S., Dani, M., Deaven, L.L., Jones, M.D., *et al.* (1988). A highly conserved repetitive DNA sequence (TTAGGG)n present at the telomeres of human chromosomes. *Proceedings of the National Academy of Sciences of the USA*, **85**, 6622–6.

Nakayasu, H. and Berezney, R. (1989). Mapping replication sites in the eukaryotic cell nucleus. *Journal of Cell Biology*, **108**, 1–11.

Palladino, F., Laroche, T., Gilson, E., Axelrod, A., Pillus, L., and Gasser, S.M. (1993). SIR3 and SIR4 proteins are required for the positioning and integrity of yeast telomeres. *Cell*, **75**, 543–55.

Rawlings, D.J., Highett, M.I., and Shaw, P.J. (1991). Localization of telomeres in plant interphase nuclei by *in situ* hybridization and 3D confocal microscopy. *Chromosoma*, **100**, 424–31.

Taruscio, D. and Manuelidis, L. (1991). Integration site preferences of endogenous retroviruses. *Chromosoma*, **101**, 141–56.

ten Hagen, K.G., Gilbert, D.M., Willard, H.F., and Cohen, S.N. (1990). Replication timing of DNA sequences associated with human centromeres and telomeres. *Molecular and Cellular Biology*, **10**, 6348–55.

Vourch, C., Taruscio, D., Boyl, A.L., and Ward, D.C. (1993). Cell cycle-dependent distribution of telomeres, centromeres and chromosome-specific subsatellite domains in the interphase nucleus of mouse lymphocytes. *Experimental Cell Research*, **205**, 142–51.

Yokota, H., van den Engh, G., Hearst, J.E., Sachs, R.K,. and Trask, B.J. (1995). Evidence for the organization of chromatin in megabase pair-sized loops arranged along a random walk path in the human GO/G1 interphase nucleus. *Journal of Cell Biology*, **130**, 1239–50.

10

Genes and transcripts in the interphase nucleus

David L. Spector

Introduction

Recent advances in fluorescence *in situ* hybridization (FISH; for a review see McNeil *et al.* 1991), including the development of probes and detection systems, have made it possible to easily detect whole chromosomes, chromosomal regions, single genes, and the RNA transcripts that the genes encode, in both cells and tissues. This approach has allowed investigators to address questions relating to the nuclear organization of chromosomes, the distribution of specific RNA species and their relationship to specific subnuclear regions, and the pathways that RNAs take from their sites of synthesis to the nuclear envelope. In this chapter I provide an overview of our current state of knowledge on the organization of genes and RNA transcripts in the interphase nucleus and the relationship of nuclear organization to gene expression.

Nuclear organization of DNA sequences

The idea that interphase chromatin may not be randomly distributed throughout the interphase nucleus, but may occupy distinct territories, was first put forth in the classical papers by Rabl (1885) and Boveri (1909). Rabl (1885) suggested that each chromosome in plant cells occupies a distinct domain throughout interphase which reflects its mitotic orientation. Boveri (1909) confirmed these studies by showing that chromosomes maintained relatively fixed positions in the nuclei of *Ascaris* eggs. Furthermore, these studies suggested that telomeres were attached to the nuclear envelope on one side of the nucleus and centromeres were attached on the opposite nuclear side. More recently, numerous studies have readdressed the initial questions asked by Rabl and Boveri in a variety of systems with more specific probes and at significantly higher resolution. In *Drosophila* polytene chromosomes have been differentially stained with vital dyes, and in conjunction with optical sectioning methods it has been possible to examine chromosomes in living cells. These studies have revealed that polytene chromosomes are closely associated with the inner surface of the nuclear membrane and contact the membrane at specific sites (Agard and Sedat 1983; Mathog *et al.* 1984; Hochstrasser and Sedat 1987). Furthermore, it was demonstrated that chromosomes occupy distinct territories within diploid and polytene nuclei and spiral with the same 'handedness' through the nucleus (Mathog *et al.* 1984; Hilliker 1985; Hochstrasser *et al.* 1986). In mammalian cells Cremer *et al.* (1982) have shown that laser UV microirradiation of particular interphase nuclear areas resulted in damage to discrete chromosomal regions, suggesting that the genome is non-randomly organized during interphase. The development of *in situ* suppression hybridization with chromosome-specific DNA libraries has allowed investigators to localize specific chromosomes in interphase cells (Plate 3). This method of 'chromosome painting' has confirmed and further established the existence of chromosome territories in normal as well as in tumour cell nuclei (Borden and Manuelidis 1988; Cremer *et al.* 1988; Lichter *et al.* 1988; Pinkel *et al.* 1988; Trask *et al.* 1988). Furthermore, Manuelidis (1985b) has examined the arrangement of individual human chromosomes in mouse–human cell hybrids and found that the human chromosome showed a reproducible position in the nuclei.

Several studies have provided data to support the concept that chromosomes are dynamic in the interphase nucleus and that their position is cell cycle dependent (for a review see Haaf and Schmid 1991; Manuelidis 1990; Spector 1993; also see Chapter 9). Using a composite probe to chromosome 8, the interphase position of this chromosome was observed to change during the cell cycle (Ferguson and Ward 1992). In G_1 cells chromosome 8 centromeres localized adjacent to the nuclear periphery and the chromosomal arms extended toward the nuclear interior. However, in G_2 the chromosome reoriented itself and the centromeres were internal and the chromosomal arms extended toward the nuclear periphery (Ferguson and Ward 1992). A similar redistribution was observed in brain tumour cells where centromeres were dispersed during G_1 and S phase and became clustered toward the nuclear interior during G_2 (Manuelidis 1985a). In a human carcinoma cell line centromeres were localized adjacent to the nucleoli or the nuclear membrane in G_1 and appeared to form necklace-like structures in S and G_2 (Haaf and Schmid 1989). These structures may result from unravelling of centromeres during DNA replication and they may represent the repetitive subunit structure of the centromere proposed by Zinkowski *et al.* (1991). In numerous plant species centromeres have been reported to be grouped together near the nuclear envelope (for a review see Hilliker and Appels 1989). Rearrangements of centromere positions have also been shown to be related to developmental stages (Manuelidis 1985b). Centromeres in post-mitotic (non-dividing) Purkinje cells were localized in three large clusters similar to what is observed in adult neurones. However, unlike the adult Purkinje neurones in which centromeres are associated with the nucleoli, centromeres in these cells were localized to various regions between the nucleoli and the nuclear envelope (Manuelidis 1985b). As nucleoli became more prominent in Purkinje cells, the centromeres began to associate with the nucleolar rim as they do in adult cells. This study clearly showed a rearrangement of centromeres as a function of development. In differentiated dorsal root ganglia, centromeres were primarily associated with the nucleolus and positions between the nucleolus and the nuclear envelope (Billia and De Boni 1991). Perhaps the most provocative study demonstrating a correlation in chromosome position and cell physiology comes from work on human epileptic foci (Borden and Manuelidis 1988). In normal male cortical neurones the X chromosome was localized to the nuclear periphery. However, when cells in an electrophysiologically defined seizure focus were observed, there was a dramatic increase from approximately 7 to 45 per cent in the number of cells exhibiting internal nuclear localization of the X chromosome (Borden and Manuelidis 1988). A similar observation was previously reported in neurones after 8 h of electrical stimulation (Barr and Bertram 1949). Taken together, these data suggest that the positioning of chromosomes in the interphase nucleus is not fixed but is dynamic and changes during the cell cycle, perhaps to reflect differentiation, physiological conditions, and/or gene expression.

Telomeres are present at the ends of chromosomes and are characterized by highly repetitive DNA sequences, which are synthesized in many organisms by a unique enzyme called telomerase (Greider and Blackburn 1987). While classically telomeres were reported to be associated with the nuclear envelope (Rabl 1885), this does not appear to be the case in all cell types examined. In two plant species telomeres were found to be associated primarily with the nuclear envelope and to a lesser extent with the nucleolar surface (Rawlins *et al.* 1991). In both cases the telomeres were clustered and in *Vicia faba* more than two-thirds of the telomeres were adjacent to an area of the nuclear envelope that represented less than 10 per cent of the total surface area, supporting the Rabl configuration. In some brain cell types telomeres have been identified in internal portions of the nucleus, whereas in some other cell types they have been shown to be associated with the nuclear membrane (Borden and Manuelidis 1988; Ferguson and Ward 1992) or in intermediate positions (Billia and De Boni 1991). However, in normal human fibroblasts (Harley 1995) and in mouse lymphocytes (Vourc'h *et al.* 1993) telomere sequences are distributed throughout the nucleoplasm and are not restricted to the nuclear periphery. Therefore, the localization of telomeres in many cell types does not follow the classical Rabl configuration observed in *Drosophila* polytene nuclei (Hochstrasser *et al.* 1986) and in nuclei of several plant species (Avivi and Feldman 1980). These differences demonstrate that the interphase organization of telomeres, and therefore chromosomes, varies in different cell lineages.

Somatic pairing of homologous chromosomes has been reported in a number of plant and Diptera species (for a review see Hilliker and Appels 1989). For example, Hiraoka *et al.* (1993) have shown that the two homologous histone loci in *Drosophila* embryo nuclei are distinct and separate through all stages of the cell cycle up to nuclear cycle 13. During interphase of cycle 14, when transcription is turned on, the two homologous clusters were found to co-localize with high frequency. Concomitant with homologue pairing at cycle 14, both histone loci were also found to move from their position near the midline of the nucleus toward the apical side (Hiraoka *et al.* 1993). This result demonstrated that there is a dramatic chromosomal reorganization coincident with the initiation of zygotic transcription. However, somatic pairing does not seem to be the case in mammalian cell systems (Cremer *et al.* 1982; Huang and Spector 1991; Lawrence and Singer 1991). Pinkel *et al.* (1986) used probes specific to the X or Y chromosomes to show that the sex chromosomes did not exhibit somatic pairing in interphase human cells. When Chinese hamster cells were UV irradiated with a laser UV microbeam (Cremer *et al.* 1974) homologous chromosomes were rarely hit together, arguing against the idea of homologous chromosome pairing in somatic cell nuclei. Furthermore, examination of the localization of homologous sequences of several single copy genes, including dystrophin, α-cardiac myosin, and the *neu* oncogene, has not identified homologous pairing of these sequences in lymphocytes or fibroblasts (Lawrence *et al.* 1990). Therefore, somatic pairing does not seem to occur in all cell systems.

Recently, Nagele *et al.* (1995) have examined the relative spatial arrangement of chromosome homologues in prometaphase rosettes of two types of normal human diploid fibroblasts and in HeLa cells with chromosome-specific DNA probes. The relative spatial location of several chromosomes was mapped and chromosome homologues were found to be positioned on opposite sides of the rosette. Examination of cells at telophase and during cytokinesis revealed that chromosomes remain arranged in a compact rosette configuration, and that the spatial positioning of homologues on opposite sides of the rosette was preserved throughout the remainder of mitosis. Based on this data it was suggested that chromosomes in the rosette are separated into two distinct groups of 23 different homologues, with each haploid chromosome set presumably derived from one parent. It will be interesting to determine if this rosette organization is maintained in interphase nuclei.

Several studies have examined the localization of individual gene sequences in interphase nuclei, and have found that genes reproducibly occupy non-random domains within the interphase nucleus. In examining the localization of single copy genes, Lawrence *et al.* (1988) showed that the Epstein Barr virus genome, which is integrated into the Namalwa cell genome on chromosome 1, was not randomly localized in the nucleus, but was positioned to the inner 50 per cent of the nuclear volume. Furthermore, both copies of the *neu* oncogene were localized to internal regions of the nucleus, whereas the dystrophin genes were found close to the nuclear periphery (Lawrence and Singer 1991). Whether individual genes occupy specific x, y, and z nuclear coordinates at specific times during the cell cycle is not yet clear. However, based on the data currently available it appears more likely that genes will occupy domains or nuclear regions rather than absolute points, and their location may change with respect to the cell cycle, cell physiology, and/or their expression. It remains to be elucidated whether the specific location of a gene in the interphase nucleus of mammalian cells directly affects its expression. However, the possibility that a gene can participate in its own regulation by being spatially localized in the cell nucleus represents an exciting possibility for the regulation of gene expression.

Distribution of specific cellular RNA transcripts

Advances in the ability to localize RNA molecules by FISH and the development of techniques to fluorescently label RNAs and then microinject them into living cells has set the stage for significant progress in our understanding of the nuclear organization of pre-mRNA metabolism. Wang *et al.* (1991) have microinjected fluorescently tagged β-globin pre-mRNA into interphase nuclei and have shown that these RNA molecules localize in a speckled distribution that is coincident with the speckled pattern enriched in pre-mRNA splicing factors (Spector 1993). In contrast, microinjection of transcripts lacking an intron or with a deleted polypyrimidine tract and 3' splice site resulted in a diffuse distribution of the injected transcripts. These results have shown that intron-containing transcripts have the ability to associate with nuclear speckles enriched in pre-mRNA splicing factors.

Numerous studies have been performed using FISH to examine the localization of endogenous pre-mRNA and mRNA transcripts in the cell nucleus. When the distribution of neurotensin (Xing *et al.* 1993) or β-actin (Zhang *et al.* 1994; Xing *et al.* 1995) RNAs was examined, the RNA localized as a dot at the site of each allele. Each dot represents the transcription site of the respective gene. Cells stably transfected with a plasmid containing the *neu* oncogene also showed a highly localized concentration of *neu* RNA. However, in many cases the distribution appeared as clusters of dots (Lawrence *et al.* 1989). A more recent examination of the distribution of collagen Iα1 RNA showed the RNA signal to be significantly larger than the dot observed for the respective DNA signal, and the RNA signal did not extend to the nuclear envelope (Xing *et al.* 1995). A dot-like distribution has also been shown for the *string* RNA within cells of a cycle-14 *Drosophila* embryo (O'Farrell *et al.* 1989). As these RNA localization signals, based upon fluorescence detection methods, for the most part do not reveal RNA contacting the nuclear envelope, it is unclear how the RNAs are transported from their sites of transcription to the cytoplasm.

In several cases the RNA distribution appeared as a track rather than a dot, raising the possibility that RNAs are vectorially transported from their sites of synthesis to the nuclear envelope. The first observation of RNA distributed in a track in mammalian cells came from a study by Lawrence *et al.* (1989) that examined Namalwa cells which contain two copies of the Epstein Barr virus (EBV) genome integrated closely together on chromosome 1. EBV *BamW* RNA transcripts were detected as a track, which averaged 5 μm in length, in the nuclei of Namalwa cells. The tracks appeared to extend between the nuclear interior and the nuclear periphery. However, as very few EBV *BamW* transcripts are transported to the cytoplasm (Dambaugh *et al.* 1986) the observed tracks may represent RNA accumulation sites rather than areas of active RNA export. Subsequently, Huang and Spector (1991) stimulated c-*fos* transcription in NIH 3T3 cells and observed nascent c-*fos* RNA to localize as two dots in interphase nuclei. Hybridization with an intron-specific probe demonstrated that pre-mRNA was present at these sites. When this distribution was examined further by confocal laser scanning microscopy and high voltage electron microscopy, the dot-like distribution of c-*fos* RNA transcripts was found to extend as an elongated path 0.75–1.5 μm through the depth of the nucleus from the site of synthesis of the RNA to the nuclear envelope. Electron microscopic *in situ* hybridization confirmed this observation, and showed these transcripts to exit the nucleus at a very limited area that could be related to a group of nuclear pores and thus support the gene gating hypothesis (Blobel 1985). More recently, Xing *et al.* (1993) have localized fibronectin mRNA by FISH and have found it to be highly concentrated at one to two sites per nucleus that frequently appeared as tracks up to 6 μm long. The gene was positioned at or near one end of the track. Since hybridization to cDNA probes produced longer tracks than with intron probes, it was suggested that splicing occurs within a portion of the tracks. Intron probes also localized as a dispersed signal throughout the nucleoplasm and did not concentrate around the nuclear periphery (Xing *et al.* 1993), as was reported previously for the acetylcholine receptor intron (Berman *et al.* 1990).

Perhaps the most intriguing case in which vectorial transport of transcripts has been suggested relates to the *Drosophila pair-rule* genes (Davis *et al.* 1993), whose striped expression in the blastoderm embryo is critical in establishing the segmentation pattern. Pair-rule transcripts accumulate exclusively on the apical side of the peripheral blastoderm nuclei. Transcript localization is determined by 3' sequences in the transcripts. Davis and Ish-Horowicz (1991) have identified several 3' pair-rule sequences that confer apical localization on reporter transcripts, whereas the 3' sequence of human α-globin was found to direct transcripts basally in the blastoderm embryo. These authors have suggested that the cytoplasmic localization of pair-rule transcripts may occur by a nuclear mechanism whereby transcripts are vectorially exported directly to the apical cytoplasm (Davis *et al.* 1993). Selective transcript exit from the nucleus could be due to export through an apical subset of nuclear pore complexes that recognize 3' localization signals. Alternatively, the RNA signals could be recognized by components of an intranuclear transport machinery and could actively direct transcripts to apical nuclear pore complexes (Davis *et al.* 1993).

Recently, Dirks *et al.* (1995) have examined the localization of three different RNA species and have suggested that the observation of nuclear RNA tracks versus dots may be related to splicing activity rather than to transport pathways. The EBV *BamW* transcripts in Namalwa cells and the human cytomegalovirus immediate early (HCMV-IE) transcripts in rat 9G cells are extensively spliced, while in the case of the luciferase transcript in X1 cells only one small intron (66 bp SV40 small-t intron) must be removed. Similar to the *BamW* transcripts (Lawrence *et al.* 1989; Dirks *et al.* 1995) (Plate 4(a)), the majority of HCMV-IE nuclear transcripts localized in a main track or elongated dot (Plate 4(b)). However, in the case of HCMV-IE transcripts many small spots of hybridization signal were also observed radiating from the main nuclear RNA signal (Dirks *et al.* 1995) (Plate 4(b)). Analysis of many tracks after double hybridization with intron- and exon-specific probes revealed complete co-localization, indicating that primary transcripts were present all along the nuclear track. This finding is in contrast to what has been observed by Xing *et al.* (1993, 1995) with regard to the fibronectin and collagen Iα1 RNAs. In these cases cDNA probes produced longer tracks than intron probes, suggesting that splicing occurs within a portion of the track. In contrast to RNAs that contain many introns (HCMV-IE, *BamW*), the nuclear RNA signal from luciferase transcripts (one small intron) mainly consisted of a bright fluorescent dot, and in addition many small fluorescent spots were observed (Dirks *et al.* 1995). The density of the small spots decreased away from the major dot and closer to the nuclear periphery. Elongated dots or tracks were never observed. The additional small spots that radiate from the transcription domains were interpreted by Dirks *et al.* (1995) to represent mRNAs in transport to the cytoplasm and spliced-out introns. Based upon these data, Dirks *et al.* (1995) suggest that when the extent for splicing is high, unspliced or partially spliced mRNAs begin to occupy elongated dot- or track-like domains in the vicinity of the gene. When the extent of splicing is low, splicing is completed co-transcriptionally, leading to a bright dot-like signal. Therefore, the nuclear RNA tracks do not represent defined RNA transport routes but, more likely, nascent transcription and/or accumulation of precursor RNA (Dirks *et al.* 1995). These authors further suggested that processed RNA transcripts radiate from this site of transcription and pre-mRNA splicing to the cytoplasm without following a specific route.

The above interpretation is consistent with several recently proposed models that argue that transcripts are transported to the cytoplasm of cells via an extrachromosomal channel network (Zachar *et al.* 1993) or interchromosome domain compartment (Zirbel *et al.* 1993; Cremer *et al.* 1993) within which transcription and pre-mRNA processing are thought to occur. Zirbel *et al.* (1993) proposed that the surfaces of chromosome territories and a space formed between them provide a network-like three-dimensional nuclear compartment for gene expression, mRNA splicing, and transport, termed the interchromosome domain compartment (ICD). More than 92 per cent of the nuclear speckles enriched in splicing factors were found at the surface of chromosome territories (Zirbel *et al.* 1993). Zachar *et al.* (1993) observed a similar nuclear compartment in *Drosophila* polytene nuclei while examining the localization of an RNA produced from a hybrid gene containing the first three introns of the suppressor-of-white-apricot gene. The RNA was localized in a small area called the primary zone that is thought to represent the gene locus (Plate 4(c)). In addition, pre-mRNA was distributed in a network-like fashion throughout the extranucleolar nucleoplasm (Plate 4(c)). As total poly(A) RNA and splicing factors were also observed in the channels, pre-mRNA metabolism is thought to occur within this region of the nucleus. Based upon the channelled distribution of RNA, Zachar *et al.* (1993) have proposed a channelled diffusion model for mRNA transport suggesting that mRNAs move from their sites of transcription to the nuclear surface at rates which are consistent with diffusion. However, it remains to be proven if RNAs are transported to the nuclear envelope by diffusion or active transport.

RNAs and splicing factors

To determine where in the cell nucleus pre-mRNA splicing occurs, Huang and Spector (1991) induced the expression of c-fos pre-mRNA in NIH 3T3 cells and examined the localization of the RNA and splicing factors in the same cells. Double labelling of c-fos RNA transcripts and the splicing factor SC35 revealed a close association between the RNA transcripts and splicing factors. Similar to what was observed for c-fos RNA (Huang and Spector 1991), fibronectin and neurotensin RNA transcripts were also usually found to be associated with splicing factors (Xing *et al.* 1993). Most recently, Xing *et al.* (1995) examined the nuclear localization of two transcriptionally active and three inactive genes. The inactive genes localized as small fluorescent dots, which were largely not associated with nuclear regions enriched in the pre-mRNA splicing factor SC35. However, RNA produced from the active β-actin gene localized as small dots only slightly larger than the DNA signal, and 89 per cent of the β-actin RNA signals were associated with SC35 domains. This finding is in sharp contrast to an earlier study (Zhang *et al.* 1994) that found only 40 per cent of the RNA signals associated with SC35 domains and which concluded that the association was random. A second transcriptionally active gene, the collagen Ia1 gene, was localized at the periphery of an SC35 domain with the RNA signal co-localizing with the domain (Xing *et al.* 1995). All of the collagen RNA signals associated with large SC35 domains and they did not extend to the nuclear envelope. Most recently, Huang and Spector (1996) have examined the nuclear localization of transiently and stably expressed nascent RNA transcripts containing introns or lacking introns (Plate 5) in order to determine if the spatial association of RNA transcripts and pre-mRNA splicing factors in nuclei is random or functionally significant. This study showed that the association between nascent RNA and splicing factors in the nucleus is intron dependent when the RNAs are either transiently (Plate 5) or stably expressed. The size and shape of the intron-containing RNA localization signal was nearly identical to the corresponding splicing factor signal at that same site. Furthermore, the fluorescence intensity of the splicing factor signal in the speckles containing the transiently expressed RNAs appeared brighter than typical speckles, suggesting a recruitment of splicing factors to the location of the nascent RNA transcripts. This finding was in agreement with the results of Jiménez-García and Spector (1993), that showed RNA polymerase II, splicing factors, and an hnRNP protein to be recruited to the sites of active adenovirus transcription. Similar to that observed in the case of β-actin and adenovirus 2 RNAs (Zhang *et al.* 1994), the presence of both pre-mRNA and mRNA at the sites of transcription of a transiently expressed template demonstrated that pre-mRNA splicing occurred at the sites of transcription (Huang and Spector 1996).

Based upon these data, Huang and Spector (1996) proposed a dynamic organization of splicing factors in the cell nucleus. Factors are thought to be present at active transcription sites (perichromatin fibrils) and storage and/or reassembly sites (interchromatin granule clusters). When transcription of intron-containing genes is activated, splicing factors are recruited to the sites of transcription from interchromatin granule clusters. The amount of splicing factors present at the sites of transcription depends upon the kinetic equilibrium of at least four parameters: the rate of transcription of the gene, the efficiency of RNA splicing, dissociation of the mRNA from the splicing complex, and the transport of the mRNA away from the transcription sites. Thus, the distribution of splicing factors between interchromatin granule clusters (storage and/or reassembly sites) and perichromatin fibrils (transcription sites) is highly regulated and dynamic in response to the transcriptional activity of the cell.

Future directions

While a significant amount of information has been recently accrued with regard to the organization of genes and RNAs in the interphase nucleus, there are still many important questions that must be addressed before we can completely understand the functional organization of the cell nucleus. Although chromosomes occupy discrete domains in the interphase nucleus, we do not yet know if this positional information is functionally significant. In addition, little information is available with regard to how rigid the three-dimensional position (x, y, z coordinates) of a specific gene must be in order for it to be properly regulated. Studies must be initiated to determine the significance of the non-random arrangement of DNA sequences in the interphase nucleus and its effect on gene expression. While the nuclear localization of many RNAs has been achieved, it is currently unclear if RNA transport from the site of transcription to the nuclear envelope occurs via active transport or a diffusion-based mechanism. Furthermore, it is still not clear if vectorial transport is utilized by a subset of nuclear RNAs. Assuming that the nucleus is an organized structure, one would expect there to be one or more architectural elements to provide the framework on which functional components reside and move. While several structural candidates have surfaced in recent years (i.e. core filaments, NuMA, actin, lamins), it is not yet known if they represent internal architectural elements (for a review see He *et al.* 1995). In looking for such elements one must not be biased by the cytoskeleton that contains several easy to recognize filament systems. The organization of the nucleus may not depend upon filaments but instead may be based on ribonucleoprotein particles or RNAs. For example, the mammalian cell nucleus is thought to contain far more snRNPs than is necessary to splice the endogenous pre-mRNA population, and recently a class of stable nuclear poly(A)$^+$ RNA has been detected (Huang *et al.* 1994), which could potentially play a role in nuclear organization. While we have learned much about nuclear organization, we are just at the very beginning and the future is sure to present answers to many of the points discussed above, as well as to raise new questions and models about the functional organization of the cell nucleus.

Summary

In summary, numerous elegant studies have been performed over the past 15 years which have provided a significant amount of support for the concept that the nucleus is an organized structure. Specific chromosomes have been found to occupy discrete domains within the interphase nucleus rather than extending throughout the nucleoplasm. The position of a chromosome can change in relation to the cell cycle, differentiation, or the physiological state of the respective cell. RNA localization studies have convincingly demonstrated for several specific RNAs that pre-mRNA splicing occurs at the site of transcription. Splicing factors appear to be recruited to the sites of intron-containing genes and the co-localization of these components is directly related to the presence of an intact intron. The dynamic organization of the genome in the interphase nucleus may be essential for the correct expression of genes during the cell cycle or for the aberrant expression of the genome during certain pathological conditions.

Acknowledgements

I would like to thank Sui Huang, Paul Mintz, and Tom Misteli for critical reading of the manuscript. This study was supported by grants to D.L.S. from the National Institutes of Health (GM42694 and 5P30 CA45508) and the Charles E. Culpeper Foundation.

References

Agard, D.A. and Sedat, J.W. (1983). Three-dimensional architecture of a polytene nucleus. *Nature*, **302**, 676–81.

Avivi, L. and Feldman, M. (1980). Arrangement of chromosomes in the interphase nucleus of plants. *Human Genetics*, **55**, 281–95.

Barr, M.L. and Bertram, E.G. (1949). A morphological distinction between neurons of the male and female, and the behavior of the nuclear satellite during accelerated nucleoprotein synthesis. *Nature*, **163**, 676–7.

Berman, S.A., Bursztajn, S., Bowen, B., and Gilbert, W. (1990). Localization of an acetylcholine receptor intron to the nuclear membrane. *Science*, **247**, 212–14.

Billia, F. and De Boni, U. (1991). Localization of centromeric satellite and telomeric DNA sequences in dorsal root ganglion neurons, *in vitro*. *Journal of Cell Science*, **100**, 219–26.

Blobel, G. (1985). Gene gating: a hypothesis. *Proceedings of the National Academy of Sciences of the USA*, **82**, 8527–9.

Borden, J. and Manuelidis, L. (1988). Movement of the X chromosome in epilepsy. *Science*, **242**, 1687–91.

Boveri, T. (1909). Die Blastomerenkerne von *Ascaris* megalocephala und die Theorie der Chromosomenindividualitat. *Archiv fuer Experimentele Zellforschung*, **3**, 181–268.

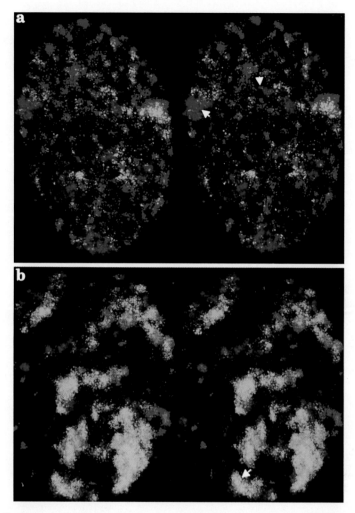

Plate 1 Replicating cells (BrdU in green) and probed with IAP sequences (chromatids labelled in red). Stereo pairs in (a) show an early replication pattern of the BrdU incorporation. IAP chromatids, known to be late replicating, are not labelled by BrdU. Note the congression of IAP chromatids at the bottom and sides that are partially in contact with the nuclear envelope. Most IAP chromatids maintain their structural identity although clustered, and many chromatid fibres are about 300 nm wide. However, they can also give rise to circular and more coiled structures with wider dimensions (about 1 μm), as shown at left and right arrows, respectively. Details of such fibres were similar to those shown for G_1 cells (see Fig. 9.3). In (b) many IAP chromatids are yellow in this late replicating cell, indicating BrdU incorporation into IAP chromatids during a 15 min pulse. The IAP chromatids are thicker than in (a) and many are still adjacent to the nuclear membrane. Arrow points to a tiny replicating loop that may be more unravelled (yellow). Note some yet to be replicated IAP chromatids in the centre and periphery of the cell, that show more delicate structures (red only). The green-only structures represent unrelated chromatids with late replicating domains. Sections were deconvoluted before reconstruction. (See pp. 156, 158.)

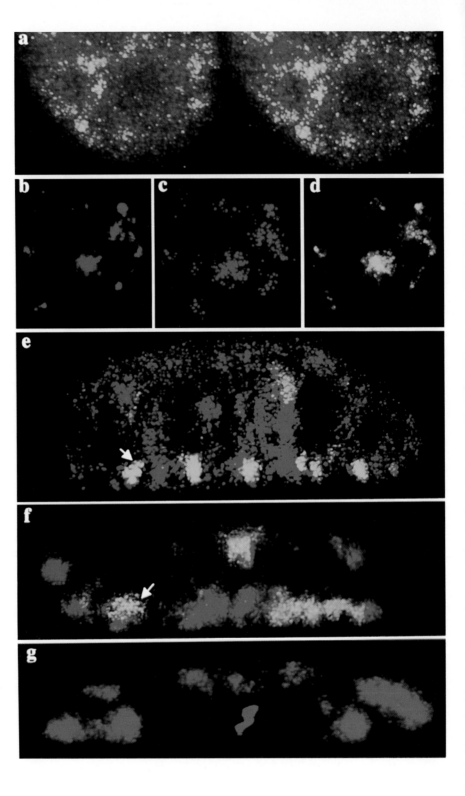

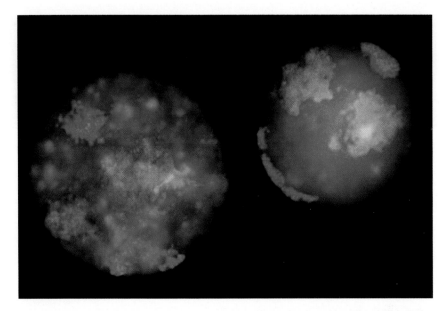

Plate 3 Visualization of the two alleles of chromosome 12 (pseudocoloured in red) and the two alleles of chromosome 13 (pseudocoloured in green) in interphase nuclei of normal human lymphocytes, visualized by dual colour fluorescence *in situ* hybridization. Probes were generated by amplification of flow-sorted chromosomes using degenerate oligonucleotide primer (DOP) polymerase chain reaction with incorporation of biotin-16-dUTP or Cy3-dUTP. The biotinylated sequences were detected with Cy2 conjugated to avidin. The nuclei were counterstained with 4,6-diamidino-2 phenylindole (DAPI). (See p. 170.) (Photo: E. Schrock and T. Ried, N.I.H.)

Plate 2 (*Facing*) In (a) IAP RNA transcripts were detected with antisense probes in non-denatured preparations. The RNA signals are yellow because total nuclear DNA was counterstained with propidium iodide (red). Note obvious compact clusters of RNA transcripts. Single corresponding confocal sections (b and c) showing hybridization to simultaneously detect DNA and RNA (green and red, respectively) in cells stimulated with Aza C. In (d) a merged image from the two confocal sections (b and c) is shown. RNA transcripts overlie the IAP chromatids both in the interior and at the periphery of the nucleus. Note that most the RNA overlies the more compact regions of the IAP chromatids at both interior and peripheral locations. In *x–z* reconstructions of early and late replicating cells (c and f, respectively) BrdU is green, with telomeres shown in red. Note replicating telomeres are present in both cells (yellow regions). Many of the telomeres in these cells are oriented close to the coverslip attachment (at bottom). Late but note synchronously replicating telomere clusters in (f) are consistent with those on termini of *β*-heterochromatic chromatids. In (e) the arrow shows fine substructure of delicate fibres in early replicating telomeres, while in (f) these fibres are markedly unravelled in some replicating telomeres (yellow, at arrow). In (g) IAP chromatids near the starting edge of the nucleus are labelled in red and begin to show up at the top of the nucleus (nearest to feeding medium). These sections also resolve the entire concatomerized heat shock locus (in green). This was one of the most extended loci found in heat shocked cells producing abundant transcripts. Notably no small fibres emanate from this locus. All transcribing heat shock DNA domains were similarly cohesive or even more compact in RNA–DNA double label studies (data not shown). Above sections were not deconvoluted. (See pp. 148, 157, 159–61.)

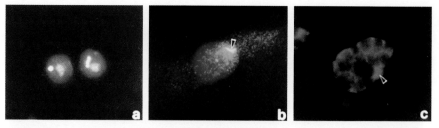

Plate 4 Tracks, dots and channels: localization patterns of RNA transcripts. (a) *Bam W* transcripts in Namalwa cells localize in a nuclear track or a dot; no cytoplasmic signal is observed. Nuclei are counterstained with DAPI (blue). (Photo: R.W. Dirks, University of Leiden.) (b) After cycloheximide induction, the human cytomegalovirus immediate early (HCMV-IE) mRNA localized in a nuclear track (arrowhead) with many small fluorescent spots radiating from the transcription site (track). Hybridization signals (dots) are also observed in the cytoplasm. The nucleus is counterstained with propidium iodide (red). (Photo: R.W. Dirks, University of Leiden.) (c) Localization of a hybrid gene containing the first three introns of the suppressor-of-white-apricot gene in a *Drosophila* polytene nucleus showing the primary zone (arrowhead) and the extrachromosomal channel network. (Photo: P. Bingham, State University of New York at Stony Brook.) (See pp. 174–5.)

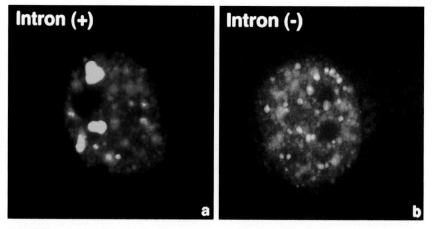

Plate 5 Pre-mRNA splicing factors localize with intron-containing and not with intron-less RNA transcripts. HeLa cells were transiently transfected with DNA templates encoding for a β-tropomyosin minigene construct (+ introns) or β-globin cDNA (– introns). Eight hours after transfection cells were hybridized with the respective probes and the SC35 splicing factor was localized in the same cells. In the case of the intron-containing RNA (a) a co-localization with splicing factors was observed (yellow regions). In addition, the speckles that were co-localized with the RNA were larger and brighter in intensity than the typical speckles (red regions). The intron-less RNA (green areas in (b)) was not localized with splicing factors (red regions in (b)) in 80 per cent of the cases examined. (See p. 176.) (Photo: S. Huang, Cold Spring Harbor Laboratory.)

Cremer, C., Zorn, C., and Cremer, T. (1974). An ultraviolet laser microbeam for 257 nm. *Microscopica Acta*, **75**, 331–7.

Cremer, T., Cremer, C., Baumann, H., Luedtke, E.-K., Sperling, K., *et al.* (1982). Rabl's model of the interphase chromosome arrangement tested in Chinese hamster cells by premature chromosome condensation and laser-UV-microbeam experiments. *Human Genetics*, **60**, 46–56.

Cremer, T., Tesin, D., Hopman, A.H., and Manuelidis, L. (1988). Rapid interphase and metaphase assessment of specific chromosomal changes in neuroectodermal tumor cells by *in situ* hybridization with chemically modified DNA probes. *Experimental Cell Research*, **176**, 199–220.

Cremer, T., Kurz, A., Zirbel, R., Dietzel, S., Rinke, B., Schrock, E., *et al.* (1993). Role of chromosome territories in the functional compartmentalization of the cell nucleus. *Cold Spring Harbor Symposia on Quantitative Biology*, **58**, 777–92.

Dambaugh, T., Hennessy, K., Fennewald, S., and Kieff, E. (1986). The virus genome and its expression in latent infection. In *The Epstein–Barr virus: recent advances* (ed. M.A. Epstein and B.G. Achong), pp. 13–45. John Wiley, New York.

Davis, I. and Ish-Horowicz, D. (1991). Apical localization of pair-rule transcripts requires 3' sequences and limits protein diffusion in the *Drosophila* blastoderm embryo. *Cell*, **67**, 927–40.

Davis, I., Francis-Lang, H., and Ish-Horowicz, D. (1993). Mechanisms of intracellular transcript localization and export in early *Drosophila* embryos. *Cold Spring Harbor Symposia on Quantitative Biology*, **58**, 793–8.

Dirks, R.W., Daniël, K.C., and Raap, A.K. (1995). RNAs radiate from gene to cytoplasm as revealed by fluorescence *in situ* hybridization. *Journal of Cell Science*, **108**, 2565–72.

Ferguson, M. and Ward, D.C. (1992). Cell cycle dependent chromosomal movement in pre-mitotic human T-lymphocyte nuclei. *Chromosoma*, **101**, 557–65.

Greider, C.W. and Blackburn, E.H. (1987). The telomere terminal transferase of *Tetrahymena* is a ribonucleoprotein enzyme with two kinds of primer specificity. *Cell*, **51**, 887–98.

Haaf, T. and Schmid, M. (1989). Centromeric association and non-random distribution of centromeres in human tumour cells. *Human Genetics*, **81**, 137–43.

Haaf, T. and Schmid, M. (1991). Chromosome topology in mammalian interphase nuclei. *Experimental Cell Research*, **192**, 325–32.

Harley, C.B. (1995). Telomeres and aging: fact, fancy, and the future. *Journal of NIH Research*, **7**, 64–8.

He, D., Zeng, C., and Brinkley, B.R. (1995). Nuclear matrixc proteins as structural and functional components of the mitotic apparatus. *International Review of Cytololgy*, **162B**, 1–73.

Hilliker, A.J. (1985). Assaying chromosome arrangement in embryonic interphase nuclei of *Drosophila melanogaster* by radiation induced interchanges. *Genetical Research*, **47**, 13–18.

Hilliker, A.J. and Appels, R. (1989). The arrangement of interphase chromosomes: structural and functional aspects. *Experimental Cell Research*, **185**, 297–318.

Hiraoka, Y., Dernburg, A.F., Parmelee, S.J., Rykowski, M.C., Agard, D.A., and Sedat, J.W. (1993). The onset of homologous chromosome pairing during *Drosophila melanogaster* embryogenesis. *Journal of Cell Biology*, **120**, 591–600.

Hochstrasser, M. and Sedat, J.W. (1987). Three-dimensional organization of *Drosophila melanogaster* interphase nuclei. II. Chromosomal spatial organization and gene regulation. *Journal of Cell Biology*, **104**, 1455–70.

Hochstrasser, M., Mathog, D., Gruenbaum, Y., Saumweber, J., and Sedat, J. W. (1986). Spatial organization of chromosomes in the salivary gland nuclei of *Drosophila melanogaster*. *Journal of Cell Biology*, **102**, 112–23.

Huang, S. and Spector, D.L. (1991). Nascent pre-mRNA transcripts are associated with nuclear regions enriched in splicing factors. *Genes and Development*, **5**, 2288–302.

Huang, S. and Spector, D.L. (1996). Intron-dependent recruitment of pre-mRNA splicing factors to sites of transcription. *Journal of Cell Biology*, **133**, 719–32.

Huang, S., Deerinck, T.J., Ellisman, M.H., and Spector, D.L. (1994). *In vivo* analysis of the stability and transport of nuclear poly(A)$^+$ RNA. *Journal of Cell Biology*, **126**, 877–99.

Jiménez-García, L.F. and Spector, D.L. (1993). *In vivo* evidence that transcription and splicing are coordinated by a recruiting mechanism. *Cell*, **73**, 47–59.

Lawrence, J.B. and Singer, R.H. (1991). Spatial organization of nucleic acid sequences within cells. *Seminars in Cell Biology*, **2**, 83–101.

Lawrence, J.B., Villnave, C.A., and Singer, R.H. (1988). Sensitive high-resolution chromatin and chromosome mapping *in situ*: presence and orientation of two closely integrated copies of EBV in a lymphoma line. *Cell*, **52**, 51–61.

Lawrence, J.B., Singer, R.H., and Marselle, L.M. (1989). Highly localized tracks of specific transcripts within interphase nuclei visualized by *in situ* hybridization. *Cell*, **57**, 493–502.

Lawrence, J.B., Singer, R.H., and McNeil, J.A. (1990). Interphase and metaphase resolution of different distances within the human dystrophin gene. *Science*, **249**, 928–32.

Lichter, P., Cremer, T., Borden, J., Manuelidis, L., and Ward, D.C. (1988). Delineation of individual human chromosomes in metaphase and interphase cells by *in situ* suppression hybridization using recombinant DNA libraries. *Human Genetics*, **80**, 224–34.

Manuelidis, L. (1985*a*). Indications of centromere movement during interphase and differentiation. *Annals of the New York Academy of Sciences*, **450**, 205–21.

Manuelidis, L. (1985*b*). Individual interphase chromosome domains revealed by *in situ* hybridization. *Human Genetics*, **71**, 288–93.

Manuelidis, L. (1990). A view of interphase chromosomes. *Science*, **250**, 1533–40.

Mathog, D., Hochstrasser, M., Gruenbaum, Y., Saumweber, H., and Sedat, J. (1984). Characteristic folding pattern of polytene chromosomes in *Drosophila* salivary gland nuclei. *Nature*, **308**, 414–21.

McNeil, J.A., Johnson, C.V., Carter, K.C., Singer, R.H., and Lawrence, J.B. (1991). Localizing DNA and RNA within nuclei and chromosomes by fluorescence *in situ* hybridization. *Genet. Analys. Tech. Appl.*, **8**, 41–58.

Nagele, R., Freeman, T., McMorrow, L., and Lee, H. (1995). Precise spatial positioning of chromosomes during prometaphase: evidence for chromosomal order. *Science*, **270**, 1831–5.

O'Farrell, P.H., Edgar, B.A., Lakich, D., and Lehner, C.F. (1989). Directing cell division during development. *Science*, **246**, 635–40.

Pinkel, D., Gray, J.W., Trask, B., van den Engh, G., Fuscoe, J., *et al.* (1986). Cytogenetic analysis by *in situ* hybridization and fluorescently labeled nucleic acid probes. *Cold Spring Harbor Symposia on Quantitative Biology*, **51**, 151–7.

Pinkel, D., Landegent, J., Collins, C., Fuscoe, J., Segraves, R., *et al.* (1988). Fluorescence *in situ* hybridization with human chromosome-specific libraries: detection of trisomy 21 and translocations of chromosome 4. *Proceedings of the National Academy of Sciences of the USA*, **85**, 9138–42.

Rabl, C. (1885). Über Zellteilung. *Morphologie Jahrbuch*, **10**, 214–330.

Rawlins, D.J., Highett, M.I., and Shaw, P.J. (1991). Localization of telomeres in plant interphase nuclei by *in situ* hybridization and 3D confocal microscopy. *Chromosoma*, **100**, 424–31.

Spector, D.L. (1993). Macromolecular domains within the cell nucleus. *Annual Review of Cell Biology*, **9**, 265–315.

Trask, B., van den Engh, G., Pinkel, D., Mullikin, J., Waldman, F., Van Dekken, H., *et al.* (1988). Fluorescence *in situ* hybridization to interphase cell nuclei in suspension allows flow cytometric analysis of chromosome content and microscopic analysis of nuclear organization. *Human Genetics*, **78**, 251–9.

Vourc'h, C., Taruscio, D., Boyle, A.L., and Ward, D.C. (1993). Cell cycle-dependent distribution of telomeres, centromeres, and chromosome-specific subsatellite domains in the interphase nucleus of mouse lymphocytes. *Experimental Cell Research*, **205**, 142–51.

Wang, J., Cao, L.-G., Wang. Y.-L., and Pederson, T. (1991). Localization of pre-messenger RNA at discrete nuclear sites. *Proceedings of the National Academy of Sciences of the USA*, **88**, 7391–5.

Xing, Y., Johnson, C.V., Dobner, P.R., and Lawrence, J.B. (1993). Higher level organization of individual gene transcription and RNA splicing: integration of nuclear structure and function. *Science*, **259**, 1326–30.

Xing, Y., Johnson, C.V., Moen, P.T. Jr., McNeil, J.A., and Lawrence, J.B. (1995). Nonrandom gene organization: structural arrangements of specific pre-mRNA transcription and splicing with SC-35 domains. *Journal of Cell Biology*, **131**, 1635–47.

Zachar, Z., Kramer, J., Mims, I.P., and Bingham, P.M. (1993). Evidence for channeled diffusion of pre-mRNAs during nuclear RNA transport in metazoans. *Journal of Cell Biology*, **121**, 729–42.

Zhang, G., Taneja, K.L., Singer, R.H., and Green, M.R. (1994). Localization of pre-mRNA splicing in mammalian nuclei. *Nature*, **372**, 809–12.

Zinkowski, R.P., Meyne, J., and Brinkley, B.R. (1991). The centromere-kinetochore complex: a repeat subunit model. *Journal of Cell Biology*, **113**, 1091–110.

Zirbel, R.M., Mathieu, U.R., Kurz, A., Cremer, T., and Lichter, P. (1993). Evidence for a nuclear compartment of transcription and splicing located at chromosome domain boundaries. *Chromosome Research*, **1**, 92–106.

Part III

Biological structure and epigenetic control of gene expression

11

DNA methylation and genomic imprinting

Marisa S. Bartolomei

Introduction

Normally in eukaryotes each allele of a gene contributes an equal amount of gene product. However, mammals exhibit an unusual phenomenon in which the expression level of the alleles of a subset of genes depends upon the parent from which the alleles were inherited. This phenomenon is called genomic imprinting and probably affects only a small number of genes (Efstratiadis 1994). One consequence of imprinted genes is the failure of uniparental mammalian embryos to develop normally. Embryos with two maternal pronuclei would be missing genes that are expressed exclusively from the paternal genome, while embryos with two paternal pronuclei lack genes expressed exclusively from the maternal genome.

Researchers who are interested in the role of genomic imprinting in development have focused their efforts on two main areas of investigation. First, extensive experimentation has centred on the identification of imprinted genes. One of the goals of these experiments is to identify genes that are responsible for the non-viability of uniparental embryos, since it has been suggested that these genes may mediate important decisions in early development. A second area of research is to characterize the mechanism by which the two parental alleles are programmed or marked with their parental origin so that they will be imprinted. Although much progress has been made in determining which genes are imprinted, less is known about the mechanism by which these genes are marked. DNA methylation of the cytosine residue in CpG dinucleotides is an excellent candidate for the imprinting mark. It is the purpose of this review to summarize the data which support methylation as the mark as well as to point out the difficulties in the definitive establishment of methylation in this role.

DNA methylation in mammals

While the evidence is largely indirect, DNA hypermethylation has been strongly associated with repression of gene expression in mammals. For example, a large majority of tissue-specific genes show a correlation between gene activity and hypomethylation of the promoter region (Yeivin and Razin 1993). Additionally, genes on the inactive X chromosome in females are hypermethylated relative to their counterparts on the active X chromosome (Riggs and Pfeiffer 1992).

Frequency of CpG methylation

Mammalian DNA can be methylated at the fifth carbon position of cytosine residues within CpG dinucleotides. These dinucleotides are found at a much lower frequency than expected. This is most likely due to the fact that most CpG dinucleotides are methylated and 5-methylcytosine is readily converted to thymine by deamination. The hypermutability at these sites accounts for approximately one-third of mutations in humans (Bestor and Coxon 1993). However, about 1 per cent of the genome does contain regions or islands rich in non-methylated CpG dinucleotides (Bird 1986). Such islands are usually associated with housekeeping genes. Occasionally CpG islands are associated with imprinted genes and in these cases the cytosine is often methylated in an allele-specific manner.

Mechanism of CpG methylation

Methyl groups are covalently added to cytosine through an enzymatic transfer reaction catalysed by DNA methyltransferase. A single DNA methyltransferase has been identified in mammalian cells (Bestor *et al.* 1988). This enzyme propagates the pre-existing methylation of CpG dinucleotides during replication by adding a methyl group to the cytosine residue on the newly replicated strand of DNA when this residue is directly across from a methylated dinucleotide on the template strand. This enzyme is essential for development since mice lacking the gene die early in gestation (Li *et al.* 1992). Since *de novo* methylation is required at several times during development, much controversy has centred over whether this initially characterized enzyme is the only DNA methyltransferase in mammalian cells. Based on the study of embryonic stem cell lines derived for the production of the mice lacking the enzyme, it appears that DNA methyltransferase identified by Bestor and colleagues serves to maintain the levels of methylation in a dividing cell (Li *et al.* 1992). Hence, either additional DNA methyltransferases or auxiliary factors for the presently characterized enzyme must exist to establish *de novo* patterns of methylation *in vivo*.

Changes in DNA methylation during development

The methylation patterns of the mammalian genome exhibit large fluctuations during development (see Fig. 11.1) (Monk *et al.* 1987; Sanford *et al.* 1987). In experiments performed with mice it was demonstrated that male and female gametes bring entirely different methylation patterns to the fertilized egg. While sperm DNA is highly methylated, oocyte DNA is generally less methylated. Immediately following fertilization the methylation patterns inherited from the gametes appear to be maintained, probably by the DNA methyltransferase

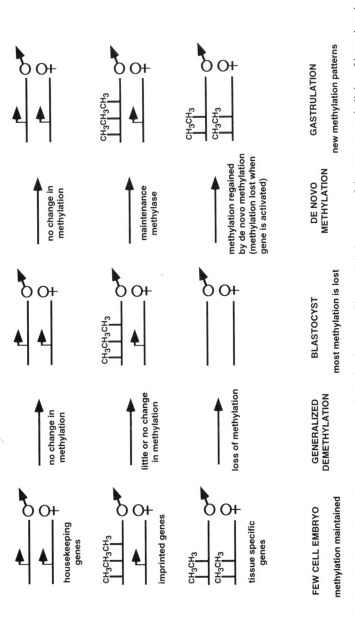

Fig. 11.1 Methylation fate of three classes of genes early in development. The methylation status of the two parental alleles of housekeeping genes (top), imprinted genes (middle), and tissue-specific genes (bottom) is shown as the embryo develops. Each line corresponds to an allele of a typical gene in the designated class. The female symbol indicates the maternally derived allele while the male symbol indicates the paternally derived allele. The bottom portion of the figure shows the time in embryogenesis and the periods of generalized demethylation and *de novo* methylation. CH₃ indicates that the allele is methylated. Alleles with the potential of being actively expressed have an arrow, while unmodified but non-expressed genes are shown as a single line. The idealized imprinted gene has a methylation imprint which survives the period of genome-wide demethylation and is subsequently maintained by DNA methyltransferase.

derived from the egg. Beginning at the eight-cell stage, much of the inherited methylation is lost as the embryo undergoes a period of massive and generalized demethylation. Following implantation at about 6.5 days of gestation, the embryonic DNA becomes slightly more methylated and within 1 day the DNA is highly methylated. As this remethylation is coincident with the onset of gastrulation, it is possible that the gene-specific methylation is essential to the differentiation of new cell lineages and organogenesis. In contrast to what is observed in the embryo, genomic DNA from extraembryonic tissues is less methylated.

It is essential to note that not all methylation is lost in the early embryo (Monk *et al.* 1987; Sanford *et al.* 1987). In order for methylation to serve as the heritable mark that distinguishes the expressed from the non-expressed alleles, it must be present throughout development, including the period of genome-wide demethylation that occurs during preimplantation. Thus, any methylated CpG dinucleotides which are implicated as the imprinting mark must remain methylated during preimplantation.

Role of DNA methylation in inhibiting gene expression

The mechanism by which DNA methylation prevents transcription of genes that are highly methylated is not known. It is widely thought that the hypermethylation of promoters restricts access to transcription factors thereby preventing transcriptional initiation (Eden and Cedar 1994). However, the evidence for this hypothesis is indirect. Another model of transcriptional repression suggests that promoter methylation may cause the assembly of a highly condensed chromatin state during replication (Meehan *et al.* 1992; see Chapter 9). Such a structure could be assembled with the help of proteins that first recognize methylated sequences and subsequently recruit repressing factors that facilitate the construction of inactive, condensed chromatin.

DNA methylation of imprinted genes

Although cytosine methylation has the potential to mediate the silencing of imprinted genes, it is difficult to establish whether methylation is the initial mark that distinguishes the maternal and paternal alleles or is simply a consequence of this distinction. The suitability of methylation serving as the imprinting mark derives from its potential to fulfil the following criteria. First, the mark must be assembled in the male and/or female germline since this is the only time when the two alleles are in separate compartments and can be differentially modified. Second, the mark must be stable and heritable so that the alleles remain differentially modified. Finally, the mark must be capable of being erased and reset. For example, a female who has inherited a maternal and paternal allele of an imprinted gene must transmit the female-specific mark to her progeny, thus causing the paternal imprint to be erased and reset as a maternal imprint.

Table 11.1 Methylation patterns of imprinted genes[a]

Imprinted gene	Expressed allele	Allele-specific methylation	Imprinting mark[b]
Igf2	Paternal	5' of gene, paternal allele	No
		3' region of gene, paternal allele[c]	No
Igf2r	Maternal	Protmoter, paternal allele	No
		Intron, maternal allele	Yes
H19	Maternal	5' of gene, paternal allele	Yes[d]
		Gene body, paternal allele	No
U2afbp-rs	Paternal	Promoter, maternal allele	No
SNRPN	Paternal	Exon α, maternal allele	n.d.[e]
		Intron 5, paternal allele	n.d.
Xist	Paternal	Promoter, maternal allele	Yes
		5' region of gene, maternal allele	Yes

[a] See text for references and details.
[b] Establishes whether experiments performed to date satisfy criteria for imprinting mark.
[c] Methylated in tissues where gene is highly expressed.
[d] Subset of CpG dinucleotides in this region are candidates.
[e] n.d. = not determined.

To determine whether methylation is involved in any of these processes, parental-specific methylation differences must first be detected. All imprinted genes examined thus far display a methylation imprint (Table 11.1). While some of these methylation imprints satisfy the criteria for the mark that designates parental identity, there has yet to be an experiment that definitively establishes methylation as the imprinting mark.

Igf2

The first imprinted gene to be identified in mouse was the *insulin-like growth factor 2* (*Igf2*) gene (DeChiara *et al.* 1991). The *Igf2* gene, which is required for normal fetal growth (DeChiara *et al.* 1990), is expressed exclusively from the paternal chromosome in mice. This gene is similarly imprinted in humans (Giannoukakis *et al.* 1993; Ohlsson *et al.* 1993; Rainier *et al.* 1993). When the mouse gene was analysed for differential methylation, a 1 kilobase pair (kb) region approximately 3 kb upstream from the start of transcription was found to be more highly methylated on the expressed paternal allele. The differential methylation was first detected at four CpG dinucleotides located within restriction enzyme sites for the methylation-sensitive enzyme *Hpa*II (Brandeis *et al.* 1993). A more extensive analysis of this region, using a technique that allows the determination of the methylation status of every CpG (bisulphite mutagenesis and sequencing (Frommer *et al.* 1992)), revealed that 13 CpG dinucleotides were generally more methylated on the paternal chromosome than on the maternal chromosome (Feil *et al.* 1994). This methylation is highly mosaic, since in no case was any CpG always methylated on the paternal allele and unmethylated on the maternal allele.

A second region of differential methylation is located in the 3' portion of the gene (Feil *et al.* 1994). This region is also more highly methylated on the expressed paternal allele, but in contrast to the upstream region, the paternal methylation is only found on genomic DNA in tissues where the gene is highly expressed. Since regions of differential methylation were found for the *Igf2* gene, it was necessary to determine if this methylation was also present in the gametes and during preimplantation. Surprisingly, the four *Hpa*II sites located in the upstream region were found to be methylated in both male and female gametes (Brandeis *et al.* 1993). Thus, although one of these sites is differentially methylated during preimplantation, the adult pattern of paternal allele-specific methylation was derived after fertilization. It remains to be determined whether the differential methylation in the 3' portion of the gene, or in any other part of the *Igf2* locus, satisfies the criteria for the methylation mark.

Igf2r

The receptor for insulin-like growth factor type 2 (*Igf2r*), which is also the cation-independent mannose-6-phosphate receptor, is crucial for regulating embryonic growth as well as lysosomal targeting (Lau *et al.* 1994; Wang *et al.* 1994). This gene is predominantly expressed from the maternal allele (Barlow *et al.* 1991). An extensive methylation analysis of the 130 kb *Igf2r* locus revealed two areas of differential methylation (Stoger *et al.* 1993). Both of these regions are CpG islands. The first region of differential methylation (designated as region 1) is approximately 0.9 kb and lies in the promoter on the inactive allele. This methylation cannot be the imprinting mark because it is acquired after fertilization. Region 2, which lies in an intron 27 kb downstream from the start of transcription, is methylated on the active maternal allele. At least a few CpG dinucleotides located within the 0.9 kb region 2 are also differentially methylated in the gametes and preimplantation embryo (Brandeis *et al.* 1993; Stoger *et al.* 1993). Thus, region 2 satisfies the previously defined criteria for methylation serving as the imprinting signal. In contrast, the promoter methylation may help to stabilize the imprint once it has been established.

It is curious that hypermethylation observed for both the *Igf2* and *Igf2r* genes resides on the active alleles. These results are opposite from what might be expected given the bias that non-expressed genes are hypermethylated. Barlow and colleagues suggest that the methylation is necessary for expression and may act by preventing the binding of a repressor on the active allele (Stoger *et al.* 1993).

H19

In contrast to what is observed for the *Igf2* and *Igf2r* genes, the inactive allele of *H19* is hypermethylated. The *H19* gene is expressed exclusively from the maternal allele in both mice and humans (Bartolomei *et al.* 1991; Rachmilewitz *et al.* 1992; Zhang *et al.* 1993). Interestingly, this gene is located about 90 kb from the oppositely imprinted *Igf2* gene. Although the function of this gene, which does not encode a protein product, is not entirely clear, it is thought that *H19* acts by regulating the imprinting of *Igf2*, since deletion of *H19* derepresses the imprinted *Igf2* gene (Leighton *et al.* 1995). Methylation analyses revealed that a 7–9 kb region encompassing the gene and a 4 kb upstream flanking sequence was methylated on the inactive paternal allele in DNA from adult somatic tissues and sperm (Bartolomei *et al.* 1993; Ferguson-Smith *et al.* 1993). When a variety of *Hpa*II sites located within the gene and promoter were tested for methylation in preimplantation embryos, it was determined that the methylation was lost (Brandeis *et al.* 1993). However, the paternal-specific methylation in the upstream region is retained in the preimplantation embryos (Tremblay *et al.* 1995). The differentially methylated CpG dinucleotides lie in a 2 kb region located about 2 kb 5' to the start of transcription. These residues satisfy the criteria for the imprinting mark of *H19* since they are differentially methylated in the gametes and this difference is maintained through development.

U2afbp-rs

The frequency at which differentially methylated regions have been described for imprinted genes has led investigators to devise assays that would detect differentially methylated fragments of genomic DNA. The assumption in these assays is that an imprinted gene would either be adjacent to or within the methylated fragment. One such scheme, restriction landmark genomic scanning with methylation-sensitive enzymes, resulted in the identification of the paternally expressed *U2afbp-rs* gene (Hatada *et al.* 1993; Hayashizaki *et al.* 1994). A single differentially methylated *Not*I site was sufficient for detection by this sensitive technique. An extensive analysis of the gene revealed that a CpG island surrounding the promoter was methylated on the inactive maternal allele (Hatada *et al.* 1995). Because the allelic methylation was not present in the parental gametes, it cannot be considered the imprinting mark for this gene.

SNRPN, IPW, and ZNF127

Prader-Willi syndrome and Angelman syndrome are two human genetic diseases which involve imprinted genes (Nicholls 1994). Many of the patients with these diseases exhibit a deletion of chromosome 15q11-13. Although the genes responsible for these syndromes are not known, two paternally expressed genes, *SNRPN* and *IPW*, have been mapped in the Prader-Willi critical region (Glenn *et al.* 1993; Reed and Leff 1994; Wevrick *et al.* 1994). A third gene, *ZNF127*, exhibits both maternal- and paternal-specific methylation imprints and lies adjacent to this region (Driscoll *et al.* 1992). *SNRPN* is the best candidate gene for Prader-Willi syndrome and is also the gene for which the most information is available. This gene encodes a developmentally regulated splicing factor which is expressed predominantly in neural tissues. It is also imprinted in mice (Leff *et al.* 1992). The *SNRPN* gene is methylated in two regions in somatic tissues. Its fifth intron is methylated on the expressed paternal allele (Glenn *et al.* 1993) and the 5' exon α is methylated on the inactive maternal allele (Sutcliffe *et al.* 1994). While these two regions are reported to be differentially methylated in gametes (Glenn *et al.* 1996), the status of preimplantation embryos is unknown.

X inactivation as a model of genomic imprinting

Dosage compensation of the X chromosome in eutherian mammals occurs when the female randomly inactivates one of her two X chromosomes during early development (Rastan 1994; see Chapter 12). In the embryo, methylation follows inactivation and most likely serves to stabilize the inactivity. X inactivation in the extraembryonic tissues is non-random and occurs exclusively on the paternally derived X chromosome. As extraembryonic tissues differentiate before embryonic tissues, non-random X inactivation occurs prior to random inactivation.

Although the mechanism by which mammalian X inactivation occurs is still poorly understood, recent investigation of *XIST*, a gene expressed exclusively from the inactive X chromosome, suggests that it may trigger the onset of inactivation (Brown *et al.* 1991; see Chapter 12). Consistent with the paternal-specific X inactivation in extraembryonic tissues, the earliest expression of *Xist* is also paternal specific (Kay *et al.* 1993). To determine whether methylation is the signal that causes the imprinting of *Xist*, the methylation status of a large region encompassing the *Xist* gene was determined. In oocytes and the early embryo, *Xist* exhibits maternally derived methylation at the promoter and 5' portion of the gene (Norris *et al.* 1994; Ariel *et al.* 1995; Zuccotti and Monk 1995). Because this methylation is absent in the sperm, the methylation is a candidate for the imprinting signal.

DNA methyltransferase

As mentioned above, only a single DNA methyltransferase has been identified thus far in mammalian cells (Bestor *et al.* 1988). Mutations were first introduced into the methyltransferase gene using gene targeting in embryonic stem cells, and mice were subsequently generated which harboured these deletions (Li *et al.*

1992). To determine if imprinting was affected in the mice that lacked DNA methyltransferase, the allelic expression of the *H19*, *Igf2* and *Igf2r* genes was analysed in homozygous null mice prior to their death at 10 days of gestation (Li *et al.* 1993). *H19*, which is normally methylated on the inactive paternal allele, was biallelically expressed and unmethylated on both alleles in the mutant mice. *Igf2* and *Igf2r* were also demethylated but neither gene was expressed. Since these genes have methylation imprints on their active alleles, it is possible that methylation is required for *Igf2* and *Igf2r* to be expressed. While these experiments offer compelling evidence that methylation is essential for determining the appropriate expression of imprinted genes, a few issues must be addressed. First, the loss of imprinting could be indirect since the entire genome is demethylated and imprinting regulators may be affected. Furthermore, these experiments address the role of maintenance methylation and do not determine whether methylation is the key to establishing the imprint.

Future directions

It remains to be determined whether methylation causes the imprinting of genes. Although the experiments performed thus far are compelling, the information should be considered as correlative. What is required is an experiment that perturbs methylation of a specific gene without disrupting the expression of it or other genes. Future experiments will undoubtedly focus on the alteration of candidate CpG dinucleotides (Table 11.1) at their endogenous loci. If such mutations were to lead to a loss of imprinting on the methylated allele without affecting the unmodified allele, then one could conclude that the residues and methylation were key to determining the imprint. It is also crucial to determine when and where the methylation imprint is made. This information will lead to a greater understanding of how imprinting is established in mammals.

Summary

DNA methylation of CpG dinucleotides plays an important role in genomic imprinting. All of the imprinted genes tested to date display allele-specific methylation. The finding that the expression of three imprinted genes is disrupted in the DNA methyltransferase mutant mice strongly suggests that methylation is, at the very least, required to maintain the imprinted expression of these genes. Whether methylation is also the initial mark that allows the parental alleles to be distinguished has yet to be determined. The differentially methylated regions detected in the *Igf2r*, *H19* and *Xist* genes are the most compelling examples of potential imprinting marks, since these regions are also differentially methylated in gametes and throughout development. Experiments that test the role of these differentially methylated regions are in progress.

References

Ariel, M., Robinson, E., McCarrey, J.R., and Cedar, H. (1995). Gamete-specific methylation correlates with imprinting of the murine *Xist* gene. *Nature Genetics*, **9**, 312–15.

Barlow, D.P., Stoger, R., Herrmann, B.G., Saito, K., and Schweifer, N. (1991). The mouse insulin-like growth factor type-2 receptor is imprinted and closely linked to the *Tme* locus. *Nature*, **349**, 84–7.

Bartolomei, M.S., Zemel, S., and Tilghman, S.M. (1991). Parental imprinting of the mouse H19 gene. *Nature*, **351**, 153–5.

Bartolomei, M.S., Webber, A.L., Brunkow, M.E., and Tilghman, S.M. (1993). Epigenetic mechanisms underlying the imprinting of the mouse H19 gene. *Genes and Development*, **7**, 1663–73.

Bestor, T.H. and Coxon, A. (1993). The pros and cons of DNA methylation. *Current Biology*, **3**, 384–6.

Bestor, T.H., Laudano, A., Mattaliano, R., and Ingram, V. (1988). Cloning and sequencing of a cDNA encoding DNA methyltransferase of mouse cells. The carboxyl-terminal domain of the mammalian enzyme is related to bacterial restriction methyltransferases. *Journal of Molecular Biology*, **203**, 971–83.

Bird, A.P. (1986). CpG-rich islands and the function of DNA methylation. *Nature*, **321**, 209–13.

Brandeis, M., Kafri, T., Ariel, M., Chaillet, J.R., McCarrey, J., Razin, A., *et al.* (1993). The ontogeny of allele-specific methylation associated with imprinted genes in the mouse. *EMBO Journal*, **12**, 3669–77.

Brown, C.J., Ballabio, A., Rupert, J.L., Lafreniere, R.G., Grompe, M., Tonlorenzi, R., *et al.* (1991). A gene from the region of the human X chromosome inactivation centre is expressed exclusively from the inactive X chromosome. *Nature*, **349**, 38–44.

DeChiara, T.M., Efstratiadis, A., and Robertson, E.J. (1990). A growth-deficiency phenotype in heterozygous mice carrying an insulin-like growth factor II gene disrupted by targeting. *Nature*, **345**, 78–80.

DeChiara, T.M., Robertson, E.J., and Efstratiadis, A. (1991). Parental imprinting of the mouse insulin-like growth factor II gene. *Cell*, **64**, 849–59.

Driscoll, D.J., Waters, M.F., Williams, C.A., Zori, R.T., Glenn, C.C., Avidano, K.M., *et al.* (1992). DNA methylation imprint, determined by the sex of the parent, distinguishes Angelman and Prader-Willi syndromes. *Genomics*, **13**, 917–24.

Eden, S. and Cedar, H. (1994). Role of DNA methylation in the regulation of transcription. *Current Opinions in Genetics and Development*, **4**, 255–9.

Efstratiadis, A. (1994). Parental imprinting of autosomal mammalian genes. *Current Opinion in Genetics and Development*, **4**, 265–80.

Feil, R., Walter, J., Allen, N.D., and Reik, W. (1994). Developmental control of allelic methylation in the imprinted mouse *Igf2* and *H19* genes. *Development*, **120**, 2933–43.

Ferguson-Smith, A.S., Sasaki, H., Cattanach, B.M., and Surani, M.A. (1993). Parental-origin-specific epigenetic modification of the mouse *H19* gene. *Nature*, **362**, 751–4.

Frommer, M., McDonald, L.E., Millar, D.S., Collis, C.M., Watt, F., Grigg, G.W., *et al.* (1992). A genomic sequencing protocol that yields a positive display of 5-methylcytosine residues in individual DNA strands. *Proceedings of the National Academy of Sciences of the USA*, **89**, 1827–31.

Giannoukakis, N., Deal, C., Paquette, J., Goodyer, C.G., and Polychronakos, C. (1993). Parental genomic imprinting of the human *IGF2* gene. *Nature Genetics*, **4**, 98–101.

Glenn, C.C., Porter, K.A., Jong, M.T.C., Nicholls, R.D., and Driscoll, D.J. (1993). Functional imprinting and epigenetic modification of the human *SNRPN* gene. *Human Molecular Genetics*, **2**, 2001–5.

Glenn, C.C., Saitoh, S., Jong, M.T.C., Filbrandt. M.M., Surti, U., Driscoll, D.J., *et al.* (1996) Gene structure, DNA methylation, and imprinted expression of the human *SNRPN* gene. *American Journal of Human Genetics* **58**, 335–46.

Hatada, I., Sugama, T., and Mukai, T. (1993). A new imprinted gene cloned by a methylation-sensitive genome scanning method. *Nucleic Acids Research*, **21**, 5577–82.

Hatada, I., Kitagawa, K., Yamaoka, T., Wang, X., Arai, Y., Hashido, K., *et al.* (1995). Allele-specific methylation and expression of an imprinted U2af1-rs1 (SP2) gene. *Nucleic Acids Research*, **23**, 36–41.

Hayashizaki, Y., Shibata, H., Hirotsune, S., Sugino, H., Okazaki, Y., Sasaki, N., *et al.* (1994). Identification of an imprinted U2af binding protein related sequence on mouse chromosome 11 using the RLGS method. *Nature Genetics*, **6**, 33–40.

Kay, G.F., Penny, G.D., Patel, D., Ashworth, A., Brockdorff, N., and Rastan, S. (1993). Expression of *Xist* during mouse development suggests a role in the initiation of X chromosome inactivation. *Cell*, **72**, 171–82.

Lau, M.M.H., Stewart, C.E.H., Liu, Z., Bhatt, H., Rotwein, P., and Stewart, C.L. (1994). Loss of the imprinted IGF2/cation-independent mannose 6-phosphate receptor results in fetal overgrowth and perinatal lethality. *Genes and Development*, **8**, 2953–63.

Leff, S.E., Brannan, C.I., Reed, M.L., Ozcelik, T., Francke, U., Copeland, N.G., *et al.* (1992). Maternal imprinting of the mouse *Snrpn* gene and conserved linkage homology with the human Prader-Willi syndrome region. *Nature Genetics*, **2**, 259–64.

Leighton, P.A., Ingram, R.S., Eggenschwiler, J., Efstratiadis, A., and Tilghman, S.M. (1995). Disruption of imprinting caused by deletion of the *H19* gene region in mice. *Nature*, **375**, 34–9.

Li, E., Bestor, T.H., and Jaenisch, R. (1992). Targeted mutation of the DNA methyltransferase gene results in embryonic lethality. *Cell*, **69**, 915–26.

Li, E., Beard, C., and Jaenisch, R. (1993). Role for DNA methylation in genomic imprinting. *Nature*, **366**, 362–5.

Meehan, R., Lewis, J., Cross, S., Nan, X., Jeppensen, P., and Bird, A. (1992). Transcriptional repression by methylation of CpG. *Journal of Cell Science Supplement*, **16**, 9–14.

Monk, M., Boubelik, M., and Lehnert, S. (1987). Temporal and regional changes in DNA methylation in the embryonic, extraembryonic and germ cell lineages during mouse embryo development. *Development*, **99**, 371–82.

Nicholls, R.D. (1994). New insights reveal complex mechanisms involved in genomic imprinting. *American Journal of Human Genetics*, **54**, 733–40.

Norris, D.P., Patel, D., Kay, G.F., Penny, G.D., Brockdorff, N., Sheardown, S.A., *et al.* (1994). Evidence that random and imprinted *Xist* expression is controlled by pre-emptive methylation. *Cell*, **77**, 41–51.

Ohlsson, R., Nystrom, A., Pfeifer-Ohlsson, S., Tohonen, V., Hedborg, F., Scholfield, P., *et al.* (1993). IGF2 is parentally imprinted during human embryogenesis and in the Beckwith–Wiedemann syndrome. *Nature Genetics*, **4**, 94–7.

Rachmilewitz, J., Goshen, R., Ariel, I., Schneider, T., de Groot, N., and Hochberg, A. (1992). Parental imprinting of the human *H19* gene. *FEBS Letters*, **309**, 25–8.

Rainier, S., Johnson, L.A., Dobry, C.J., Ping, A.J., Grundy, P.E., and Feinberg, A.P. (1993). Relaxation of imprinted genes in human cancer. *Nature*, **362**, 747–9.

Rastan, S. (1994). X chromosome inactivation and the *Xist* gene. *Current Opinion in Genetics and Development*, **4**, 292–7.

Reed, M.L. and Leff, S.E. (1994). Maternal imprinting of human *SNRPN*, a gene deleted in Prader-Willi syndrome. *Nature Genetics*, **6**, 163–7.

Riggs, A.D. and Pfeifer, G.P. (1992). X-chromosome inactivation and cell memory. *Trends in Genetics*, **8**, 169–74.

Sanford, J.P., Clark, H.J., Chapman, V.M., and Rossant, J. (1987). Differences in DNA methylation during oogenesis and spermatogenesis and their persistance during early embryogenesis in the mouse. *Genes and Development*, **1**, 1039–46.

Stoger, R., Kubicka, P., Liu, C.-G., Kafri, T., Razin, A., Cedar, H., *et al.* (1993). Maternal-specific methylation of the imprinted mouse *Igf2r* locus identifies the expressed locus as carrying the imprinting signal. *Cell*, **73**, 61–71.

Sutcliffe, J.S., Nakao, M., Christian, S., Orstavik, K.H., Tommerup, N., Ledbetter, D.H., *et al.* (1994). Deletions of a differentially methylated CpG island at the *SNRPN* gene define a putative imprinting control region. *Nature Genetics*, **8**, 52–8.

Tremblay, K.D., Saam, J.R., Ingram, R.S., Tilghman, S.M., and Bartolomei, M.S. (1995). A paternal-specific methylation imprint marks the alleles of the mouse *H19* gene. *Nature Genetics*, **9**, 407–13.

Wang, Z., Fung, M.R., Barlow, D.P., and Wagner, E.F. (1994). Regulation of embryonic growth and lysosomal targeting by the imprinted *Igf2/Mpr* gene. *Nature*, **372**, 464–7.

Wevrick, R., Kerns, J.A. and Francke, U. (1994). Identification of a novel paternally expressed gene in the Prader-Willi syndrome region. *Human Molecular Genetics*, **3**, 1877–82.

Yeivin, A. and Razin, A. (1993). Gene methylation patterns and expression. In *DNA methylation: molecular biology and biological significance* (ed. J.P. Jost and H.P. Saluz), pp. 523–68. Birkhauser Verlag, Basel.

Zhang, Y., Shields, T., Crenshaw, T., Hao, Y., Moulton, T., and Tycko, B. (1993). Imprinting of human H19; allele-specific CpG methylation, loss of the active allele in Wilms tumor, and potential for somatic allele switching. *American Journal of Human Genetics*, **53**, 113–24.

Zuccotti, M. and Monk, M. (1995). Methylation of the mouse *Xist* gene in sperm and eggs correlates with imprinted *Xist* expression and paternal X-inactivation. *Nature Genetics*, **9**, 316–20.

12

X chromosome inactivation and heritable chromatin structure

Arthur D. Riggs and Maty Hershkovitz

Introduction

X chromosome inactivation results in the heterochromatinization and genetic silencing of one of the two X chromosomes in female mammalian cells and is a premier example of an epigenetic phenomenon. Transcriptional silencing by X chromosome inactivation is efficient, stable, and somatically heritable, yet genes on the inactive X chromosome (the Xi) function unaltered in the next generation. By definition, epigenetic changes do not involve irreversible alteration of base sequence. Instead, phenomena such as X chromosome inactivation are assumed to depend on heritable chromatin structure for the mitotic or meiotic transfer of information from parent to progeny. Chromatin-dependent epigenetic silencing is also seen in other organisms. Yeast mating type is an excellent example of alternate, heritable chromatin states (see Chapters 11 and 15), and the maintenance of chromatin activity states is also part of *Drosophila* development (see Chapters 13 and 14).

Monoallelic expression, factor immunity, and cell memory

As first noted by Harris (1982), epigenetic variants sometimes masquerade as mutations in cultured mammalian cells (reviewed by Riggs and Jones 1983; Holliday 1987, 1989), but the most studied epigenetic phenomena are X chromosome inactivation and genomic imprinting (also termed parental imprinting or genetic imprinting; see Chapter 11). Both of these phenomena are part of normal mammalian development, involve heritable chromatin structure, and display a very important additional feature not found in *Drosophila*, i.e. monoallelic expression in diploid cells. Even though two potentially functional alleles are present in the same nucleus, only one is active and the other is efficiently silenced. The chromatin structure of the inactive allele presumably renders it immune or inaccessible to the transcription factors present in levels sufficient to support full-level transcription from the active allele. The regulation of chromatin accessibility seems often to be an important part of gene control, cell memory, and development. Alberts *et al.* (1983) introduced the term 'cell memory' to describe the faithful transmission of determined states to progeny cells. Good cell memory is essential for *Drosophila* (Paro 1990, 1993) and even

Table 12.1 Epigenetic mechanisms and X chromosome inactivation[a]

I		*Stable feedback loops*
	Assumptions	Free diffusion; mass action; equilibrium
	Comments	Several examples exist, including Lac, lambda, and *Drosophila* sex determination
	Main difficulty	Does not adequately explain monoallelic expression such as seen for X chromosome inactivation and imprinted genes
II		*DNA cytosine methylation*
	Assumptions	Maintenance methylase system; methylation affects protein binding
	Comments	Proven essential for mammals; apparently not used by *Drosophilia, C. elegans* or yeast
III		*Nuclear compartmentalization*
	Assumptions	Spatial separation into chemically different compartments
	Comments	How maintained through mitosis? Require one active compartment (entity) per diploid set of autosomes
IV		*Temporal separation/replication timing*
	Assumptions	Early and late replicating chromatin are different
	Comments	Needs experimental testing
V		*Heritable chromatin structure*
	Assumptions	Replication- and mitosis-resistant nucleoprotein complexes
	Comments	More molecular detail needed to explain heritability

[a] For more detailed discussion and references, see text and Rigg and Pfeifer (1992), Wolffe (1994) and Hendrich and Willard (1995).

more so for mammalian development (Riggs 1989; Riggs and Pfeifer 1992). In general, the response of a mammalian cell to new signals during ontogeny depends on developmental history. For example, mammalian muscle cells bear a cell-autonomous, heritable memory of their anterior–posterior position and this positional memory, which is established during embryogenesis, can be maintained even after long-term growth in culture (Donoghue *et al.* 1992). Recently, Stanworth *et al.* (1995) reported that somatically heritable epigenetic modifications determine the expression of developmentally regulated human globin genes. Although this overview will focus on X chromosome inactivation, evolution is very conservative, so other somatically heritable phenomena which are dependent on chromatin structure are likely to use similar mechanisms. What are these mechanisms? Some mechanisms that have been proposed are listed in Table 12.1 and will be discussed later, after review of the fundamentals of X chromosome inactivation and DNA methylation.

Key features of X chromosome inactivation

Figure 12.1 schematically illustrates X chromosome inactivation, which is a chromosome-wide phenomenon unique to mammals. Both X chromosomes are active soon after fertilization, at least to the extent that several housekeeping

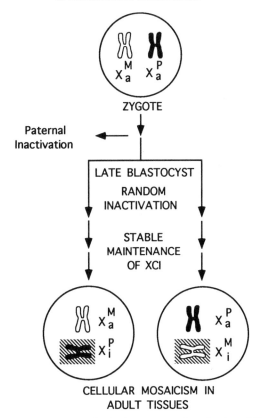

Fig. 12.1 Schematic overview of X chromosome inactivation. X^M and X^P depict maternally and paternally derived X chromosomes, respectively; X_a and X_i depict active and inactive X chromosomes, respectively. The X^M is white; the X^P is black. The shaded box represents the heterochromatinization which correlates with genetic inactivation.

genes are expressed in female cleavage-stage embryos from both the paternal and maternal X chromosomes (see Singer-Sam *et al.* 1992a). Biallelic expression continues until differentiation from totipotency, at which time cytodifferentiation signals trigger the genetic silencing of most genes on one X chromosome. The resulting monoallelic expression thus compensates for gene dosage differences between XX female and XY male cells. For a comprehensive summary and discussion of the earlier literature the reader is referred to reviews by Gartler and Riggs (1983) and Grant and Chapman (1988). X chromosome inactivation has also been the subject of several recent reviews (Riggs and Pfeifer 1992; Lyon 1993; Singer-Sam and Riggs 1993; Willard *et al.* 1993; Gartler and Goldman 1994; Migeon *et al.* 1994; Rastan 1994; Hendrich and Willard 1995).

Cytodifferentiation and X chromosome inactivation take place earlier in the extraembryonic lineages. In these lineages the paternal X chromosome is preferentially silenced, providing a clear example of genetic imprinting. In the

Table 12.2 Differences between the Xa and Xi

Property or feature	Xa/Xi differences[a]
Transcriptional silence	Most genes on Xi are inactive, a few are not
Appearence in interphase	Xi is more condensed, like heterochromatin
Chromatin structure	Little known. Positioned nucleosomes likely on PGK promoter of Xi. General DNAase I sensitivity not much different, but see text
Mitotic chromosomes	Histone H4 is relatively hypoacetylated on Xi. Transcription factors absent from Xa and Xi
Xist/XIST	Only transcribed from Xi
Replication timing	Xa replicates in first half of S phase, Xi in second half
Cytosine methylation	Many CpG islands are heavily methylated only on Xi. *Xist* promoter is opposite, methylated on Xa

[a] See text for more details and references.

embryonic lineage which forms the adult mouse, X chromosome inactivation does not take place until late blastocyst, at about the time of placental implantation at 5.5–6.5 days. By this time the imprint marking the paternal X as different from the maternal X has been erased, thus inactivation is random with respect to paternal source in the progenitors of all somatic cells. Once the chromosomal differentiation is established at the late blastocyst stage, the inactive X becomes a somatically heritable entity. The same X is maintained inactive in all progeny of a given cell and as a result mammalian females are cellular mosaics with respect to X-linked genes. Macroscopic monoclonal patches are seen in most tissues, including the skin (Lyon 1961) and bladder epithelium (Tsai *et al.* 1995). The stability of the Xi is truly remarkable. For example, the reactivation of the *HPRT* gene on an Xi is undetectable (less than one reactivant in 10^8 cells) in normal human cells in culture. There is no evidence for reactivation in the course of a human lifetime (Migeon *et al.* 1988), during which roughly 10^6 cells are dividing every second. In transformed cells or in cross-species hybrid cells the human Xi is significantly less stable (reviewed by Grant and Chapman 1988; Gartler and Goldman 1994), but still the spontaneous reactivation rate is 10^{-6} or less. X chromosome inactivation in the mouse may be less stable for some genes without CpG islands, as both ornithine carbamoyl transferase (*Otc*) and mottled coat colour (*Mo*[blo]) genes show detectable reactivation with age (Wareham *et al.* 1987; Brown and Rastan 1988). Also, the coat colour gene *albino*, when inserted by translocation into the X chromosome, shows slow reactivation during ageing (Cattanach 1974; Deol *et al.* 1986).

Table 12.2 lists some of the differences between the active and inactive X chromosomes (the Xa and Xi, respectively). Cytologically and functionally the Xa is very similar to autosomes and is considered a normal euchromatic chromosome. In humans the Xi is cytologically distinguishable in interphase and is called the Barr body. In mice the Barr body is also present, but is difficult to distinguish from constitutive heterochromatin. Since the Xi is genetically silent and looks similar to heterochromatin, heterochromatinization is said to occur as part of the inactivation process and the Xi is considered to be facultative or conditional heterochromatin. The Xa and Xi replicate asynchronously: in somatic cells the Xi replicates later in S phase than the Xa.

With regard to chromatin structure–function relationships two important features of X chromosome inactivation should be kept in mind. The first is factor immunity. As mentioned in the Introduction, the Xi is in the same nucleus with the Xa and yet housekeeping genes such as phosphoglycerate kinase (*PGK1*) and *HPRT* are actively transcribed only from the Xa. At least two other phenomena in mammals display monoallelic factor immunity. First, the hallmark feature of imprinted autosomal genes is monoallelic expression in diploid cells (Efstratiadis 1994), and second, evidence is emerging that chromatin structure limits accessibility of the immunoglobulin VDJ recombination machinery to only one allele (Hsieh and Lieber 1992).

It is assumed that monoallelic immunity to factors is a result of differential chromatin packaging which limits access to one allele. The importance of chromatin repression and restricted accessibility is becoming increasingly recognized as a dominant aspect of gene control (Croston and Kadonaga 1993; Kornberg and Lorch 1995) and seems likely to be a key factor for control of genes that are monoallelically expressed. A second important feature of X chromosome inactivation is that it is *cis*-limited. In order to be subject to inactivation DNA must be in physical continuity with a locus called the X inactivation centre (XIC), located at human Xq13 (Brown *et al.* 1991a,b). A similar locus (Xic or Xce) has been mapped on the mouse X chromosome (Simmler *et al.* 1993). X-linked genes separated from the XIC are not inactivated and autosomal genes translocated to the X chromosome are subject to inactivation. The inactive chromatin configuration is able to 'spread' into autosomal chromatin, although the extent of spreading is limited, variable, and probably not as stable as for X chromosomal chromatin (Cattanach 1974; Schanz and Steinbach 1989). Recently, a gene that maps within the XIC region has been identified. In contrast to all other known X-linked genes, it is expressed only from the inactive X chromosome (Brown *et al.* 1991a). For this reason the gene has been named *XIST* (*Xist* in mouse) for 'X inactivation-specific transcript'. The transcript is large (15–17 kb) and seems not to be translated. Though its mechanism of action is completely unknown, the *XIST/Xist* gene is likely to be involved in, and necessary for, the establishment of X chromosome inactivation (Kay *et al.* 1993; Migeon *et al.* 1994; Jani *et al.* 1995), although it seems not to be necessary for the maintenance of X chromosome inactivation (Brown and Willard 1994).

DNA methylation and X chromosome inactivation

The notion that DNA modification by enzymatic methylation could provide a molecular mechanism for the inheritance of chromatin states and some key

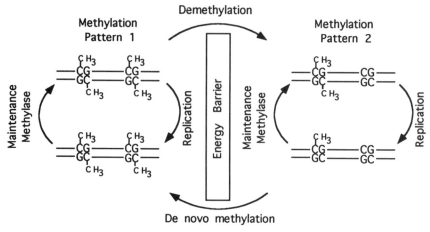

Fig. 12.2   Model for cell memory and epigenetic inheritance by DNA methylation. Alternate chromatin states survive DNA replication because of a maintenance DNA methyltransferase that acts, to a first approximation, only on the hemi-methylated sites that are formed by the replication of a symmetrically methylated site. As a result of this mechanism the methylation pattern is self-templating. The pattern of methylated cytosines can, however, be altered either by *de novo* methylation of unmethylated sites or by demethylation of methylated sites. Changes are likely to be mediated or guided by cell- and sequence-specific proteins, not the maintenance methyltransferase which by itself is specific only for CpG sites.

aspects of X chromosome inactivation was first proposed by Riggs (1975) and Holliday and Pugh (1975) (see Chapter 11). The epigenetic mechanism, illustrated in Fig. 12.2, depends on 5-methylcytosine at CpG sites and an enzyme, often termed a maintenance methylase, to maintain pre-existing methylation patterns. This model is supported by a large body of experimental evidence (reviewed in Jost and Saluz 1993) and is now generally accepted as being a component mechanism of X chromosome inactivation. Mammalian cells have a DNA (cytosine-5-)-methyltransferase enzyme that post-replicationally converts certain cytosines to 5-methylcytosine. This enzyme preferentially methylates cytosine in CpG sites and *in vitro* shows up to a 100-fold preference for hemi-methylated sites, such as are generated by the replication of a CpG site symmetrically methylated in both strands (Fig. 12.2). All mammalian cells have cell type-specific patterns of methylated and unmethylated CpG sites. On average, about 80 per cent of CpG sites are methylated in both strands. Hemi-methylated sites are rare and usually not detected in normal genomic DNA.

The maintenance methylase system will tend to maintain a pre-existing methylation pattern on progeny chromosomes. However, although persistent in the absence of specific perturbation, a methylation pattern can be changed during development either by specific *de novo* methylation of unmethylated sites, or by specific demethylation of already methylated sites. The methylation system can

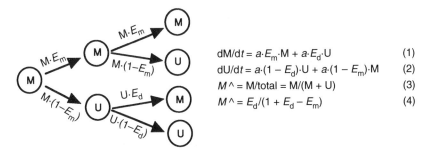

$$dM/dt = a \cdot E_m \cdot M + a \cdot E_d \cdot U \qquad (1)$$

$$dU/dt = a \cdot (1 - E_d) \cdot U + a \cdot (1 - E_m) \cdot M \qquad (2)$$

$$M^\wedge = M/\text{total} = M/(M + U) \qquad (3)$$

$$M^\wedge = E_d/(1 + E_d - E_m) \qquad (4)$$

Fig. 12.3 Methylation dynamics in cell culture. Methylation level at a specific site in a growing population of cells is considered. Each circle represents a cell. M represents the methylated state of a specific CpG site in a specific gene, whereas U represents the unmethylated state of the same site. Since M is the starting state, U states arise in the population by a failure of maintenance. After some U states arise they can be converted back to M states by *de novo* methylation. With time a steady-state level of methylation of the site will be attained. The equations shown are from Pfeiffer *et al.* (1990a) and describe the fraction methylation, *M*, as a function of two primary parameters: E_m, the efficiency of maintenance, and E_d, the efficiency of *de novo* methylation. It is assumed that the growth rate constant, *a*, is the same for M or U cells. The first three equations can be used to model *M* as a function of either time or population doublings (see Pfeifer *et al.* 1990a). For the special case where a steady state is reached *M* is given by equation (4), which can be used to estimate E_m and E_d. (Adapted from Singer-Sam and Riggs 1993.)

be considered as providing an energy barrier between two alternate states (Fig. 12.2). Energy is required to change one state to the other, a property that seems ideally suited for developmental switches. It is now known that mechanisms for the specific removal of methylated cytosines exist (Paroush *et al.* 1990; Frank *et al.* 1991; Jost 1993; Kafri *et al.* 1993; Jost and Jost 1994; Lichtenstein *et al.* 1994), as do mechanisms for the addition of methyl groups to unmethylated sites (*de novo* methylation). Methylation patterns are dynamic during development (reviewed by Razin and Shemer 1995). The emerging picture is that *de novo* methylation occurs during late meiosis and/or gametogenesis, global demethylation shortly after fertilization, global remethylation at the time of implantation, and then gene-specific demethylation during later cellular differentiation (Razin and Shemer 1995).

In tissue culture previously unmethylated sites occasionally become *de novo* methylated, including CpG islands. For example, tumours often have hypermethylated CpG islands (Counts and Goodman 1995; Gonzalez-Zulueta *et al.* 1995; Merlo *et al.* 1995). In culture and most likely also *in vivo* the equilibrium methylation level of any given CpG (and thus methylation pattern) is dependent on the efficiency of *de novo* methylation (E_d in Fig. 12.3) relative to the efficiency of maintenance of methylation (E_m in Fig. 12.3) (Pfeifer *et al.* 1990a; Singer-Sam and Riggs 1993). Pfeifer *et al.* (1990a) found that *de novo* methylation at the CpG island of *PGK* on the Xi can be as high as 5 per cent per CpG site per cell generation. We have recently found that *de novo* methylation of an *Hpa*II site in the differentially methylated region 3 kb upstream of *Igf2*, an imprinted autosomal gene, is also about 3 per cent per generation (Z. Xiong and A.D. Riggs, unpublished). Thus, though the methylation system is stable enough to aid inheritance, it is to a considerable extent dynamic with stochastic as well as cell-specific and tissue-specific aspects. The methylation level of any given CpG will depend on (i) the previous methylation state, (ii) the level and activity of DNA methyltransferase, (iii) the level of general and specific demethylation activity, and (iv) local chromatin structure, which will include the influence of nucleosomes, specific and non-specific protein factors, higher order chromatin structure, and perhaps DNA structure such as superhelical tension.

A second postulate necessary to relate the epigenetic mechanism of Fig. 12.2 to X chromosome inactivation is that cytosine methylation affects chromatin function. This postulate is now very well supported as both *in vitro* and *in vivo* studies have shown that methylation inhibits transcription either directly or indirectly (Boyes and Bird 1991, 1992). Adding to these observations are over 20 reports of proteins whose DNA binding is affected by cytosine methylation (see Ehrlich and Ehrlich 1993). Most methylation-sensitive protein factors are sequence specific and inhibited by methylation at key sites, but some proteins, such as MeCP1 and MeCP2, are not sequence specific but do require methylated cytosines (see Tate and Bird 1993). Sp1 was the first transcription factor studied *in vitro* for the effect of methylation on specific DNA binding and, although initially reported to be methylation insensitive, more recent reports indicate that at least some members of the consensus binding sequence bind Sp1 more strongly if methylated (Jane *et al.* 1993; Joel *et al.* 1993). The binding of histone H1 and its effect on *in vitro* transcription is sensitive to methylation (Levine *et al.* 1993), and a protein first identified as binding to a methylated site in the chicken vitellogenin gene was found to be a member of the histone H1 family (Jost and Hofsteenge 1992).

Methylation and chromatin structure

Little is know about the effect of methylation on chromosome structure other than inhibiting or enhancing the binding of some protein factors as mentioned above. Methylation stabilizes Z-form DNA (Zacharias *et al.* 1988). Thus the potential for influencing alternate DNA structures exists, but the significance of this remains to be determined. With regard to methylation and chromatin structure at an X-linked gene, one of the best studied promoters is that of human *PGK1*. Studies done at nucleotide-level resolution by ligation-mediated polymerase chain reaction (LMPCR) (Mueller and Wold 1989; Pfeifer *et al.* 1989) have established that the active promoter on the Xa is completely unmethylated and shows clear footprints for several transcription sites (Pfeifer *et al.* 1990b; Pfeifer and Riggs 1991; Riggs and Pfeifer 1992). In contrast, the inactive *PGK1* promoter on the Xi is covered by two positioned nucleosomes or nucleosome-sized complexes and is methylated at 60 out of 61 CpG sites in an 800 bp region. The large number of CpG sites is because the *PGK1* promoter is a typical CpG island. A hallmark feature of autosomal CpG islands is that they are maintained methylation free in spite of containing numerous potentially methylatable sites (Antequera and Bird 1993). The heavy methylation of the *PGK1* island is thus in dramatic contrast to most CpG islands and exemplifies what seems to be a key feature of X chromosome inactivation: methylation of X-linked CpG islands (Norris *et al.* 1991). Differential methylation of CpG islands is also a candidate for the primary mark which influences transcriptional control of imprinted genes (Bartolomei *et al.* 1993; Efstratiadis 1994; Surani 1994).

It is now generally accepted that methylation of CpG islands is essential for the efficient maintenance of X chromosome inactivation. Evidence for this was comprehensively reviewed by Grant and Chapman (1988) and includes the following observations.

1. Xi-derived DNA functions less well for DNA-mediated transfectionthan does Xa-derived DNA.

2. DNA cytosine methylation inhibitors, such as 5-azacytidine or 5-azadeoxycytidine, are the only known exogenous agents that efficiently cause reactivation of X-linked genes, as much as 10^4-fold over the spontaneous rate.

3. Demethylation and then changes in chromatin nuclease sensitivity precede transcription during the reactivation of *HPRT* by 5-azacytidine (Sasaki *et al.* 1992).

4. In marsupials, X-linked CpG islands at genes so far tested are not heavily methylated at CpG islands, and these genes are not kept stably inactivated either in culture or *in vivo* (Migeon *et al.* 1989).

Reactivation of X-linked genes by methylation inhibitors is piecemeal, not chromosome-wide, suggestive of regional or domain-wide control. X chromosome inactivation also can be reversed by fusion of mouse lymphocytes with teratocarcinoma cells (Takagi *et al.* 1983) or by fusing transformed mouse somatic cells with human chorionic villi cells (Migeon *et al.* 1986; Luo *et al.* 1995). In these cases reactivation is thought to be chromosome-wide.

Is methylation a primary or secondary event in X chromosome inactivation?

The complete answer to this question is not yet known. Though accepting that methylation is involved in and may be sufficient for the maintenance of X chromosome inactivation, many question whether methylation is involved in the initial steps for establishment of X chromosome inactivation in the early embryo. The two primary reasons for considering methylation only as a locking mechanism are the following.

1. Methylation is less extensive at CpG islands in marsupials (which are
    assumed to have a more primitive X inactivation system) and also in the
    extraembryonic tissues of placental mammals, yet inactivation and asyn-
    chronous replication are seen (Kaslow and Migeon 1987).

2. Some methylation-sensitive restriction sites in the first exon of the mouse
    Hprt gene become methylated well after X chromosome inactivation takes
    place (Lock *et al.* 1987).

Despite these data it is nevertheless premature to draw any conclusions with
respect to the role of methylation in the initial establishment of X chromosome
inactivation. First, with regard to gene-specific control, most methylation is
likely to be irrelevant, representing just a background, which tends to obscure
methylation at key sites. There are reported cases where methylation at a single
CpG site is crucial for the maintenance of transcriptional silence (Graessmann *et
al.* 1994; Robertson *et al.* 1995) and the elimination of protein factor binding (see
Gaston and Fried 1995). Second, the time of X chromosome inactivation
establishment is a time of major changes in genome-wide methylation, going
from very hypomethylated in the early blastocyst to heavy methylation in the late
blastocyst (Razin and Shemer 1995). In contrast to the earlier study of intron sites
(Lock *et al.* 1987), methylation changes at the promoter regions of the X-linked
genes *Pgk-1*, *G6pd*, and *Xist* take place concomitant with genetic silencing
(Singer-Sam *et al.* 1990; Grant *et al.* 1992; Norris *et al.* 1994; Beard *et al.*
1995).

Of most direct relevance to the establishment of X chromosome inactivation
are recent studies on methylation of the *Xist* gene in DNA methyltransferase-
deficient embryos (Beard *et al.* 1995). *Xist* is differentially methylated in the
gametes, with sperm DNA unmethylated and oocyte DNA methylated (Norris *et
al.* 1994; Ariel *et al.* 1995); thus the embryo starts with methylation-marked *Xist*
genes. In the extraembryonic tissues the unmethylated, paternal *Xist* gene is
preferentially transcribed and the methylated, maternal *Xist* gene is transcriptionally silent (Kay *et al.* 1994; Norris *et al.* 1994). Recalling that *Xist* is expressed
only from the Xi, this result is in keeping with methylation silencing, the
correlation seen for most autosomal and X-linked genes. In the embryonic
lineages, which differentiate later in development, the methylation imprint is lost
and now the paternal and maternal X chromosomes are equally likely to be
inactivated (Kay *et al.* 1994; Norris *et al.* 1994). Beard *et al.* (1995) studied the
expression of *Xist* in transgenic mouse embryos with two levels of DNA
methyltransferase deficiencies, moderate and severe. Homozygous deficiency of
DNA methyltransferase is embryonic lethal (Li and Jaenisch 1993), but the
embryos do survive long enough to permit studies of *Xist* expression. The key
finding of Beard *et al.* (1995) is that the *Xist* gene in severely methyltransferase-
deficient embryos is hypomethylated and expressed even in males, which have
only an Xa and normally have a methylated and transcriptionally silent *Xist* gene.
The involvement of methylation in controlling the expression of *Xist* in the

embryo is made even more likely by the finding that only the severe, not the moderate, methylation-deficient embryos are hypomethylated at the *Xist* promoter and also transcribe *Xist*.

An important emerging generalization is that sites vary in their resistance to demethylation and the sites most resistant to the extensive demethylation that occurs during the morula and early blastocyst stages of development are the best candidates for primary imprint sites. Thus for imprinted genes the leading model has become that methylation at these limited number of primary sites survives through the morula stage to later guide the establishment of the final methylation pattern of the gene or larger chromosomal domain. This notion is also relevant for the maintenance of X chromosome inactivation in germ cells. Primordial germ cells have an Xi, but the methylation level of pre-meiotic and early-stage-meiotic germ cells is very low (Driscoll and Migeon 1990; Ghazi *et al.* 1992; Singer-Sam *et al.* 1992b). As in methylase-deficient mice, though, demethylation-resistant sites could be maintained.

Before methylation-pattern correlations can be used to argue for or against the involvement of methylation in a process, the key DNA elements controlling the process must first be identified and this is often difficult. In this regard it should be noted that though *Xist* does map to the Xce, the mouse X inactivation centre, mapping resolution is limited, and Courtier *et al.* (1995) have recently reported that a region lying 15 kb distal to *Xist* contains several sites that show hypermethylation specifically associated with the Xa. Furthermore, analysis of this region in various Xce strains has revealed a correlation between the strength of the Xce allele carried and the methylation status of this region. It seems likely that genes or elements other than or in addition to *Xist* influence the choice of which of the two X chromosomes is inactivated.

Methylation-independent mechanisms

Overview of mechanisms

The various proposed models listed in Table 12.1 are not mutually exclusive. On the contrary, the use of several mechanisms is to be fully expected, especially for genes that need tight control and complete, stable shut-down in most tissues. For example, the first model listed, feedback loops, is almost certain to be used for the switching of some genes in mammals as it is in *Drosophila* and *Caenorhabiditis elegans* (Hodgkin 1994). A particularly well-characterized example of a stable, positive feedback loop is provided in the sex determination system of *Drosophila*, where positive autoregulation of *Sex-lethal* by alternative splicing maintains the female-determined state (Bell *et al.* 1991). However, something more definitely is needed to explain monoallelic expression and *cis*-limited action such as seen for X chromosome inactivation. Despite the solid evidence that cytosine methylation is one epigenetic system essential for mammalian development, there is no reason to think that it is the only epigenetic system. This

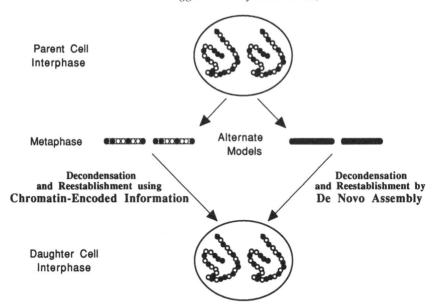

Fig. 12.4   Metaphase chromosome structure and cell memory. Information about the alternate activity states (active: open circles; inactive: filled circles) of a gene, chromosomal domain, or an entire chromosome (in the case of the X chromosome) can be either preserved (left pathway) or erased (right pathway) from the chromatin of a metaphase chromosome when a cell undergoes mitosis. If no information is retained on the metaphase chromosome, then a very complex pattern of active and inactive genes and chromosomal domains must be re-established *de novo* each generation in the daughter cells (from Hershkovitz and Riggs 1995).

is especially so because both *Drosophila* and yeast display cell memory and, for yeast mating type, chromatin-state heritability and factor immunity, yet neither of these organisms has any detectable 5-methylcytosine. Moreover, the fourth model, replication timing, has been suggested to be the ancestral mechanism for X chromosome inactivation (Riggs 1990a,b), as it is the most obviously conserved feature (Graves 1987). If the Xa and Xi replicate in different parts of S phase, then they could be temporally separated into a different environment during DNA synthesis and the reformation of mature chromatin (see Riggs and Pfeifer 1992). A similar model was earlier proposed to explain the switch from oocyte to somatic 5S gene transcription in *Xenopus* (Gottesfeld and Bloomer 1982; Wormington *et al.* 1982). Though no longer thought necessary for the 5S genes (Bouvet *et al.* 1994), the temporal separation model is still an attractive model for X chromosome inactivation. The key chromosomal elements for this model are likely to be replication origins. The third model, nuclear compartmentalization mediated by binding to a single entity per nucleus, was one of the first models proposed (Comings 1968) and continues to be attractive. Recently, Schlossher *et al.* (1994) have discussed specific sequestration into

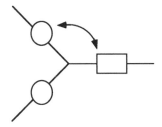

Fig. 12.5   Maintenance of interacting but separated nucleoprotein complexes. Two different complexes (indicated by a circle and a rectangle) near a replication fork are shown sufficiently separated so that the machinery of the replication fork does not disrupt both at the same time. Interactions between the two elements can preserve chromatin states (Wolffe 1994). As an example, the circle could represent an enhancer located several kb upstream of its promoter.

separate nuclear territories as a possible mechanism for position effect variegation (PEV) and cell memory in *Drosophila*. Nuclear compartmentalization could aid maintenance by keeping the Xa and Xi in different environments with respect to transcription factors and chromosomal proteins. A model for the replication of nuclear compartments has not been discussed, however, except in the context of needing one chromosome binding entity per diploid set of autosomes (Gartler and Riggs 1983).

Heritable chromatin structure

The last class of models listed in Table 12.1 is heritable chromatin structure, and includes models that propose 'heterochromatinization' as the first step of X chromosome inactivation (e.g. Gartler *et al.* 1985). This class of models depends on different nucleoprotein complexes being established on the Xa and Xi and the differences somehow surviving DNA replication. It has been proposed that replication does not disrupt nucleoprotein complexes as the DNA is reeled through the replication machinery (Brown 1984), but experimental evidence is that basal transcription complexes are removed from DNA by passage of a fork (Wolffe and Brown 1986). Nucleosomes also are at least transiently removed and are segregated randomly to daughter strands, but may stay in the general vicinity of their previous location (reviewed by Wolffe 1994). Thus pre-existing nucleoprotein complexes, though 50 per cent new, may nevertheless be able to 'seed' the assembly of similar structures within a domain. Wolffe (1994) has discussed several protein modification models for epigenetic inheritance and suggested that separated bipartite structures can aid the re-establishment of active or inactive chromatin states following replication. A key aspect of this interesting model is illustrated in Fig. 12.5. An element which is downstream with respect to replication fork movement can direct the re-formation of an already duplicated upstream element. Then, when the replication fork passes and disrupts the downstream element, the previously re-established upstream element can guide the re-formation of the proper activity state of the downstream element. This mechanism could also apply to any multipartite, adequately large nucleoprotein complex. The model also is relevant to, and provides good rationale for, studies on metaphase chromosomes.

Metaphase chromatin structure as an approach to the study of epigenetic mechanisms

Promoters and other control elements of the Xi are presumed to be packaged in chromatin so as to be inaccessible to the transcription factors for housekeeping genes, because these must be constantly present in the nucleus. For example, *PGK1* is expressed at moderately high levels in virtually all cells, yet transcription from the Xi is undetectable, i.e. less than 0.1 per cent of the Xa (LeBon *et al.* 1995). Thus it is especially difficult to imagine how resistance is stably preserved if the complex pattern of active and inactive nucleoprotein complexes and/or chromosomal domains is created completely *de novo* for the entire genome each cell generation. More likely, some kind of chromatin structural information passes to progeny cells and helps guide the re-formation of the complex pattern of active and inactive genes that determine cell type. In order for chromatin-based information to pass to progeny cells it must survive not only DNA replication, but also mitosis (left pathway in Fig. 12.4).

Metaphase chromosomes have an extremely compacted chromatin fibre, approximately 10 000-fold condensed relative to extended DNA (Van Holde 1988; Manuelidis 1990). What information remains on compacted metaphase chromosomes that can help progeny cells faithfully maintain X chromosome inactivation and monoallelic expression from imprinted genes? To investigate this question for X-linked genes we isolated metaphase cells by mitotic shake off, treated them with dimethylsulphate (DMS), and assayed for *in vivo* footprints by LMPCR (Mueller and Wold 1989; Pfeifer *et al.* 1989). We found that footprints for eight transcription factors on the constitutively active promoter of the human *PGK1* gene are absent from the chromatin of metaphase cells (Hershkovitz and Riggs 1995). Similar results have been obtained using DNAase I on permeabilized cells (M. Hershkovitz and A.D. Riggs, unpublished results). Clear footprints are seen on the promoter of interphase chromatin, but these are absent from metaphase chromatin. Martínez-Balbás *et al.* (1995) did genomic footprinting of HeLa cells arrested at mitosis by nacodozole and also found that several transcription factors seen in interphase cells on the induced promoter of the heat shock gene *hsp70* are absent from metaphase chromatin. They also showed that the transcription factors are still present in the cell, just not bound to mitotic chromatin. Thus the multiprotein transcriptional complexes on the constitutive *PGK1* gene, the inducible *hsp70* gene, and presumably a myriad of other transcription complexes in the genome must form *de novo* each cell generation.

An interesting requirement of the bipartite control element model of Wolffe (1994) (Fig. 12.5) is that at least one of the protein complexes must remain on the DNA of metaphase chromosomes, otherwise this mechanism cannot be used to guide the chromosomal decondensation and re-establishment of transcription complexes in progeny cells. The absence of transcription factors from the promoters so far examined by *in vivo* footprinting does not mean that all nucleoprotein structures are stripped from metaphase chromatin, although only a few proteins, such as topoisomerase II, have been firmly identified so far as components of metaphase chromosomes (Manuelidis 1990; Saitoh and Laemmli 1994). Immunocytochemical studies indicate that out of eight transcription factors studied, only AP-2 is detectable on metaphase chromosomes of HeLa cells (Martínez-Balbás *et al.* 1995). Raff *et al.* (1994) found that GAGA factor, a transcription factor that binds to GA/AT-rich sequences found in *Drosophila* heterochromatin, can be detected by immunostaining on chromosomes at all stages of the cell cycle, indicating that it is possible for sequence-specific proteins to remain on tightly condensed chromatin. We have recently studied the matrix attachment site described by Mielke *et al.* (1990) that is near the β-interferon gene. We find in this case that the structure of metaphase chromatin is not significantly different from interphase chromatin, although both are different from the pattern seen for naked DNA. Sites of enhanced reactivity are seen *in vivo* (M. Hershkovitz and A.D. Riggs, unpublished). Whether the altered reactivity is due to DNA structure or a nucleoprotein complex is not yet known, but matrix attachment sites clearly are promising elements to study as key elements for the epigenetic inheritance of chromatin.

Xa/Xi differences other than methylation

Histone acetylation

A very strong positive correlation has emerged between histone acetylation and the active chromatin state (O'Neill and Turner 1995; see Chapter 3). This correlation holds for the differences between the chromatin of the active Xa and the facultative heterochromatin of the Xi. Jeppesen and Turner (1993) did immunofluorescent staining of metaphase mammalian cells using antibodies specific for acetylated histone H4 and found that the Xa is similar to other chromosomes, but the Xi is barely visible, like heterochromatin. Since histone acetylation is preserved in metaphase chromosomes and affects nucleosomal and chromatin function, acetylation is attractive as a component of the information transfer system(s) that guides chromatin decondensation and reactivation in progeny cells. It should be noted though that inhibitors of deacetylation, such as sodium butyrate or trichostatin, which increase histone acetylation levels, have not been reported to reactivate X-linked or imprinted genes, and our attempts to reactivate *HPRT* in Chinese hamster–human hybrid cells by treatment with trichostatin were negative (M. Hershkovitz and A.D. Riggs, unpublished).

Chromatin condensation and accessibility to enzyme and protein factors

A more condensed state during interphase is almost universally accepted as a key difference between active euchromatin and inactive heterochromatin and the long-standing cytological evidence does seem incontrovertible. The cytological evidence interpreted as evidence for continued condensation during interphase usually is based on differential staining, e.g. the Barr body. Some regions, such as centromeric heterochromatin and G-bands, stain differently at metaphase than do the R-bands, which replicate earlier than the G-bands and have a higher density of genes (Holmquist 1992). By the usual staining methods, however, the Xi at metaphase is not distinguishable from the Xa. Replication banding, which depends on differential incorporation of BrdU due to asynchronous replication, is usually used to distinguish the Xi from the Xa. Of the non-immunological procedures for staining metaphase chromosomes only the Kanda procedure, which involves incubation of cells in hypotonic KCl prior to staining with Giemsa (Kanda 1973), distinguishes the two X chromosomes.

Active chromatin has long been known to have increased general sensitivity to DNAase I as well as localized hypersensitive sites (Wolffe 1992; Elgin 1995). Southern blot studies have demonstrated increased sensitivity of X-linked genes to DNAase I in interphase nuclei (Riley *et al.* 1986; Yang and Casky 1987; Lin and Chinault 1988). In view of the above discussion of metaphase chromatin it is noteworthy that differences in DNAase I sensitivity are preserved in metaphase chromatin, as evidenced both by cytogenetic studies and treatment of isolated chromatin (Gazit and Cedar 1982; Kuo *et al.* 1982; Kerem *et al.* 1983; Martínez-Balbás *et al.* 1995). However, despite the seemingly obvious fact that heterochromatin is more condensed, there is a body of data that does not support this notion.

In a study of nuclear compartments, Cremer *et al.* (1993) used *in situ* hybridization methods and careful analysis of serial sections to compare the volumes occupied by the Xa and Xi in interphase nuclei. The surprising result is that, though the shape of the Xi is significantly different from the Xa, the volume occupied is not. Schlossher *et al.* (1994), using a sensitive, quantitative variation on LMPCR (Steigerwald *et al.* 1990), have come to similar conclusions in a study comparing active and inactive chromatin in *Drosophila*. Genes silenced by heritable chromatin alteration, either by PEV or by the *polycomb* gene product (Paro 1993; see Chapter 14), do not show altered restriction enzyme sensitivity (but see Wallrath and Elgin 1995). Schlossher *et al.* (1994) also found that mutations affecting PEV and *polycomb* silencing have no effect on enzyme accessibility. Even though DNAase I-detectable differences are observed between the Xa and Xi in mammalian cells, the differences are only about two-fold, as measured by Southern blot experiments (Riley *et al.* 1986; Yang and Casky 1987; Lin and Chinault 1988). Interphase chromatin has been thought to be coiled into a 30 nm solenoid, which is then compacted by looping to form a

250 nm fibre. Further compaction takes place during mitosis to form the 700 nm chromatid of a metaphase chromosome (Van Holde 1988; Manuelidis 1990; see Chapter 11). The solenoid model has been called into question by recent data and a looser three-dimensional zigzag model has been proposed (Woodcock *et al.* 1993; Leuba *et al.* 1994; Woodcock 1994). A tight solenoid would be expected to exclude protein factors and enzymes such as DNAase I, but perhaps the zigzag structure would not. *In vivo* footprinting experiments done by Pfeifer and Riggs (1991), which involved treatment of permeabilized cells and had a good internal control, indicated little difference between the Xa and Xi in susceptibility to DNAase I nicking. These experiments, which have been confirmed for mouse cells (L. Wang and A.D. Riggs, unpublished), are consistent with and confirm the Southern blot experiments, which are done at relatively high nick densities. However, our recent preliminary studies indicate that Xa/Xi differences are larger in chromatin that is only occasionally nicked, suggesting that the special Xi chromatin structure is differentially sensitive to DNA strand nicking, thereby raising the question whether superhelical tension could be involved in compaction of the Xi.

Future directions

The key question addressed in this review is how the differences between the Xa and the Xi can be maintained through DNA replication and mitosis in spite of the fact that the underlying primary DNA sequences are the same and the nucleus contains all the factors needed to maintain full activity of the active alleles. Recent studies on the *Xist* gene in DNA methyltransferase-deficient mice (Beard *et al.* 1995) have provided strong evidence that methylation is involved in maintaining the silence of this gene, which is likely to be a key gene at least in the establishment of X chromosome inactivation in the early embryo. However, the mechanism of action of the *Xist* gene is completely unknown and additional studies are needed. With regard to maintenance, perhaps the clearest difference between the Xa and the Xi is the full methylation on the Xi of 60 CpG sites in the CpG island that encompasses the *PGK* promoter and no methylation on the Xa. *De novo* methylation of the Xi CpG island and promoter is known to occur at about 5 per cent per cell generation, so this observation, along with the likelihood of multiple key sites affected by methylation, essentially explains the stability of the inactive state (Pfeifer *et al.* 1990a). *De novo* methylation will repair any loss due to maintenance failure. These observations, however, just change the question to: 'how is the promoter-containing CpG island of the Xa maintained free of methylation?' CpG islands are not resistant to *de novo* methylation *in vivo* (Antequera *et al.* 1990; Jones *et al.* 1990; Pfeifer *et al.* 1990b). The involvement of Sp1 should be considered, as it has been reported to keep nearby DNA free of methylation (Brandeis *et al.* 1994; Macleod *et al.* 1994). In this context, it should be noted that Sp1 binding is either not affected or is increased by methylation of its binding site, yet Sp1 is not bound to the methylated *PGK* promoter of the Xi either in interphase chromatin (Pfeifer and Riggs 1991) or at metaphase (Hershkovitz and Riggs 1995). In this case, the action of methylation on Sp1 binding seems necessarily to be indirect. Why is Sp1 on the Xa but not on the Xi?

Greater understanding of chromatin structure both at the nucleosomal and higher order level will obviously be helpful in order to more fully understand the details of X chromosome inactivation. Areas of crucial importance are histone acetylation (see Chapter 3) and studies of metaphase chromosomes. Nucleoprotein complexes that survive on metaphase chromosomes should be the focus of more detailed analyses because, as illustrated in Fig. 12.3 and discussed in an earlier section, these are likely also to be involved in the transfer to progeny of information needed to maintain monoallelic expression of X-linked and imprinted genes. Replication origins are particularly strong candidates, because of conceptually attractive replication-timing models for X chromosome inactivation (Riggs 1990a,b), and also because origin of replication (ORC) complexes are known to be involved in the chromatin-based silencing of yeast mating-type genes (Pillus and Grunstein 1995; see Chapter 15). It is of interest that a region of high density methylation was found near a putative mammalian replication origin (Tasheva and Roufa 1994). This study also found methylation in non-CpG sites by use of a method for methylation analysis dependent on the deamination of 5-methylcytosine by bisulphite treatment (Frommer *et al.* 1992). Results obtained using this bisulphite method, which is an important technical advance, have generally confirmed that methylation is only in CpG sites, but there are clearly some exceptional non-CpG sites (Clark *et al.* 1995) and the function of these sites needs to be determined. This is especially true in light of studies by Laayoun and Smith (1995) who found that the mammalian DNA methyltransferase preferentially *de novo* methylates certain unusual DNA structures such as slipped duplexes and cruciforms.

Summary

Most X-linked genes and at least 13 imprinted genes are monoallelically expressed in diploid cells. In addition to monoallelic expression, all of these genes share two additional features: asynchronous replication and differential methylation of CpG islands. In this overview we have focused on the features of X chromosome inactivation that are relevant to the differential treatment of sequence-identical genes and the stable epigenetic inheritance of alternate chromatin states. The now solid evidence that DNA methylation is one component mechanism of X chromosome inactivation is reviewed, but other models for heritable chromatin structure are also considered, such as replication timing, nuclear compartmentalization, and multipartite nucleoprotein complexes. A hierarchy of mechanisms is likely to be needed to explain the tight control seen for most X-linked genes. The study of metaphase chromosome structure is emphasized as a new experimental approach to elucidating some of the mechanisms of epigenetic inheritance.

Acknowledgement

This work was supported by a National Institutes Health grant GM50575 to A.D.R.

References

Alberts, B., Bray, D., Lewis, J., Roberts, K., and Watson, J.D. (1983). *Molecular biology of the cell*. Garland Publishing, New York.

Antequera, F. and Bird, A. (1993). CpG Islands. In *DNA methylation: molecular biology and significance* (ed. J.P. Jost and H.P. Saluz), pp. 169–85. Birkhauser Verlag, Berlin.

Antequera, F., Boyes, J., and Bird, A. (1990). High levels of *de novo* methylation and altered chromatin structure at CpG islands in cell lines. *Cell*, **62**, 503–14.

Ariel, M., Robinson, E., McCarrey, J.R., and Cedar, H. (1995). Gamete-specific methylation correlates with imprinting of the murine Xist gene. *Nature Genetics*, **9**, 312–15.

Bartolomei, M.S., Webber, A.L., Brunkow, M.E., and Tilghman, S.M. (1993). Epigenetic mechanisms underlying the imprinting of the mouse H19 gene. *Genes and Development*, **7**, 1663–73.

Beard, C., Li, E., and Jaenisch, R. (1995). Loss of methylation activates Xist in somatic but not in embryonic cells. *Genes and Development*, **9**, 2325–34.

Bell, L.R., Horabin, J.L., Schedl, P., and Cline, T.W. (1991). Positive autoregulation of Sex-lethal by alternative splicing maintains the female determined state in *Drosophila*. *Cell*, **65**, 229–39.

Bouvet, P., Dimitrov, S., and Wolffe, A.P. (1994). Specific regulation of *Xenopus* chromosomal 5S gene transcription *in vivo* by histone H1. *Genes and Development*, **8**, 1147–59.

Boyes, J. and Bird, A. (1991). DNA methylation inhibits transcription indirectly via a methyl-CpG binding protein. *Cell*, **64**, 1123–34.

Boyes, J. and Bird, A. (1992). Repression of genes by methylation depends on CpG density and promoter strength: evidence for involvement of a methyl-CpG binding protein. *EMBO Journal*, **11**, 327–33.

Brandeis, M., Frank, D., Keshet, I., Siegfried, Z., Mendelsohn, M., Nemes, A., *et al.* (1994). Sp1 elements protect a CpG island from *de novo* methylation. *Nature*, **371**, 435–8.

Brown, D.D. (1984). The role of stable complexes that repress and activate eucaryotic genes. *Cell*, **37**, 359–65.

Brown, S. and Rastan, S. (1988). Age-related reactivation of an X-linked gene close to the inactivation centre of the mouse. *Genetical Research*, **52**, 151–4.

Brown, C.J. and Willard, H.F. (1994). The human X-inactivation centre is not required for maintenance of X-chromosome inactivation. *Nature*, **368**, 154–6.

Brown, C.J., Ballabio, A., Rupert, J.L., Lafreniere, R.G., Grompe, M., Tonlorenzi, R., *et al.* (1991a). A gene from the region of the human X inactivation centre is expressed exclusively from the inactive X chromosome. *Nature*, **349**, 38–44.

Brown, C.J., Lafreniere, R.G., Powers, V.E., Sebastio, G., Ballabio, A., Pettigrew, A.L., *et al.* (1991b). Localization of the X inactivation centre on the human X chromosome in Xq13. *Nature*, **349**, 82–4.

Cattanach, B.M. (1974). Position effect variegation in the mouse. *Genetical Research*, **23**, 291–306

Clark, S.J., Harrison, J., and Frommer, M. (1995). CpNpG methylation in mammalian cells. *Nature Genetics*, **10**, 20–7.

Comings, D.E. (1968). The rationale for an ordered arrangement of chromatin in the interphase nucleus. *American Journal of Human Genetics*, **20**, 440–60.

Counts, J.L. and Goodman, J.I. (1995). Alterations in DNA methylation may play a variety of roles in carcinogenesis. *Cell*, **83**, 13–15.

Courtier, B., Heard, E., and Avner, P. (1995). Xce haplotypes show modified methylation in a region of the active X chromosome lying 3' to Xist. *Proceedings of the National Academy of Sciences of the USA*, **92**, 3531–5.

Cremer, T., Kurz, A., Zirbel, R., Dietzel, S., Rinke, B., Schrock, E., *et al.* (1993). Role of chromosome territories in the functional compartmentalization of the cell nucleus. *Cold Spring Harbor Symposia on Quantitative Biology*, **58**, 777–92.

Croston, G.E. and Kadonaga, J.T. (1993). Role of chromatin structure in the regulation of transcription by RNA polymerase II. *Current Opinion in Cell Biology*, **5**, 417–23.

Deol, M.S., Truslove, G.M., and McLaren, A. (1986). Genetic activity at the albino locus in Cattenachs insertion in the mouse. *Journal of Embryology and Experimental Morphology*, **96**, 295–302.

Donoghue, M.J., Morris-Valero, R., Johnson, Y.R., Merlie, J.P., and Sanes, J.R. (1992). Mammalian muscle cells bear a cell-autonomous, heritable memory of their rostocaudal position. *Cell*, **69**, 67–77.

Driscoll, D.J. and Migeon, B.R. (1990). Sex difference in methylation of single-copy genes in human meiotic germ cells: implications for X chromosome inactivation, parental imprinting, and origin of CpG mutations. *Somatic Cell and Molecular Genetics*, **16**, 267–82.

Efstratiadis, A. (1994). Parental imprinting of autosomal mammalian genes. *Current Opinion in Genetics and Development*, **4**, 265–80.

Ehrlich, M. and Ehrlich, K.C. (1993). Effect of DNA methylation on the binding of vertebrate and plant proteins to DNA. In *DNA methylation: molecular biology and biological significance* (ed. J.P. Jost and H.P. Saluz), pp. 145–68. Birkhauser Verlag, Berlin.

Elgin, C. R. (ed.) (1995). *Chromatin structure and gene expression*. Oxford University Press, New York.

Frank, D., Keshet, I., Shani, M., Levine, A., Razin, A., and Cedar, H. (1991). Demethylation of CpG islands in embryonic cells. *Nature*, **351**, 239–41.

Frommer, M., McDonald, L.E., Millar, D.S., Collis, C.M., Watt, F., Grigg, G.W., *et al.* (1992). A genomic sequencing protocol that yields a positive display of 5-methylcytosine residues in individual DNA strands. *Proceedings of the National Academy of Sciences of the USA*, **89**, 1827–31.

Gartler, S.M. and Goldman, M.A. (1994). Reactivation of inactive X-linked genes. *Developmental Genetics*, **15**, 504–14.

Gartler, S.M. and Riggs, A.D. (1983). Mammalian X-chromosome inactivation. *Annual Review of Genetics*, **17**, 155–90.

Gartler, S.M., Dyer, K.A., Graves, J.A.M., and Rocchi, M. (1985). A two step model for mammalian X chromosome inactivation. In *Biochemistry and biology of DNA methylation* (ed. G.L. Cantoni and A. Razin), pp. 223–35. Alan R. Liss, New York.

Gaston, K. and Fried, M. (1995). CpG methylation has differential effects on the binding of YY1 and ETS proteins to the bi-directional promoter of the Surf-1 and Surf-2 genes. *Nucleic Acids Research*, **23**, 901–9.

Gazit, B. and Cedar, H. (1982). Active genes are sensitive to deoxyribonuclease during metaphase. *Science*, **217**, 648–50.

Ghazi, H., Gonzales, F., and Jones, P. A. (1992). Methylation of CpG-island-containing genes in human sperm, fetal and adult tissues. *Gene*, **114**, 203–10.

Gonzalez-Zulueta, M., Bender, C.M., Yang, A.S., Nguyen, T.D., Beart, R.W., Van Tornout, J.M., *et al.* (1995). Methylation of the 5' CpG island of the p16/CDKN2 tumor suppressor gene in normal and transformed human tissues correlates with gene silencing. *Cancer Research*, **55**, 4531–5.

Gottesfeld, J. and Bloomer, L.S. (1982). Assembly of transcriptionally active 5S RNA gene chromatin *in vitro. Cell*, **28**, 781–91.

Graessmann, A., Sandberg, G., Guhl, E., and Graessmann, M. (1994). Methylation of single sites within the herpes simplex virus tk coding region and the simian virus 40 T-antigen intron causes gene inactivation. *Molecular and Cellular Biology*, **14**, 2004–10.

Grant, S.G. and Chapman, V.M. (1988). Mechanisms of X-chromosome inactivation. *Annual Review of Genetics*, **22**, 199–233.

Grant, M., Zuccotti, M., and Monk, M. (1992). Methylation of CpG sites of two X-linked genes coincides with X-inactivation in the female mouse embryo but not in the germ line. *Nature Genetics*, **2**, 161–6.

Graves, J.A.M. (1987). The evolution of mammalian sex chromosomes and dosage compensation: clues from marsupials and monotremes. *Trends in Genetics*, **3**, 252–6.

Harris, M. (1982). Induction of thymidine kinase in enzyme-deficient Chinese-hamster cells. *Cell*, **29**, 483–92.

Hendrich, B.D. and Willard, H.F. (1995). Epigenetic regulation of gene expression: the effect of altered chromatin structure from yeast to mammals. *Human Molecular Genetics*, **4**, 1765–77.

Hershkovitz, M. and Riggs, A.D. (1995). Metaphase chromosome analysis by ligation-mediated PCR: heritable chromatin structure and a comparison of active and inactive X chromosomes. *Proceedings of the National Academy of Sciences of the USA*, **92**, 2379–83.

Hodgkin, J. (1994). Epigenetics and the maintenance of gene activity states in *Caenorhabiditis elegans. Developmental Genetics*, **15**, 471–7.

Holliday, R. (1987). The inheritance of epigenetic defects. *Science*, **238**, 163–70.

Holliday, R. (1989). A different kind of inheritance. *Scientific American*, **260**, 60–73.

Holliday, R. and Pugh, J.E. (1975). DNA modification mechanisms and gene activity during development. *Science*, **187**, 226–32.

Holmquist, G.P. (1992). Chromosome bands, their chromatin flavors, and their functional features. *American Journal of Human Genetics*, **51**, 17–37.

Hsieh, C.-L. and Lieber, M.R. (1992). CpG methylated minichromosomes become inaccessible for V(D)J recombination after undergoing replication. *EMBO Journal*, **11**, 315–25.

Jane, S.M., Gumucio, D.L., Ney, P.A., Cunningham, J.M., and Nienhuis, A.W. (1993). Methylation-enhanced binding of Sp1 to the stage selector element of the human gamma-globin gene promoter may regulate developmental specificity of expression. *Molecular and Cellular Biology*, **13**, 3272–81.

Jani, M.M., Torchia, B.S., Pai, G.S., and Migeon, B.R. (1995). Molecular characterization of tiny ring X chromosomes from females with functional X chromosome disomy and lack of cis X inactivation. *Genomics*, **27**, 182–8.

Jeppesen, P. and Turner, B.M. (1993). The inactive X chromosome in female mammals is distinguished by a lack of histone H4 acetylation, a cytogenetic marker for gene expression. *Cell*, **74**, 281–9.

Joel, P., Shao, W., and Pratt, K. (1993). A nuclear protein with enhanced binding to methylated Sp1 sites in the AIDS virus promoter. *Nucleic Acids Research*, **21**, 5786–93.

Jones, P.A., Wolkowicz, M.J., Rideout, W.M., Gonzales, F.A., Marziasz, C.M., Coetzee, G.A., *et al.* (1990). *De novo* methylation of the MyoD1 CpG island during the establishment of immortal cell lines. *Proceedings of the National Academy of Sciences of the USA*, **87**, 6117–21.

Jost, J.P. (1993). Nuclear extracts of chicken embryos promote an active demethylation of DNA by excision repair of 5-methyldeoxycytidine. *Proceedings of the National Academy of Sciences of the USA*, **90**, 4684–8.

Jost, J.P. and Hofsteenge, J. (1992). The repressor MDBP-2 is a member of the histone H1 family that binds preferentially *in vitro* and *in vivo* to methylated nonspecific DNA sequences. *Proceedings of the National Academy of Sciences of the USA*, **89**, 9499–503.

Jost, J.P. and Jost, Y.C. (1994). Transient DNA demethylation in differentiating mouse myoblasts correlates with higher activity of 5-methyldeoxycytidine excision repair. *Journal of Biological Chemistry*, **269**, 10040–3.

Jost, J.P. and Saluz, H.P. (ed.) (1993). *DNA methylation: molecular biology and biological significance*. Birkhauser Verlag, Boston.

Kafri, T., Gao, X., and Razin, A. (1993). Mechanistic aspects of genome-wide demethylation in the preimplantation mouse embryo. *Proceedings of the National Academy of Sciences of the USA*, **90**, 10558–62.

Kanda, N. (1973). A new differential technique for staining the heteropycnotic X chromosome is female mice. *Experimental Cell Research*, **80**, 463–7.

Kaslow, D.C. and Migeon, B.R. (1987). DNA methylation stabilizes X-chromosome inactivation in eutherians but not in marsupials: evidence for multistep maintenance of mammalian X dosage compensation. *Proceedings of the National Academy of Sciences of the USA*, **84**, 6210–14.

Kay, G.F., Penny, G.D., Patel, D., Ashworth, A., Brockdorff, N., and Rastan, S. (1993). Expression of Xist during mouse development suggests a role in the initiation of X chromosome inactivation. *Cell*, **72**, 171–82.

Kay, G.F., Barton, S.C., Surani, M.A., and Rastan, S. (1994). Imprinting and X chromosome counting mechanisms determine Xist expression in early mouse development. *Cell*, **77**, 639–50.

Kerem, B.S., Goitein, R., Richler, C., Marcus, M., and Cedar, H. (1983). *In situ* nick-translation distinguishes between active and inactive X chromosomes. *Nature*, **304**, 88–90.

Kornberg, R.D. and Lorch, Y. (1995). Interplay between chromatin structure and transcription. *Current Opinion in Cell Biology*, **7**, 371–5.

Kuo, M.T., Iyer, B., and Schwarz, R.J. (1982). Condensation of chromatin into chromosomes preserves an open configuration but alters the DNase I hypersensitive cleavage sites of the transcribed gene. *Nucleic Acids Research*, **10**, 4565–79.

Laayoun, A. and Smith, S.S. (1995). Methylation of slipped duplexes, snapbacks and cruciforms by human DNA(cytosine-5)methyltransferase. *Nucleic Acids Research*, **23**, 1584–9.

LeBon, J.M., Tam, P.P.L., Singer-Sam, J., Riggs, A.D., and Tan, S.S. (1995). Mouse endogenous X-linked genes do not show lineage-specific delayed inactivation during development. *Genetical Research*, **65**, 223–7.

Leuba, S.H., Yang, G., Robert, C., Samori, B., Van Holde, K., Zlatanova, J., *et al.* (1994). Three-dimensional structure of extended chromatin fibers as revealed by tapping-mode scanning force microscopy. *Proceedings of the National Academy of Sciences of the USA*, **91**, 11621–5.

Levine, A., Yeivin, A., Ben-Asher, E., Aloni, Y., and Razin, A. (1993). Histone H1-mediated inhibition of transcription initiation of methylated templates *in vitro*. *Journal of Biological Chemistry*, **268**, 21754–9.

Li, E.B.C. and Jaenisch, R. (1993). Role of DNA methylation in genomic imprinting. *Nature*, **366**, 362–5.

Lichtenstein, M., Keini, G., Cedar, H., and Bergman, Y. (1994). B cell-specific demethylation: a novel role for the intronic kappa chain enhancer sequence. *Cell*, **76**, 913–23.

Lin, D. and Chinault, A.C. (1988). Comparative study of DNase I sensitivity at the X-linked human HPRT locus. *Somatic Cell and Molecular Genetics*, **14**, 261–72.

Lock, L.F., Takagi, N., and Martin, G.R. (1987). Methylation of the Hprt gene on the inactive X occurs after chromosome inactivation. *Cell*, **48**, 39–46.

Luo, S., Torchia, B.S., and Migeon, B.R. (1995). XIST expression is repressed when X inactivation is reversed in human placental cells: a model for study of XIST regulation. *Somatic Cell and Molecular Genetics*, **21**, 51–60.

Lyon, M.F. (1961). Gene action in the X-chromosome of the mouse. *Nature*, **190**, 372–3.

Lyon, M.F. (1993). Epigenetic inheritance in mammals. *Trends in Genetics*, **9**, 123–8.

Macleod, D., Charlton, J., Mullins, J., and Bird, A.P. (1994). Sp1 sites in the mouse aprt gene promoter are required to prevent methylation of the CpG island. *Genes and Development*, **8**, 2282–92.

Manuelidis, L. (1990). A view of interphase chromosomes. *Science*, **250**, 1533–40.

Martínez-Balbás, M.A., Dey, A., Rabindran, S.K., Ozato, K., and Wu, C. (1995). Displacement of sequence-specific transcription factors from mitotic chromatin. *Cell*, **83**, 29–38.

Merlo, A., Herman, J.G., Mao, L., Lee, D.J., Gabrielson, E., Burger, P.C., *et al.* (1995). 5' CpG island methylation is associated with transcriptional silencing of the tumour suppressor p16/CDKN2/MTS1 in human cancers. *Nature Medicine*, **1**, 686–92.

Mielke, C., Kohwi, Y., Kohwi, S.T., and Bode, J. (1990). Hierarchical binding of DNA fragments derived from scaffold-attached regions: correlation of properties *in vitro* and function *in vivo*. *Biochemistry*, **29**, 7475–85.

Migeon, B.R., Schmidt, M., Axelman, J., and Cullen, C.R. (1986). Complete reactivation of X chromosomes from human chorionic vili with a switch to early DNA replication. *Proceedings of the National Academy of Sciences of the USA*, **83**, 2182–6.

Migeon, B.R., Axelman, J., and Beggs, A.H. (1988). Effect of ageing on reactivation of the human X-linked HPRT locus. *Nature*, **335**, 93–6.

Migeon, B.R., de Beur, S.J., and Axelman, J. (1989). Frequent derepression of G6PD and HPRT on the marsupial inactive X chromosome associated with cell proliferation *in vitro*. *Experimental Cell Research*, **182**, 597–609.

Migeon, B.R., Luo, S., Jani, M., and Jeppesen, P. (1994). The severe phenotype of females with tiny ring X chromosomes is associated with inability of these chromosomes to undergo X inactivation. *American Journal of Human Genetics*, **55**, 497–504.

Mueller, P.R. and Wold, B. (1989). *In vivo* footprinting of a muscle specific enhancer by ligation mediated PCR. *Science*, **246**, 780–6.

Norris, D.P., Brockdorff, N., and Rastan, S. (1991). Methylation status of CpG-rich islands on active and inactive mouse X cromosomes. *Mammalian Genome*, **1**, 78–83.

Norris, D.P., Patel, D., Kay, G.F., Penny, G.D., Brockdorff, N., Sheardown, S.A., *et al.* (1994). Evidence that random and imprinted Xist expression is controlled by pre-emptive methylation. *Cell*, **77**, 41–51.

O'Neill, L.P. and Turner, B.M. (1995). Histone H4 acetylation distinguishes coding regions of the human genome from heterochromatin in a differentiation-dependent but transcription-independent manner. *EMBO Journal*, **14**, 3946–57.

Paro, R. (1990). Imprinting a determined state into the chromatin of *Drosophila*. *Trends in Genetics*, **6**, 416–21.

Paro, R. (1993). Mechanisms of heritable gene repression during development of *Drosophila*. *Current Opinion in Cell Biology*, **5**, 999–1005.

Paroush, Z., Keshet, I., Yisraeli, J., and Cedar, H. (1990). Dynamics of demethylation and activation of the alpha-actin gene in myoblasts. *Cell*, **63**, 1299–37.

Pfeifer, G.P. and Riggs, A.D. (1991). Chromatin differences between active and inactive X chromosomes revealed by genomic footprinting of permeabilized cells using DNase I and ligation-mediated PCR. *Genes and Development*, **5**, 1102–13.

Pfeifer, G.P., Steigerwald, S.D., Mueller, P.R., Wold, B., and Riggs, A.D. (1989). Genomic sequencing and methylation analysis by ligation mediated PCR. *Science*, **246**, 810–13.

Pfeifer, G.P., Steigerwald, S.D., Hansen, R.S., Gartler, S.M., and Riggs, A.D. (1990*a*). Polymerase chain reaction-aided genomic sequencing of an X chromosome-linked CpG island: methylation patterns suggest clonal inheritance, CpG site autonomy, and an explanation of activity state stability. *Proceedings of the National Academy of Sciences of the USA*, **87**, 8252–6.

Pfeifer, G.P., Tanguay, R.L., Steigerwald, S.D., and Riggs, A.D. (1990*b*). *In vivo* footprint and methylation analysis by PCR-aided genomic sequencing: comparison of active and inactive X chromosomal DNA at the CpG island and promoter of human PGK-1. *Genes and Development*, **4**, 1277–87.

Pillus, L. and Grunstein, M. (1995). Chromatin structure and epigenetic regulation in yeast. In *Chromatin structure and gene expression* (ed. C.R. Elgin), pp. 123–46. Oxford University Press, New York.

Raff, J.W., Kellum, R., and Alberts, B. (1994). The *Drosophila* GAGA transcription factor is associated with specific regions of heterochromatin throughout the cell cycle. *EMBO Journal*, **13**, 5977–83.

Rastan, S. (1994). X chromosome inactivation and the Xist gene. *Current Opinion in Genetics and Development*, **4**, 292–7.

Razin, A. and Shemer, R. (1995). DNA methylation in early development. *Human Molecular Genetics*, **4**, 1751–5.

Riggs, A.D. (1975). X chromosome inactivation, differentiation and DNA methylation. *Cytogenetics and Cellular Genetics*, **14**, 9–25.

Riggs, A.D. (1989). DNA methylation and cell memory. *Cell Biophysics*, **15**, 1–13.

Riggs, A.D. (1990*a*). DNA methylation and late replication probably aid cell memory, and type I DNA reeling could aid chromosome folding and enhancer function. *Philosophical Transactions of the Royal Society of London*, **326**, 285–297.

Riggs, A.D. (1990*b*). Marsupials and mechanisms of X chromosome inactivation. *Australian Journal of Zoology*, **37**, 419–41.

Riggs, A.D. and Jones, P.A. (1983). Methylcytosine, gene regulation, and cancer. *Advances in Cancer Research*, **40**, 1–30.

Riggs, A.D. and Pfeifer, G.P. (1992). X chromosome inactivation and cell memory. *Trends in Genetics*, **8**, 169–74.

Riley, D.E., Goldman, M.A. and Gartler, S.M. (1986). Chromatin structure of active and inactive human X-linked phosphoglycerate kinase gene. *Somatic Cell and Molecular Genetics*, **12**, 73–80.

Robertson, K.D., Hayward, S.D., Ling, P.D., Samid, D., and Ambinder, R.F. (1995). Transcriptional activation of the Epstein–Barr virus latency C promoter after 5-azacytidine treatment: evidence that demethylation at a single CpG site is crucial. *Molecular and Cellular Biology*, **15**, 6150–9.

Saitoh, Y. and Laemmli, U.K. (1994). Metaphase chromosome structure: bands arise from a differential folding path of the highly AT-rich scaffold. *Cell*, **76**, 609–22.

Sasaki, T., Hansen, R.S., and Gartler, S.M. (1992). Hemimethylation and hypersensitivity are early events in transcriptional reactivation of human inactive X-linked genes in a hamster × human somatic cell hybrid. *Molecular and Cellular Biology*, **12**, 3819–26.

Schanz, S. and Steinbach, P. (1989). Investigation of the 'variable spreading' of X inactivation into a translocated autosome. *Human Genetics*, **82**, 244–8.

Schlossher, J., Eggert, H., Paro, R., and Cremer, S. (1994). Gene inactivation in *Drosophila* mediated by the polycomb gene product or by position-effect variegation does not involve major changes in the chromatin fibre. *Molecular and General Genetics*, **243**, 453–62.

Simmler, M.C., Cattanach, B.M., Rasberry, C., Rougeulle, C., and Avner, P. (1993). Mapping the murine Xce locus with (CA)n repeats. *Mammalian Genome*, **4**, 523–30.

Singer-Sam, J. and Riggs, A.D. (1993). X chromosome inactivation and DNA methylation. In *DNA methylation: molecular biology and biological significance* (ed. J. Jost and H. Saluz), pp. 358–84. Birkhauser Verlag, Berlin.

Singer-Sam, J., Grant, M., LeBon, J.M., Okuyama, K., Chapman, V., Monk, M., *et al.* (1990). Use of a HpaII-polymerase chain reaction assay to study DNA methylation in the Pgk-1 CpG island of mouse embryos at the time of X-chromosome inactivation. *Molecular and Cellular Biology*, **10**, 4987–9.

Singer-Sam, J., Chapman, V., LeBon, J. M., and Riggs, A.D. (1992*a*). Parental imprinting studied by allele-specific primer extension after PCR: paternal X chromosome-linked genes are transcribed prior to preferential paternal X chromosome inactivation. *Proceedings of the National Academy of Sciences of the USA*, **89**, 10469–73.

Singer-Sam, J., Goldstein, L., Dai, A., Gartler, S.M., and Riggs, A.D. (1992*b*). A potentially critical Hpa II site of the X-linked PGK-1 gene is unmethylated prior to the onset of meiosis of human oogenic cells. *Proceedings of the National Academy of Sciences of the USA*, **89**, 1413–17.

Stanworth, S.J., Roberts, N.A., Sharpe, J.A., Sloane-Stanley, J.A., and Wood, W.G. (1995). Established epigenetic modifications determine the expression of developmentally regulated globin genes in somatic cell hybrids. *Molecular and Cellular Biology*, **15**, 3969–78.

Steigerwald, S.D., Pfeifer, G.P., and Riggs, A.D. (1990). Ligation-mediated PCR improves the sensitivity of methylation analysis by restriction enzymes and detection of specific DNA strand breaks. *Nucleic Acids Research*, **18**, 1435–9.

Surani, M. A. (1994). Genomic imprinting: control of gene expression by epigenetic inheritance. *Current Opinion in Cell Biology*, **6**, 390–5.

Takagi, N., Yoshida, M.A., Sugawara, O., and Sasaki, M. (1983). Reversal of X-inactivation in female mouse somatic cells hybridized with murine teratocarcinoma stem cells *in vitro*. *Cell*, **34**, 1053–62.

Tasheva, E.S. and Roufa, D.J. (1994). Densely methylated DNA islands in mammalian chromosomal replication origins. *Molecular and Cellular Biology*, **14**, 5636–44.

Tate, P.H. and Bird, A. (1993). Effects of DNA methylation on DNA-binding proteins and gene expression. *Current Opinion in Genetics and Development*, **3**, 226–31.

Tsai, Y.C., Simoneau, A.R., Spruck, C.H.R., Nichols, P.W., Steven, K., Buckley, J.D., *et al.* (1995). Mosaicism in human epithelium: macroscopic monoclonal patches cover the urothelium. *Journal of Urology*, **153**, 1697–700.

Van Holde, K.E. (1988). *Chromatin*. Springer-Verlag, New York.

Wallrath, L.L. and Elgin, S.C.R. (1995). Position effect variegation in *Drosophila* is associated with an altered chromatin structure. *Genes and Development*, **9**, 1263–77.

Wareham, K.A., Lyon, M.F., Glenister, P.H., and Williams, E.D. (1987). Age related reactivation of an X-linked gene. *Nature*, **327**, 725–7.

Willard, H.F., Brown, C.J., Carrel, L., Hendrich, B., and Miller, A.P. (1993). Epigenetic and chromosomal control of gene expression: molecular and genetic analysis of X chromosome inactivation. *Cold Spring Harbor Symposia on Quantitative Biology*, **58**, 315–22.

Wolffe, A.P. (1992). *Chromatin structure and function*. Academic Press, New York.

Wolffe, A.P. (1994). Inheritance of chromatin states. *Developmental Genetics*, **15**, 463–70.

Wolffe, A.P. and Brown, D.D. (1986). DNA replication *in vitro* erases a *Xenopus* 5S RNA gene transcription complex. *Cell*, **47**, 217–27.

Woodcock, C.L. (1994). Chromatin fibers observed *in situ* in frozen hydrated sections. *Journal of Cell Biology*, **125**, 11–19.

Woodcock, C.L., Grigoryev, S.A., Horowitz, R.A., and Whitaker, N. (1993). A chromatin folding model that incorporates linker variability generates fibers resembling the native structures. *Proceedings of the National Academy of Sciences of the USA*, **90**, 9021–5.

Wormington, J.L., Taylor, I.C.A., and Brown, D.D. (1982). Developmental regulation of *Xenopus* 5S RNA genes. *Cold Spring Harbor Symposia on Quantitative Biology*, **47**, 879–84.

Yang, T.P. and Casky, C.T. (1987). Nuclease sensitivity of the mouse HPRT gene promoter region: differential sensitivity on the active and inactive X chromosome. *Molecular and Cellular Biology*, **7**, 2994–8.

Zacharias, W., Caserta, M., O'Connor, T.R., Larson, J.E., and Wells, R.D. (1988). Cytosine methylation as an effector of right-handed to left-handed DNA structural transitions. *Gene*, **74**, 221–4.

13

Drosophila heterochromatin: retreats for repeats

Amy K. Csink, Georgette L. Sass, and Steven Henikoff

Introduction

The ultimate goal of the Human Genome Project is to obtain the entire sequences of human and several model genomes, which will greatly facilitate numerous lines of genetic research. The gene-rich regions are most amenable to analysis, however a very large fraction of higher eukaryotic genomes consists of enormous blocks of repeats, often simple oligomers reiterated hundreds of thousands of times. Such a level of repetitiveness must be dismaying to even the most enthusiastic participants in the Project: highly repetitive DNA is difficult to clone and sequence, and non-informational sequence is of questionable relevance to mainstream genetic research. Nevertheless, the finding that many human diseases are associated with simple sequence repeat expansions (Bates and Lehrach 1994) has excited renewed interest in repetitive sequence blocks, which have long been known to lie in heterochromatin.

Heterochromatin was originally defined visually as the heavily staining region seen in both the mitotic and interphase nucleus (Heitz 1928). The contrast between the openness of euchromatin and the compactness of the heterochromatin implied that the euchromatin is the location of most gene functions and led to the neglect of heterochromatin. However, it has become increasingly clear that the nuclear genome is more than a collection of autonomous DNA sequences and that a deeper knowledge of the structure of the nucleus and chromatin, including the heterochromatin, will contribute to our understanding of gene expression.

In *Drosophila melanogaster* heterochromatin accounts for approximately 30 per cent of the genome. Its properties have been studied since the advent of *Drosophila*'s use as an experimental organism. Heterochromatin in this species is found surrounding the centromeres of the X chromosome and the two large autosomes. Additionally, the entirety of the Y chromosome is heterochromatic, as well as much of the small fourth chromosome. In diploid metaphase chromosome preparations heterochromatin can be mapped using differential staining patterns (Gatti *et al.* 1976; Dimitri 1991), analogous to the classic, distinctive banding patterns of the euchromatic arms in polytene chromosomes. The chromosomes in most larval tissues in dipterans undergo endoreplication, leading to polytene chromosomes that consist of more than 1000 copies of a chromatid in a rope-like strand. However, heterochromatin is exceptional in that it is mostly underrepresented in polytene nuclei. Furthermore, the hetero-chromatic regions of polytene nuclei associate into a single chromocentre, so that when squashed, the euchromatic arms of all the chromosomes appear to radiate out from a single point.

A phenomenon in *Drosophila* called position effect variegation (PEV) has been intensively studied with the anticipation of gaining insight into the proteins that interact with heterochromatin and are responsible for its distinctive proper-ties. Classical PEV is observed when a euchromatic gene is placed next to heterochromatin by a chromosomal rearrangement. The aberrant proximity of heterochromatin leads to the silencing of the gene. This silencing is thought to be an all-or-nothing event, so that one sees patches of tissue where the gene is expressed at its normal levels interspersed with tissue where it is totally off, hence variegation. A classic example is *Inversion(1)white-mottled-4* (w^{m4}), an allele of the gene that allows the deposition of eye pigments, whose absence leads to a white eye. In the case of w^{m4}, the *white* gene is brought within 20 kb of heterochromatin, resulting in varying sized patches of red and white (Tartof *et al.* 1984). Many second-site modifier genes that either enhance (make more white patches in the eye) or suppress (make more red patches) the w^{m4} phenotype have been isolated and a few have been sequenced in the hope that these would encode constituents of heterochromatin. In general, modifiers of PEV isolated using w^{m4} similarly affect other chromosomal rearrangements that cause variegation of *white* or other loci (Spofford 1976).

In this overview we will describe the types of sequences and structures of *Drosophila* heterochromatin and point out the connection between repetitiveness and heterochromatic behaviour. We will then discuss the relationship between the structure of heterochromatin and its properties and functions. Finally, we will comment on the nature of the proteins that have been implicated in the control of chromatin structure based on their modification of PEV. We hope that our discussion of these subjects will motivate the reader to further interest in this field. More comprehensive treatments of heterochromatin (John 1988; Gatti and Pimpinelli 1992) and PEV (Spofford 1976; Weiler and Wakimoto 1995) can be found in other reviews.

Components of *Drosophila* heterochromatin

Satellite sequences

Like other eukaryotic organisms, *Drosophila* contains a significant fraction (about 20 per cent in *D. melanogaster*) of simple, highly repetitive DNA sequences that can be isolated as 'satellites' of the main band by equilibrium centrifugation. When these sequences are cloned and used as *in situ* probes to diploid metaphase chromosome spreads, they hybridize to specific bands in the heterochromatin (Lohe *et al.* 1993). The sequences of the satellites give rise to the classic cytogenetic pattern of bright fluorescent bands (AT satellites) alternating with N-bands (GA satellites). Some satellite sequences hybridize to a specific subset of chromosomes, while others are found on all chromosomes. In addition to the very simple 5–12 bp repeat units, there are larger units, such as the 359 bp repeats found exclusively on the X chromosome. Simple satellite sequences are present in the heterochromatin in blocks of 50–900 kb (Le *et al.* 1995). Indeed, many of the blocks are interrupted by islands of unique or moderately repetitive sequence DNA. Interestingly, these satellite sequences can differ greatly in their abundance and chromosomal locations even among sibling species of *Drosophila* (Lohe and Brutlag 1987; Carmena *et al.* 1993).

Middle repetitive sequences

About 10 per cent of the *Drosophila* genome consists of middle repetitive elements, probably of transposable element origin. Insertions of transposable elements are scattered throughout the euchromatic arms at sites that vary greatly between populations. Additionally, many transposable element families are concentrated in heterochromatin, but the elements in the heterochromatin differ from their euchromatic siblings in a number of ways. Heterochromatic transposable elements are often truncated and/or degenerate and incapable of further transposition. Also, many heterochromatic transposable element sequences are tandemly repeated, or at least clustered in specific regions of the heterochromatin, and the location of these clusters is conserved among different populations of *D. melanogaster* (Carmena and Gonzalez 1994; Pimpinelli *et al.* 1995). It was previously thought that the majority of the heterochromatic transposable elements are near the euchromatin–heterochromatin boundary, a region referred to as β-heterochromatin. This misimpression was probably due to the reliance on polytene chromosomes for *in situ* localization and the cloning of β-heterochromatic genes that were riddled with middle repetitive DNA (see below). However, examination of diploid mitotic chromosomes (Pimpinelli *et al.* 1995) and dissection of the molecular structure of centric and Y heterochromatin (Le *et al.* 1995; Zhang and Spradling 1995) have revealed patches of middle repetitive DNA deep in heterochromatin.

The presence of high concentrations of defective transposable elements in heterochromatin is mysterious. Speculations for the observation have included: (i) an inability to remove the elements because meiotic recombination is absent from heterochromatic regions, and (ii) a relaxation of selection against specific insertions due to the low number of genes in the heterochromatin (Charlesworth and Langley 1989). A third possibility is that heterochromatin is a more likely target for transposons, perhaps because of a sequence preference for insertion. A fourth explanation reflects the location of the heterochromatin at the nuclear periphery during interphase; many transposable elements may originally enter the genome by horizontal transmission (Daniels *et al.* 1990; Maruyama and Hartl 1991) or contain a cytoplasmic stage of their life cycle (i.e. reverse transcription), so that heterochromatin might well be the first DNA encountered by a transposable element entering from the cytoplasm.

Speculations also abound concerning consequences and functions of transposable elements in heterochromatin. The simplest idea is that these elements are trapped, as mentioned above, and perhaps stochastically amplified and rearranged with no impact on the organism. Alternatively, at least some of these elements could be influencing or providing an essential function for the genome. One idea is that the activity of mobile elements may be regulated by the accumulation of amplified bits of themselves, perhaps by titrating out a specific *trans*-acting factor necessary for some part of the transposition cycle of the element (Pimpinelli *et al.* 1995). Another suggestion is that these elements insert into satellite sequences and provide origins of replication within very long stretches of simple sequence repeats that are hypothesized to lack origins (Le *et al.* 1995).

Heterochromatic genes

The heterochromatin of *D. melanogaster* harbours at least 40 conventional genetic loci, some nestled deep in centric heterochromatin (Gatti and Pimpinelli 1992). The heterochromatic genes that have been cloned contain numerous insertions of middle repetitive DNA surrounding the gene and within the introns (Devlin *et al.* 1990; Hackstein and Hochstenbach 1995). Interestingly, these genes can show a heterochromatin-dependent distance effect that can be thought of as an inverse of standard euchromatic PEV (Eberl *et al.* 1993; Howe *et al.* 1995). Rearrangements that move a heterochromatic gene from its normal position to a more distal position in euchromatin have been found to decrease gene expression. Additionally, this repression is often enhanced by mutations that have been shown to suppress classical PEV. Taken together, these observations imply that there is some special feature of heterochromatin that is required for expression, perhaps a specific location in the nucleus (Hilliker 1976), a certain abundance of proteins associated with heterochromatin, or the repetitive nature of the surrounding sequences (Wakimoto and Hearn 1990).

Centromeres

The location of the centromeres within large masses of heterochromatin indicates that heterochromatin may be important in centromere function. The centromere enables proper chromosome disjunction by providing a site for both spindle attachment via the kinetochore and cohesion between sister chromatids. The sequence requirements that define a functional centromere have been determined in both *Saccharomyces cerevisiae* and *Schizosaccharomyces pombe* yeasts

(Clarke *et al.* 1993; Hegemann and Fleig 1993; Baum *et al.* 1994) and more recently in *Drosophila melanogaster* (Le *et al.* 1995; Murphy and Karpen 1995). In the case of a centromere located on a mini-chromosome from *Drosophila*, the sequences needed for minimal centromere function were delimited to a 220 kb region that contains both single copy and middle repetitive DNA. An additional about 200 kb of flanking, highly repetitive DNA was found to be required for efficient transmission of the mini-chromosome. It has been proposed that the essential core is the site of kinetochore formation, and the flanking sequences play a role in stabilizing this structure and/or mediating sister chromatid adhesion (Murphy and Karpen 1995).

Specific sequence requirements have yet to be identified for centromere function in other higher eukaryotes, although it is clear that large regions of repetitive DNA are involved (Tyler-Smith *et al.* 1993). In fact, with the exception of *S. cerevisiae*, all centromeres are associated with repetitive DNA that can be packaged into heterochromatin. The specific sequence composition of repetitive DNA associated with centromeric heterochromatin is variable not only between eukaryotic organisms (Vig 1994), but even between centromeres of the same organism (Clarke *et al.* 1993). The lack of sequence similarity among centromeres leads to the question of how all these diverse centromeres mediate the same process of chromosome segregation. It is of interest to note that repeats of *S. pombe* centromeres have highly phased nucleosomes (Polizzi and Clarke 1991), reminiscent of the ordered nucleosome arrangement typical of heterochromatin in higher eukaryotes (Doshi *et al.* 1991; Wallrath and Elgin 1995). Additionally, *S. pombe* centromeres can exhibit silencing analogous to that caused by heterochromatin in PEV (Allshire *et al.* 1994). These observations hint that the underlying structural similarity among eukaryotic centromeres may be heterochromatin. It has been suggested that centromere function is related to a unique physical property of centromeric heterochromatin, such as DNA curvature rather than sequence composition *per se* (Vig 1994; Willard 1990). Recognition of such a unique structure by centromere-specific proteins could then contribute to centromere function.

In *Drosophila* the requirement for heterochromatin in centromere function is inferred from studies with a mitotically unstable chromosome (Wines and Henikoff 1992). Mitotic instability is thought to result from deletion of pericentric heterochromatin and is observed as lack of sister chromatid adhesion and frequent non-disjunction. Genetic elements such as the Y chromosome that generally affect PEV also modify the mitotic instability of this chromosome.

Arrangement and localization

The common attribute of all the sequences described is the repetitive nature of the sequences themselves or the sequence milieu in which they are found. The prediction that reiteration is the defining property of heterochromatin can be traced to the heyday of *Drosophila* genetics (Ephrussi and Sutton 1944). This was later confirmed with the discovery that highly repetitive sequences are associated with heterochromatic structures (Mayfield and Ellison 1975). Attention was refocused on heterochromatin repetitiveness with the observation that tandem repeats of a transgene in *Drosophila* acquired increasingly heterochromatic properties with increasing copy number. When flies were transformed with a *white* transgene and the copy number of that gene was increased at the same location, flies showed patchy inactivation of the transgene, much like classical PEV. Indeed, this effect could be suppressed and enhanced by genic modifiers of PEV, despite the fact that there was no rearrangement of the chromosome and the transgene arrays were well within the euchromatin. This led to the conclusion that heterochromatic DNA silencing could be produced *de novo* by simply repeating a sequence (Dorer and Henikoff 1994).

Heterochromatin is thought to be arrayed around the periphery in the interphase nucleus. Electron micrographs of cross-sections of interphase nuclei often show a concentration of more densely staining material on the inner edge of the nuclear envelope. However, there seems to be very little known about the specific sequence contents of the DNA at the nuclear periphery. Indeed, the location of the heterochromatin seems to vary with cell type and cell cycle. In a *Drosophila* species where 75 per cent of the satellite sequences consist of a single repeat, it has been shown that instead of coating the interior of the nuclear envelope, the heterochromatin was found in two chromocentres that associate with the nuclear envelope, the nucleolus, and with each other (Mayfield and Ellison 1975). One chromocentre seemed to contain the predominant satellite class, while the other type consisted of the rest of the minor satellite classes. There was no densely stained chromatin on the remainder of the periphery of the nucleus. This extreme case may illustrate that there are two forces controlling the localization of heterochromatin: (i) the tendency of repetitive DNA to self-associate and (ii) the affinity of heterochromatin for the periphery of the nucleus, perhaps through an interaction with a component of the nuclear envelope. Alternatively, the position of the individual chromocentres at the end of telophase might be maintained as the nucleus cycles into interphase, and so the presence of heterochromatin on the periphery would be merely a historical relic (Lifschytz and Hareven 1982). In *D. melanogaster*, where no single species of satellite predominates, there would be less tendency to associate into distinct chromocentres, leading to a more even distribution of satellites along the periphery of the nucleus.

Structure

Other chapters in this book deal with the specifics of nucleosomes and chromatin structure and their relationship to gene expression (see Chapters 1–4). An example in *Drosophila* in which heterochromatin has been associated with altered chromatin structure was seen for a set of variegating transgenes inserted into various heterochromatic sites (Wallrath and Elgin 1995). Nuclear preparations from *Drosophila* lines containing these heterochromatic transgene insertions showed that the transgene DNA was less accessible to restriction digestion and showed a more ordered nucleosome structure than the same transgene inserted into a non-variegating, euchromatic site.

Repetitive sequences may have special structural properties. The likely self-association of repetitive sequences in heterochromatin might involve pairing of complementary sequences from adjacent repeats. This pairing could take the form of intrastrand hairpins or pairing between similar nearby sequences, or perhaps even unlinked sequences in the same region of the nucleus. In fact, electron microscopy of triplet repeat sequences has revealed that such hairpins do indeed spontaneously form (Hopkin 1995). The idea that transgene arrays can form paired structures is supported by the observation that more extreme heterochromatic silencing occurs upon reversal of a single repeat unit within a direct repeat array of transgenes (Dorer and Henikoff 1994). The resulting inverted repeats should pair more easily than direct repeats, because they would fold into a hairpin (Gubb *et al.* 1990). Such promiscuously paired sequences might be recognized by specific proteins that would tag them as potential heterochromatin and promote their interaction with other similar sequences (Dorer and Henikoff 1994).

Some heterochromatic sequences from *Drosophila* seem to be selectively resistant to transfer from agarose gels during Southern blotting, suggesting that there is some structural property of heterochromatin that is retained during this procedure (Glaser and Spradling 1994). This observation could reflect either DNA modification, although base modifications are not detectable in *Drosophila* (Urieli-Shoval *et al.* 1982), or a very tightly associated protein (Pfutz *et al.* 1992). Alternatively, DNA secondary structure, such as hairpin association between repeats, might have been maintained during the procedure.

Heterochromatic proteins

Modifiers of position effect variegation

One motivation for the intensive study of PEV is the hope that PEV modifier loci encode proteins directly involved in chromatin structure. It is supposed that *Suppressors-of-PEV* (*Su(var)*s) are mutations in genes encoding structural components of heterochromatin, whereas *Enhancers-of-PEV* (*E(var)*s) are mutations in genes encoding components of euchromatin. Many mutagenic screens for dominant modifiers of PEV have been performed and at least 50 such loci have been identified (for reviews see Wustmann *et al.* 1989; Sinclair *et al.* 1992). Some have been cloned and sequenced. Contrary to expectations not all modifiers appear to be structural components of chromatin. Although *Su(var)* loci that encode heterochromatin structural proteins have been identified, e.g. histones and heterochromatin protein 1 (HP1), molecular analyses of other modifiers of PEV have revealed a variety of proteins, including a member of the

DNA replication complex, a subunit of protein phosphatase I, and several proteins with intriguing motifs (for review see Weiler and Wakimoto 1995). It is important to keep in mind that there may be many genes with a pleiotropic effect of modifying PEV, yet without direct structural effects on heterochromatin (Michailidis *et al.* 1988). Indeed, the number of actual structural components of PEV may well be a small subset of the more than 50 loci that have been shown to modify PEV. In light of this it is relevant to point out that diverse cell cycle mutants in *S. cerevisiae* have been shown to modify yeast mating-type silencing (Laman *et al.* 1995), a phenomenon analogous to PEV in *Drosophila*.

Heterochromatin protein 1

In polytene chromosomes HP1 localizes primarily to the chromocentre. The protein contains a region referred to as the chromodomain that has been identified in other regulatory genes thought to exert gene silencing effects by modifying chromatin structure (Paro and Hogness 1991; Saunders *et al.* 1993; see Chapter 14). The chromodomain mediates protein–protein interactions. In the case of HP1 it is responsible for targeting to the heterochromatin (Platero *et al.* 1995). The protein or proteins that the HP1 chromodomain interacts with are unknown. HP1 was originally identified by screening monoclonal antibodies to hetero-chromatin-specific proteins from early *Drosophila* embryos (James and Elgin 1986). Despite cloning and characterization of many other *Su(var)*s, HP1 remains the only clear-cut example of a heterochromatin-specific protein. Perhaps this tells us that few of the *Su(var)*s have direct effects on chromatin structure.

HP1 is thought to be involved in proper chromosome segregation. The embryonic lethal phenotype of flies deficient in HP1 is due to defective chromosome condensation that may in turn affect chromosome segregation (Kellum and Alberts 1995). It is probable that the function of the centromere is severely compromised in the absence of heterochromatin condensation. Additionally, the general failure of heterochromatin condensation could have an indirect effect on mitosis.

GAGA protein

GAGA protein is involved in redistribution of nucleosomes in actively tran-scribed regions. It binds at the sequence GAGA in promoter regions of genes and results in a more open nucleosome structure (Lu *et al.* 1993). The loss of function of this gene has been shown to act as an *E(var)*, leading to the proposal that reduced levels of GAGA protein result in chromatin condensation (Farkas *et al.* 1994). GAGA protein also prominently binds heterochromatin in mitotic cytological preparations (Raff *et al.* 1994). It is rather surprising that a protein shown to be responsible for opening up chromatin should also be a constituent of the highly condensed heterochromatin. This localization of GAGA protein likely results from the high abundance of GA-rich satellites in *D. melanogaster.* The correspondence of a binding site for a transcription factor and the repeat unit of a satellite could be a coincidence and similar coincidences may be common in heterochromatin. For example, topoisomerase II preferentially binds to its consensus site found in the 359 bp satellite (Käs and Laemmli 1992; Swedlow *et al.* 1993). Perhaps each of the various DNA binding proteins increases the stability of its bound repeats, thereby fostering their accumulation in the genome.

Future directions

The study of PEV has provided a useful genetic assay for heterochromatin. To date much of the research has been directed towards the identification and characterization of *trans*-acting modifiers of PEV, which are of uncertain relevance to heterochromatin. Therefore, a more direct approach is warranted if the molecular mechanisms underlying heterochromatin function are to be understood. Such an approach may involve biochemical isolation of hetero-chromatin components. For instance, the *Drosophila* homologue of a human centromere binding protein, CENP-B, was recently isolated in this way (Avides and Sunkel 1994). Also, *in situ* assays can identify macromolecular interactions in heterochromatin (Platero *et al.* 1995).

An increasingly comprehensive description of heterochromatin has emerged from the characterization of its structure and organization in *Drosophila*; this in spite of the unpopularity of heterochromatin in genome sequencing circles. In the case of heterochromatin analysis other strategies are more appropriate than complete sequence determination. For example, the localization of specific heterochromatic components by fluorescent *in situ* techniques promises to elucidate the relationship between heterochromatic compartmentalization and gene expression (Csink and Henikoff 1996).

Summary

In *Drosophila* heterochromatin is found mostly in the pericentric regions, where there are few genetic loci. The DNA sequences of heterochromatin consist mainly of simple satellite sequence repeats alternating with moderately repetitive elements. The repetitive nature of these sequences appears to confer hetero-chromatic properties. We discuss possible structural differences between euchro-matin and heterochromatin, including altered nucleosome structure and secondary structures caused by promiscuous pairing of repeats. The juxtaposition of heterochromatin and euchromatic genes can silence these genes. This implies that heterochromatic properties can be conferred on other regions of chromatin.

Modifiers of heterochromatin silencing are discussed in light of their possible role as structural components of chromatin.

References

Allshire, R.C., Javerzat, J.P., Redhead, N.J., and Cranston, G. (1994). Position effect variegation at fission yeast centromeres. *Cell*, **76**, 157–69.

Avides, M.C. and Sunkel, C.E. (1994). Isolation of chromosome-associated proteins from *Drosophila melanogaster* that bind a human centromeric DNA sequence. *Journal of Cell Biology*, **127**, 1159–71.

Bates, G. and Lehrach, H. (1994). Trinucleotide repeat expansions and human genetic disease. *BioEssays*, **16**, 277–284.

Baum, M., Ngan, V.K., and Clarke, L. (1994). The centromeric K-type repeat and the central core are together sufficient to establish a functional *Schizosaccharomyces pombe* centromere. *Molecular Biology of the Cell*, **5**, 747–61.

Carmena, M. and Gonzalez, C. (1994). Transposable elements map in a conserved pattern of distribution extending from beta-heterochromatin to centromeres in *Drosophila melanogaster*. *Chromosoma*, **103**, 676–84.

Carmena, M., Abad, J.P., Villasante, A. and Gonzalez, C. (1993). The *Drosophila melanogaster* dodecasatellite sequence is closely linked to the centromere and can form connections between sister chromatids during mitosis. *Journal of Cell Science*, **105**, 41–50.

Charlesworth, B. and Langley, C.H. (1989). The population genetics of *Drosophila* transposable elements. *Annual Review of Genetics*, **23**, 251–87.

Clarke, L., Baum, M., Marschall, L.G., Ngan, V.K., and Steiner, N.C. (1993). Structure and function of *Schizosaccharomyces pombe* centromeres. *Cold Spring Harbor Symposia on Quantitative Biology*, **58**, 687–95.

Csink, A.K. and Henikoff, S. (1996). Genetic modification of heterochromatic association and nuclear organization in *Drosophila*. *Nature*, **381**, 529–31.

Daniels, S.B., Peterson, K.R., Strausbaugh, L.D., Kidwell, M.G., and Chovnick, A. (1990). Evidence for horizontal transmission of the P transposable element between *Drosophila* species. *Genetics*, **124**, 339–55.

Devlin, R.H., Bingham, B., and Wakimoto, B.T. (1990). The organization and expression of the *light* gene, a heterochromatic gene of *Drosophila melanogaster*. *Genetics*, **125**, 129–40.

Dimitri, P. (1991). Cytogenetic analysis of the second chromosome heterochromatin of *Drosophila melanogaster*. *Genetics*, **127**, 553–64.

Dorer, D.R. and Henikoff, S. (1994). Expansions of transgene repeats cause heterochromatin formation and gene silencing in *Drosophila*. *Cell*, **77**, 993–1002.

Doshi, P., Kaushal, S., Benyajati, C., and Wu, C.I. (1991). Molecular analysis of the responder satellite DNA in *Drosophila melanogaster*: DNA bending, nucleosome structure, and Rsp-binding proteins. *Molecular Biology and Evolution*, **8**, 721–41.

Eberl, D.F., Duyf, B.J., and Hilliker, A.J. (1993). The role of heterochromatin in the expression of a heterochromatic gene, the rolled locus of *Drosophila melanogaster*. *Genetics*, **134**, 277–92.

Ephrussi, B. and Sutton, E. (1944). A reconsideration of the mechanism of position effect. *Proceedings of the National Academy of Sciences of the USA*, **30**, 183–97.

Farkas, G., Gausz, J., Galloni, M., Reuter, G., Gyurkovics, H., and Karch, G. (1994). The *Trithorax-like* gene encodes the *Drosophila* GAGA factor. *Nature*, **371**, 806–8.

Gatti, M. and Pimpinelli, S. (1992). Functional elements in *Drosophila melanogaster* heterochromatin. *Annual Review of Genetics*, **26**, 239–75.

Gatti, M., Pimpinelli, S., and Santini, G. (1976). Characterization of *Drosophila* heterochromatin. I. Staining and decondensation with Hoechst 33258 and quinacrine. *Chromosoma*, **75**, 351–75.

Glaser, R.L. and Spradling, A.C. (1994). Unusual properties of genomic DNA molecules spanning the euchromatic–heterochromatic junction of a *Drosophila* minichromosome. *Nucleic Acids Research*, **22**, 5068–75.

Gubb, D., Ashburner, M., Roote, J., and Davis, T. (1990). A novel transvection phenomenon affecting the *white* gene of *Drosophila melanogaster*. *Genetics*, **126**, 167–76.

Hackstein, J.H.P. and Hochstenbach, R. (1995). The elusive fertility genes of *Drosophila*: the ultimate haven for selfish genetic elements. *Trends in Genetics*, **11**, 195–200.

Hegemann, J.H. and Fleig, U.N. (1993). The centromere of budding yeast. *Bioessays*, **15**, 451–60.

Heitz, E. (1928). Das Heterochromatin der Moose. *Jahrbücher für Wissenschaftliche Botanik*, **69**, 762–818.

Hilliker, A.J. (1976). Genetic analysis of the centromeric heterochromatin of chromosome 2 of *Drosophila melanogaster*: deficiency mapping of EMS-induced lethal complementation groups. *Genetics*, **83**, 765–82.

Hopkin, K. (1995). Hairpins and heterochromatin: how triplet repeats may lead to disease. *Journal of NIH Research*, **7**, 45–8.

Howe, M., Dimitri, P., Berloco, M., and Wakimoto, B.T. (1995). Cis-effects of heterochromatin on heterochromatic and euchromatic gene activity in *Drosophila melanogaster*. *Genetics*, **140**, 1033–45.

James, T.C. and Elgin, S.C.R. (1986). Identification of a nonhistone chromosomal protein associated with heterochromatin in *Drosophila melanogaster* and its gene. *Molecular and Cellular Biology*, **6**, 3862–72.

John, B. (1988). The biology of heterochromatin. In *Heterochromatin: molecular and structural aspects* (ed. R.S. Verma), pp. 1–147. Cambridge University Press, Cambridge.

Käs, E. and Laemmli, U.K. (1992). *In vivo* topoisomerase II cleavage of the *Drosophila* histone and satelite repeats: DNA sequence and structural characteristics. *EMBO Journal*, **11**, 705–15.

Kellum, R. and Alberts, B.M. (1995). Heterochromatin protein 1 is required for correct chromosome segregation in *Drosophila* embryos. *Journal of Cell Science*, **108**, 1419–31.

Laman, H., Balderes, D., and Shore, D. (1995). Disturbance of normal cell cycle progression enhances the establishment of transcriptional silencing in *Saccharomyces cerevisiae*. *Molecular and Cellular Biology*, **15**, 3608–17.

Le, M.-H., Duricka, D., and Karpen, G.H. (1995). Islands of complex DNA are widespread in *Drosophila* centric heterochromatin. *Genetics*, **141**, 283–303.

Lifschytz, E. and Hareven, D. (1982). Heterochromatin markers: arrangement of obligatory heterochromatin, histone genes and multisite gene families in the interphase nucleus of *Drosophila melanogaster*. *Chromosoma*, **86**, 443–55.

Lohe, A.R. and Brutlag, D.L. (1987). Identical satellite DNA sequences in sibling species of *Drosophila*. *Journal of Molecular Biology*, **194**, 161–70.

Lohe, A.R., Hilliker, A.J., and Roberts, P.A. (1993). Mapping simple repeated DNA sequences in heterochromatin of *Drosophila melanogaster*. *Genetics*, **134**, 1149–74.

Lu, Q., Wallrath, L.L., Granok, H., and Elgin, S.C.R. (1993). (CT)-n (GA)-n repeats and heat shock elements have distinct roles in chromatin structure and transcriptional activation of the *Drosophila hsp26* gene. *Molecular and Cellular Biology*, **13**, 2802–14.

Maruyama, K. and Hartl, D.L. (1991). Evidence for interspecific transfer of the transposable element mariner between *Drosophila* and *Zaprionus*. *Journal of Molecular Evolution*, **33**, 514–24.

Mayfield, J.E. and Ellison, J.R. (1975). The organization of interphase chromatin in *Drosophilidae*. The self adhesion of chromatin containing the same DNA sequences. *Chromosoma*, **52**, 37–48.

Michailidis, J., Murray, N.D., and Graves, J.A.M. (1988). A correlation between development time and variegated position effect in *Drosophila melanogaster*. *Genetical Research*, **52**, 119–23.

Murphy, T.D. and Karpen, G.H. (1995). Localization of centromere function in a *Drosophila* minichromosome. *Cell*, **82**, 599–609.

Paro, R. and Hogness, D.S. (1991). The Polycomb protein shares a homologous domain with a heterochromatin-associated protein of *Drosophila*. *Proceedings of the National Academy of Sciences of the USA*, **88**, 263–7.

Pfutz, M., Gileadi, O., and Werner, D. (1992). Identification of human satellite DNA sequences associated with chemically resistant nonhistone polypeptide adducts. *Chromosoma*, **101**, 609–17.

Pimpinelli, S., Berloco, M., Fanti, L., Dimitri, P., Bonaccorsi, S., Marchetti, E., *et al.* (1995). Transposable elements are stable structural components of *Drosophila melanogaster* heterochromatin. *Proceedings of the National Academy of Sciences of the USA*, **92**, 3804–8.

Platero, J.S., Hartnett, T., and Eissenberg, J.C. (1995). Functional analysis of the chromo domain of HP1. *EMBO Journal*, **14**, 3977–86.

Polizzi, C. and Clarke, L. (1991). The chromatin structure of centromeres from fission yeast: differentiation of the central core that correlates with function. *Journal of Cell Biology*, **112**, 191–201.

Raff, J.W., Kellum, R., and Alberts, B. (1994). The *Drosophila* GAGA transcription factor is associated with specific regions of heterochromatin throughout the cell cycle. *EMBO Journal*, **13**, 5977–83.

Saunders, W.S., Chue, C., Goebl, M., Craig, C., Clark, R.F., Powers, J.A., *et al.* (1993). Molecular cloning of a human homologue of *Drosophila* heterochromatin protein HP1 using anti-centromere autoantibodies with anti-chromo specificity. *Journal of Cell Science*, **104**, 573–82.

Sinclair, D.A., Ruddell, A.A., Brock, J.K., Clegg, N.J., Lloyd, V.K., and Grigliatti, T.A. (1992). A cytogenetic and genetic characterization of a group of closely linked second chromosome mutations that suppress position-effect variegation in *Drosophila melanogaster*. *Genetics*, **130**, 333–44.

Spofford, J.B. (1976). Position-effect variegation in *Drosophila*. In *The genetics and biology of Drosophila* (ed. M. Ashburner and E. Novitiski), pp. 955–1018. Academic Press, New York.

Swedlow, J.R., Sedat, J.W., and Agard, D.A. (1993). Multiple chromosomal populations of topoisomerase II detected *in vivo* by time-lapse, three dimensional wide-field microscopy. *Cell*, **73**, 97–108.

Tartof, K.D., Hobbs, C., and Jones, M. (1984). A structural basis for variegating position effects. *Cell*, **37**, 869–78.

Tyler-Smith, C., Oakey, R.J., Larin, Z., Fisher, R.B., Crocker, M., Affara, N.A., *et al.*

(1993). Localization of DNA sequences required for human centromere function through an analysis of rearranged Y chromosomes. *Nature Genetics*, **5**, 368–75.

Urieli-Shoval, S., Gruenbaum, Y., Sedat, J., and Razin, A. (1982). The absence of detectable methylated bases in *Drosophila melanogaster* DNA. *FEBS Letters*, **146**, 148–52.

Vig, B.K. (1994). Do specific nucleotide bases constitute the centromere? Review. *Mutation Research*, **309**, 1–10.

Wakimoto, B.T. and Hearn, M.G. (1990). The effects of chromosome rearrangements on the expression of heterochromatic genes in chromosome 2L of *Drosophila melanogaster*. *Genetics*, **125**, 141–54.

Wallrath, L.L. and Elgin, S.C. (1995). Position effect variegation in *Drosophila* is associated with an altered chromatin structure. *Genes and Development*, **9**, 1263–77.

Weiler, K.S. and Wakimoto, B.T. (1995). Heterochromatin and gene expression in *Drosophila*. *Annual Review of Genetics* **9**, 577–605.

Willard, H.F. (1990). Centromeres of mammalian chromosomes. *Trends in Genetics*, **6**, 410–16.

Wines, D.R. and Henikoff, S. (1992). Somatic instability of a *Drosophila* chromosome. *Genetics*, **131**, 683–91.

Wustmann, G., Szidonya, J., Taubert, H., and Reuter, G. (1989). The genetics of position-effect variegation modifying loci in *Drosophila melanogaster*. *Molecular and General Genetics*, **217**, 520–7.

Zhang, P. and Spradling, A.C. (1995). The *Drosophila* salivary gland chromocenter contains highly polytenized subdomains of mitotic heterochromatin. *Genetics*, **139**, 659–70.

14

Mechanisms of heritable gene silencing during development of *Drosophila*

Achim Breiling and Renato Paro

Introduction

An unsolved problem in the field of developmental biology is how a determined state can be faithfully maintained over many cell generations. Genetic analysis in *Drosophila* uncovered two classes of genes which play a role in such a process. The *Polycomb* group (PcG) and *trithorax* group (trxG) genes were first identified as transcriptional regulators of homeotic selector (HOM) genes. They maintain the differential expression pattern of these important determining factors by keeping either the inactive (PcG) or the active state (trxG) during development. The two groups are not directly involved in the process of pattern formation. They are responsible for 'freezing' a developmental decision, such that during the differentiation process appropriate structures can be generated based on the committed programmes of the cells. In *Drosophila* it was found that the mechanisms of pattern formation rely heavily on differentially localized molecules (morphogens, region-specific transcription factors, etc.). On the other hand, the process of maintaining a determined state, also called 'cellular memory', appears to rely on molecules that are expressed in every cell of the developing fly. How can such a set of homogeneously distributed factors memorize differentially determined states of cells? There is increasing evidence that the proteins encoded by the PcG and trxG genes exert such a regulatory role at the level of chromatin structure. Differential patterns of gene expression are encoded in higher order chromatin structures and are epigenetically transmitted from one cell generation to the next.

Genetics of PcG and trxG genes

Mutations in PcG genes result in ectopic expression of their target genes. This is most dramatically observed with the HOM genes (McKeon and Brock 1991), which control aspects of segmental identity. These HOM genes exhibit finely tuned expression patterns, and an abnormal redistribution at any time of development, as observed in PcG mutants, results in homeotic body transformations. However, ectopic HOM expression is only observed late in embryogenesis. The early pattern, as established by the patterning functions of the maternal and segmentation genes, is not disturbed in PcG mutants. This indicates that the PcG/ trxG system is involved in memorizing programmed expression states. Indeed, if HOM genes are patterned incorrectly, this aberrant state will also be maintained by the PcG/trxG. A hallmark of the PcG is the transformation of the second and third legs into the first leg, which is normally characterized in males by the 'sex combs'. Hence the 'multiple sex combs' phenotype was used as a very sensitive means to identify and group these genes together. In the meantime, however, it has become clear that the PcG not only regulate HOM genes, but also a large set of other genes. Interestingly, all the targets identified so far appear to encode developmentally important regulatory factors. This suggests that this type of transcriptional control is exerted on genes whose products have to be strictly regulated regionally. Indeed, the process of cellular memory is required for different developmental processes (segmentation, oogenesis, etc.) and once set up, is used to maintain the specifically determined state for the rest of development.

Around 15 PcG genes have been genetically described so far (Landecker *et al.* 1994; for a review see Kennison 1995). Based on calculations of genetic material the entire group is expected to consist of 30–40 members. All identified genes show a strong maternal effect, such that a substantial amount of the gene products is deposited in the oocyte during oogenesis and subsequently used for embryogenesis. Most of the genes are haplo-insufficient, indicating that the amount of gene product is crucial for function. In addition, it appears that the PcG proteins interact in some manner. Various combinations of mutant alleles show a synergistic enhancement of the embryonic mutant phenotype (Jürgens 1985; Cheng *et al.* 1994; Campbell *et al.* 1995). Indeed, molecular analyses indicate that several PcG proteins are physically linked in multimeric complexes (see below).

The trxG genes are required for maintaining the active state of homeotic gene expression, as mutations show loss of HOM gene function, which results in anteriorly directed homeotic transformations. Arguably, every factor involved in transcriptional activation could be grouped into the trxG. However, the particular genetic screens used to identify such elements appear to have restricted the candidate genes to a particular level of regulation. Several members were identified in a screen searching for mutations that suppress the phenotype caused by PcG-induced transformations. Their products are expected to antagonize the repressive nature of PcG complexes. Interestingly, several of the molecularly characterized members of the trxG encode factors that were found *in vitro* to open chromatin in an energy-dependent fashion. The molecular nature of the trxG will not be discussed further, as recently several excellent original papers and reviews have been published on this topic (Kennison 1995; Kingston *et al.* 1996; Paro and Harte 1996). This chapter will focus on the regulatory role of the PcG, the molecular nature of its constituents, and discuss the underlying molecular mechanisms needed to stably and heritably maintain inactive transcriptional states.

Molecular nature of PcG proteins

Seven PcG genes have been cloned and characterized so far (see Fig. 14.1). Not surprisingly, all were found to encode nuclear proteins. However, this appears to be the only obvious common denominator. Unlike the HOM proteins, whose common basic function as sequence-specific transcription factors is earmarked by the presence of a homeo domain, the PcG proteins are, with one exception, not characterized by a shared protein motif. This suggest that these proteins have different partner molecules and/or that their site of function might be found at different levels of the silencing process.

Several of the conserved protein motifs found in PcG proteins indicate functions involving protein–protein interactions (see Fig. 14.1). Indeed, there is molecular evidence that at least four PcG proteins interact in large (2.5 MDa) multiprotein complexes (Franke *et al.* 1992; H. Strutt and R. Paro, unpublished results). At the chromosomal level many of these proteins co-localize at a common set of target genes (Lonie *et al.* 1994). In addition, it appears that the stability of the complex is dependent on the appropriate stoichiometry and constitution of the various components. Mutations in the proteins Enhancer of zeste (E(z)) and Polycomb (PC) were also found to dislodge other proteins of the PcG complex from their target sites (Rastelli *et al.* 1993; Franke *et al.* 1995). Identification of the PcG protein sequences has revealed several new as well as previously identified conserved motifs.

SPM domain

The recent cloning of the *Sex comb on midleg* (*Scm*) gene identified a C-terminal sequence encoding 60 amino acids with strong homology to the Polyhomeotic (PH) protein (Bornemann *et al.* 1996). This domain, termed SPM (because of its presence in the three proteins SCM, PH and MBT), is also found in the corresponding murine PH homologue (RAE 28; Nomura *et al.* 1994), as well as in the gene product of a *Drosophila* tumour suppressor gene, *1(3) malignant brain tumour* (*mbt*) (Wismar *et al.* 1995). The SPM domain depicts a substructure reminiscent of helix-loop-helix domains and was suggested to mediate homomeric SPM association. Alternatively, this domain could allow heteromeric association of SCM with PH, particularly in the light of the finding that *Scm* mutations exhibit extragenic non-complementation with *ph* mutations (Cheng *et al.* 1994). In addition, the SCM and MBT proteins contain, respectively, two and three repeat units of about 100 amino acids in length (MBT domain). The function of these conserved sequence motifs is unknown.

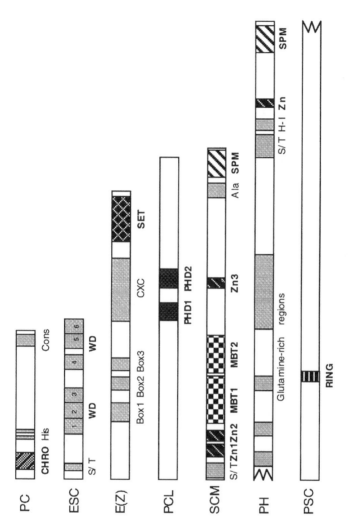

Fig. 14.1 Characteristic protein domains of PcG members. Some relevant domains of the PcG proteins Polycomb (PC, 390aa), Extra sex combs (ESC, 425aa), Enhancer of zeste (E(z), 760aa), Polycomblike (PCL, 857aa), Sex comb on midleg (SCM, 877aa), Polyhomeotic (PH, 1589aa) and Posterior sex combs (PSC, 1603aa). Grey boxes refer to domains not particularly mentioned in the text (mostly containing homopolymeric amino acid repeats). Abbreviations: Ala, alanine-rich region; Box, conserved regions in E(z); CHRO, chromo domain; Cons, conserved C-terminus in PC; CXC, cysteine-rich region with multiple CXC motifs; H-I, homology domain shared by PH and RAE-28; His, histidine repeats; HTH, conserved region in PSC containing a helix-turn-helix motif; MBT, MBT repeats; PHD, PHD finger; RING, RING finger; SET, domain found in E(z), Su(var)3–9 and TRX; SPM, homology domain found in PH and SCM; S/T, serine/threonine-rich regions; WD, WD-40 repeats; Zn, zinc finger.

Zinc fingers

Several PcG proteins carry zinc finger motifs of various types. SCM and PH contain a single zinc finger (Bornemann *et al.* 1996). Posterior sex combs (*PSC*) contains zinc fingers of the C3HC4 RING finger type and Polycomblike (*PCL*) of the C4HC3 PHD type (Brunk *et al.* 1991; van Lohuizen *et al.* 1991; Lonie *et al.* 1994). These motifs are found in a variety of other proteins. The exact molecular function of the domains in the PcG proteins is unknown. It remains to be seen whether the interaction known best for zinc finger motifs, nucleic acid binding, also applies as a function for the various PcG motifs. This would be particularly important to determine, as so far no PcG protein has been reported to exhibit sequence-specific DNA binding.

SET domain

The SET domain is found in both the PcG gene *Enhancer of zeste* (*E(z)*) and a suppressor of position effect variegation (PEV), *Su(var)3–9* (Jones and Gelbart 1993; Tschiersch *et al.* 1994). Interestingly, a SET domain is also present in the Trithorax protein (TRX), an important member of the activating trxG (Paro and Harte 1996). Thus, the SET domain (named from the first letters of Su(var)3–9, Enhancer of zeste and Trithorax) might give these functionally diverse proteins the ability to interact with a common partner. Indeed, there is accumulating evidence that some PcG proteins and the TRX protein are co-localized on chromosomes and bind to a similar set of target sequences, termed PRE (see below).

WD motifs

The Extra sex combs protein (ESC) contains six WD-40 repeats (Gutjahr *et al.* 1995; Sathe and Harte 1995; Simon *et al.* 1995). The WD repeat has this name because the motif usually ends with the sequence W–D (Trp–Asp). Mapped mutations in the WD-40 repeats indicate that this motif is necessary for the ESC repressive function. This motif is found in a variety of other proteins involved in different biological roles in the cell and is suggested to function as a protein–protein or protein–nucleic acid interface. The yeast Tup1p protein is the most intriguing counterpart in this class as it also acts as a transcriptional repressor in cooperation with the α2 homeodomain protein (Komachi *et al.* 1994) and was found to directly interact with histones H3 and H4 (Edmondson *et al.* 1996).

Chromo domain

The Polycomb protein (PC) contains a chromo domain. This conserved 50 amino acid motif was first identified in a comparison between PC and the heterochromatin-associated protein, HP1 (Paro and Hogness 1991). The chromo domain coding sequence was used to isolate related genes in a variety of organisms (Singh *et al.* 1991; Reijnen *et al.* 1995). In addition, recent database searches indicate that this protein motif has been conserved from yeast to humans (Koonin *et al.* 1995). In all the cases tested proteins carrying the chromo domain were found to be nuclear and associated with chromosomes (James and Elgin 1986; Zink and Paro 1989; Saunders *et al.* 1993; Wreggett *et al.* 1994; Ekwall *et al.* 1995), but the exact functional role remains unclear. The chromo domain was however found to be crucial for the specific targeting of proteins to the correct chromosomal location (see below). Interestingly, a chromo domain was also found in the *Drosophila* male-specific lethal protein (MSL-3) involved in X-chromosome dosage compensation in males (Koonin *et al.* 1995). MSL-3 is involved in the hyperactivation of transcription from the male X chromosome rather than in repression. A second, slightly alternate form, termed 'chromo shadow domain', is often found in a subset of chromo domain-containing proteins (Aasland and Stewart 1995).

The chromo domain: a link to chromatin

HP1 is encoded by a suppressor of PEV, *Su(var)205* (Eissenberg *et al.* 1990). Another heterochromatin-regulating gene identified in a PEV screen, *Su(var)3–9*, was also found to encode a chromo domain protein (Tschiersch *et al.* 1994). These genes belong to a class collectively called modifiers of PEV. Members of this group are considered to encode structural and/or regulatory elements of chromatin (see Chapter 13). This similarity suggested an interesting mode of action for the PcG regulation of homeotic genes, supported by additional common features that have been found during the investigation of the repressive action of the PcG and the phenomenon of PEV. It was suggested that PcG proteins could form heterochromatin-like structures to maintain genes in a permanently silenced state. However, what is heterochromatin?

At the light microscopic level heterochromatin can be recognized as a late replicating part of the genome, that stays condensed during interphase and is transcriptionally inert, although some expressed genes embedded in hetero-chromatin regions have been identified. Everything below this level of resolution remains an enigma (see Chapters 9 and 13). As such, the finding of a similarity between PcG regulation and PEV pointed to a possible new level of transcrip-tional regulation. However, little was known of the molecular processes involved. Heterochromatin is considered to consist of tightly packaged chroma-tin. Multiprotein complexes, encoded by the modifiers of PEV, have been invoked to organize a particular higher order structure. However, the lack of appropriate tools prevented a demonstration of the existence of these structures. It is hoped that through the analysis of the genetically identified regulatory components of this particular type of chromatin, understanding of the molecular structures may be gained. In this respect the analyses of HP1 and PC are at the most advanced stage. On polytene chromosomes HP1 shows a tight association with centromeric heterochromatin, telomeric sequences, and with sequences on the mostly heterochromatic chromosome 4 and at some sites on the euchromatic arms (James and Elgin 1986; James *et al.* 1989). The PC protein, on the other hand, binds to over 100 different sites on the euchromatic arms (Paro and Zink 1992). Many of the binding sites coincide with genetically identified target genes of the PcG (like the HOM clusters).

The chromo domain was found to play an important role in the differential targeting of the two proteins. Mutations in the PC chromo domain eliminate the binding of the protein to its target sequences (Messmer *et al.* 1992). Similarly, both the HP1 chromo domain and the C-terminal chromo shadow domain can target HP1 to heterochromatin, as demonstrated with HP1–β-galactosidase fusion proteins (Platero *et al.* 1995). When the HP1 chromo domain is exchanged with the PC chromo domain, the corresponding chimeric protein is targeted to heterchromatin and to the PC binding sites. In addition, the endogenous proteins become attracted to the corresponding ectopic sites. This suggests that these domains are involved, possibly through protein–protein interactions, in generating protein complexes which are characteristic for the respective chromatin domains. The finding that the PC–HP1 chimeric protein can rescue *Su(var)205* mutations additionally suggests that the two chromo domains, albeit having different site specificity, do exert their function in a similar molecular context.

Starting silencing at PcG-responsive elements

None of the PcG proteins analysed so far shows a sequence-specific DNA binding ability *in vitro*, but by observing the distribution of the proteins on polytene chromosomes it is quite obvious that there must be an inherent sequence specificity in the system. Indeed, several DNA elements which attract PcG proteins have been identified. These have been termed PcG response elements (PREs) (Simon *et al.* 1993). The role of PREs can be tested in lines of transgenic flies containing reporter gene constructs. PREs cooperate with region-specific enhancers. For several PREs of the HOM clusters it was shown that they are necessary to maintain the repression of genes in appropriate domains (Busturia and Bienz 1993; Simon *et al.* 1993; Chan *et al.* 1994). In addition, PREs can induce silencing on exogenous promoters (Zink and Paro 1995). Silencing is always accompanied by a recruitment of PcG proteins to the PREs on the construct (see Fig. 14.2). Two possible models have been proposed for the particular chromatin structures imposed in PcG-silenced domains. The apparent tight packaging of heterochromatin could be induced by multiprotein complexes, which organize the chromatin fibre into a regular higher order structure. Alternatively, heterochromatin could be compartmentalized in the nucleus, thus forming particular protein environments which are accessible only to a particular set of regulators.

It is becoming increasingly clear that many PcG proteins interact in multiprotein complexes. Indeed, the formation of these complexes is crucial for the functioning of silencing (Franke *et al.* 1995). However, the subnuclear organiza-

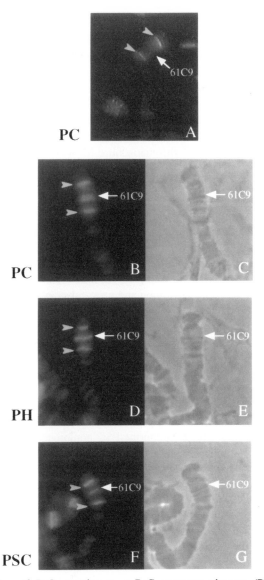

Fig. 14.2 Binding of PcG proteins to a PcG response element (PRE). Polytene chromosomes from the transgenic fly line 24F6 15,5 (Zink and Paro 1995) carrying a transgene on a P-element at the chromosomal position 61C9. This transgenic construct contains a PRE from the bithorax HOM complex. Chromosomes were stained with antibodies against the three PcG proteins PC, PH, and PSC. (a) Immunostaining with PC antibodies to a chromosome carrying no transgene at 61C9. Two endogenous signals are detected at position 61B3 and 61F2–3 (arrowheads), while no PC binding is observed at position 61C9. (b, d, f) Immunostainings at the tip of chromosome 3L of the line 24F6 15,5 with PC, PH, or PSC antibodies, respectively. In all three cases a new signal at 61C9 (marked with an arrow) is due to the binding of the respective PcG proteins to the inserted PRE. (c, e, g). Phase contrast pictures of the chromosome tips stained in (b), (d) and (f). (Photographs were kindly provided by Giacomo Cavalli.)

tion of these complexes, and their role in structuring the chromatin fibre, is still unclear. PcG protein complexes seem to act over extended chromosomal domains (Orlando and Paro 1993). However, they seem not to simply 'coat' or package an entire gene as experiments analysing the accessibility of DNA in such domains have demonstrated (Schlossherr *et al.* 1994; McCall and Bender 1996). PcG proteins appear rather to bind to relevant regulatory DNA elements and package these into a structure which prevents binding of transactivators and thus an unwanted opening of the chromatin fibre for transcriptional activation.

PcG silencing has been conserved during evolution

In recent years various homologues to the *Drosophila* PcG proteins have been identified through directed homology screens or by database searches (Brunk *et al.* 1991; van Lohuizen *et al.* 1991; Pearce *et al.* 1992; Nomura *et al.* 1994; Reijnen *et al.* 1995; Hobert *et al.* 1996). Utilizing the sequences encoding the PC chromo domain, two highly related genes in mouse and *Xenopus* were identified through low stringency hybridization and were subsequently used to demonstrate also a functional relationship. Even though the murine PC protein (M33) is not extremely well conserved in sequence when compared to the fly protein, it was found to be able to rescue the *Polycomb* mutant phenotype to a large extent when expressed in flies (Müller *et al.* 1995). The *Xenopus* PC homologue was found in biochemical assays to interact with itself and with the vertebrate homologue of the PSC protein, the BMI-1 protein (Reijnen *et al.* 1995). This supports the notion that also in vertebrates PcG proteins interact in multiprotein complexes.

The most convincing proof of the functional conservation of the PcG system was given by the analysis of the murine PSC homologues. Knock-out mutations of *bmi-1* reveal multiple posteriorly directed homeotic transformations (Van der Lugt *et al.* 1994). Similarly, the murine gene *mel-18*, a close relative to *bmi-1*, was found to exhibit homeotic transformations in null mutants. These transformations were found to be caused by ectopic expression of the genes of the HOX clusters (Akasaka *et al.* 1996). Interestingly, *mel-18* was found to have tumour suppresser activity in mouse (Kanno *et al.* 1995), while *bmi-1* had originally been identified as a proto-oncogene.

Future directions

There is increasing evidence that the proteins of the PcG and trxG interact in multimeric complexes to exert their respective repressive and activating roles. How are they organized within the nuclear context? In this respect molecular interactions at two levels of resolution need to be analysed.

The first is the detection of a possible functional interaction with the nucleosomal fibre. Biochemical evidence seems to indicate that certain members of the trxG can 'open' the nucleosomal fibre in an energy-dependent manner, probably by counteracting PcG-induced chromatin structures (see Kingston *et al.*

1996; for a review). No such interactions with the basic nucleosomal structures have yet been reported for PcG complexes. Do these proteins organize the nucleosomal fibre into a regular higher order structure as has been suggested for heterochromatin, and what is the extent of such structures? Recent genetic and biochemical studies of silencing complexes in yeast seem to indicate that a eukaryotic cell utilizes such organized chromatin structures to generate stable and heritable gene inactivity (Hecht *et al.* 1995; see Chapter 15). Comparisons with this work may give clues as to the basic organizing features of the PcG silencing complexes.

A second important issue that will need clarification is the possible organization of PcG silencing complexes in the context of the nuclear architecture. Is a subnuclear positioning required for the stable maintenance of the silenced states? Recent results on the *brown* gene (*bw*), subjected to PEV in the *bw*^{D} mutation, suggest that the stable inactivity of the gene is correlated with its positioning close to the centromeric heterochromatic territory within the nucleus (Dernburg *et al.* 1996; Csink and Henikoff 1996). Since both the tools for visualizing the PcG proteins and their target DNAs are available, it should be possible to identify potential functional features connecting silencing to subnuclear positioning (see Chapter 13).

As a major role of the PcG complexes is to maintain inactive transcriptional states throughout many cell divisions, it will be important to identify the molecular events occurring during DNA replication and mitosis. In particular, it will be important to elucidate how protein complexes, expressed homogeneously in all cells, can 'tag' only a subset of genes and maintain their pattern during cell division.

Genetically accessible organisms like yeast and *Drosophila* will be most useful in elucidating the fundamental mechanisms of chromatin silencing. The identification of homologues in vertebrates will furthermore substantiate the basic role this type of regulation plays in eukaryotic organisms. Within vertebrate systems the organization and functional relevance of HOX genes has been quite well established, but their transcriptional control is far less understood. In this respect, the analysis of the corresponding PcG and trxG proteins should certainly lead to some interesting new insights. In addition to these fundamental issues of development, it will also be interesting to identify why in mammalian systems PcG and trxG counterparts are connected to cancer processes. This could indicate that the loss of 'cellular memory' might be an obligatory route through which a cancer cell has to pass. As such, mutations in PcG/trxG components might represent an early event in tumorigenesis.

Summary

In *Drosophila* the genes of the *Polycomb* group and the *trithorax* group are part of the cellular memory system, maintaining the differential expression patterns of genes necessary for defining the determined states of cells. Both groups seem to be acting at the level of chromatin. The repressor PcG appears to induce heterochromatin-like structures on genes that need to be stably and heritably inactivated. The role of the trxG is to counteract these special chromatin domains to render genes accessible to activating transcription factors.

PcG proteins interact in multiprotein complexes which become tethered to their target genes at particular PREs. An important question to be resolved concerns the molecular nature of the higher order chromatin structures involved in this type of regulation.

Heritable gene silencing plays an important but little understood part in pattern formation, as well as in other biological phenomena that range from the inactivation of the mating-type loci in yeast to X chromosome inactivation in mammals. The *Drosophila* PcG system appears to present excellent tools to unravel the molecular mechanisms whereby transcriptional inactivity is maintained by higher order chromatin structures.

Acknowledgements

We would like to thank Helen Strutt and Giacomo Cavalli for a critical review of the manuscript, and the members of the group for many interesting discussions. The work of R.P. is supported by grants from the Deutsche Forschungsgemeinschaft and from the Fond der Chemischen Industrie.

References

Aasland, R. and Stewart, F.A. (1995). The chromo shadow domain, a second chromodomain in heterochromatin-binding protein, HP1. *Nucleic Acids Research*, **23**, 3168–73.

Akasaka, T., Kanno, M., Balling, R., Mieza, M.A., Taniguchi, M., and Koseki, H. (1996). A role for *mel-18*, a Polycomb group related vertebrate gene, during the anteroposterior specification of the axial skeleton. *Development*, **122**, 1513–22.

Bornemann, D., Miller, E., and Simon, J. (1996). The *Drosophila Polycomb* group gene *Sex comb on midleg* (*Scm*) encodes a zinc finger protein with similarity to Polyhomeotic protein. *Development*, **122**, 1621–30.

Brunk, B.P., Martin, E.C., and Adler, P.N. (1991). *Drosophila* genes *Posterior sex combs* and *Suppressor (2) of zeste* encode proteins with homology to the murine *bmi-1* oncogene. *Nature*, **353**, 351–3.

Busturia, A. and Bienz, M. (1993). Silencers in *abdominal-B*, a homeotic *Drosophila* gene. *EMBO Journal*, **12**, 1415–25.

Campbell, R.B., Sinclair, D.A., Couling, M., and Brock, H.W. (1995). Genetic interactions and dosage effects of *Polycomb* group genes of *Drosophila*. *Molecular and General Genetics*, **246**, 291–300.

Chan, C.S., Rastelli, L., and Pirrotta, V. (1994). A *Polycomb* response element in the *Ubx* gene that determines an epigenetically inherited state of repression. *EMBO Journal*, **13**, 2553–64.

Cheng, N.N., Sinclair, D.A., Campbell, R.B., and Brock, H.W. (1994). Interactions of *Polyhomeotic* with *Polycomb* group genes of *Drosophila melanogaster*. *Genetics*, **138**, 1151–62.

Csink, A.K., and Henikoff, S. (1996). Genetic modification of heterochromatin association and nuclear organization in *Drosophila*. *Nature*, **381**, 529–31.

Dernburg, A.F., Broman, K.W., Fung, J.C., Marshall, W.F., Philips, J., Agard, D.A., *et al.* (1996). Perturbation of nuclear architecture by long-distance chromosome interactions. *Cell*, **85**, 745–60.

Edmondson, G.G., Smith, M.M., and Roth, S.Y. (1996). Repression domain of the yeast global repressor Tup1 interacts directly with histones H3 and H4. *Genes and Development*, **10**, 1247–59.

Eissenberg, J.C., James, T.C., Foster-Hartnett, D.M., Hartnett, T., Ngan, V., and Elgin, S. (1990). Mutation in a heterochromatin-specific chromosomal protein is associated with suppression of position-effect variegation in *Drosophila melanogaster*. *Proceedings of the National Academy of Sciences of the USA*, **87**, 9923–7.

Ekwall, K., Javerzat, J.P., Lorentz, A., Schmidt, H., Cranston, G., and Allshire, R. (1995). The chromodomain protein Swi6: a key component at fission yeast centromeres. *Science*, **269**, 1429–31.

Franke, A., DeCamillis, M., Zink, D., Cheng, N., Brock, H.W., and Paro, R. (1992). *Polycomb* and *Polyhomeotic* are constituents of a multimeric protein complex in chromatin of *Drosophila melanogaster*. *EMBO Journal*, **11**, 2941–50.

Franke, A., Messmer, S., and Paro, R. (1995). Mapping functional domains of the Polycomb protein of *Drosophila melanogaster*. *Chromosome Research*, **3**, 351–60.

Gutjahr, T., Frei, E., Spicer, C., Baumgartner, S., White, R.A., and Noll, M. (1995). The *Polycomb*-group gene, *extra sex combs*, encodes a nuclear member of the WD-40 repeat family. *EMBO Journal*, **14**, 4296–306.

Hecht, A., Laroche, T., Bolsinger, S.S., Gasser, S.M., and Grunstein, M. (1995). Histone H3 and H4 N termini interact with the silent information regulators Sir3 and Sir4: a molecular model for the formation of heterochromatin in yeast. *Cell*, **80**, 583–92.

Hobert, O., Sures, I., Ciossek, T., Fuchs, M., and Ullrich, A. (1996). Isolation and developmental expression analysis of *Enx*-1, a novel mouse *Polycomb* group gene. *Mechanisms of Development*, **55**, 171–84.

James, T.C. and Elgin, S.C.R. (1986). Identification of a nonhistone chromosomal protein associated with heterochromatin in *Drosophila melanogaster* and its gene. *Molecular and Cellular Biology*, **6**, 3862–72.

James, T.C., Eissenberg, J.C., Craig, C., Dietrich, V., Hobson, A., and Elgin, S.C.R. (1989). Distribution patterns of HP1, a heterochromatin-associated nonhistone chromosomal protein of *Drosophila*. *European Journal of Cell Biology*, **50**, 170–80.

Jones, R.S. and Gelbart, W.M. (1993). The *Drosophila Polycomb*-group gene *Enhancer of zeste* contains a region with sequence similarity to *Trithorax*. *Molecular and Cellular Biology*, **13**, 6357–66.

Jürgens, G. (1985). A group of genes controlling the spatial expression of the *bithorax* complex in *Drosophila*. *Nature*, **316**, 153–5.

Kanno, M., Hasegawa, M., Ishida, A., Isono, K., and Taniguchi, M. (1995). *mel-18*, a *Polycomb* group-related mammalian gene, encodes a transcriptional negative regulator with tumor suppressive activity. *EMBO Journal*, **14**, 5672–8.

Kennison, J.A. (1995). The *Polycomb* and *trithorax* group proteins of *Drosophila*: trans-regulators of homeotic gene function. *Annual Review of Genetics*, **29**, 289–303.

Kingston, R.E., Bunker, C.A., and Imbalzano, A.N. (1996). Repression and activation by multiprotein complexes that alter chromatin structure. *Genes and Development*, **10**, 905–20.

Komachi, K., Redd, M.J., and Johnson, A.D. (1994). The WD repeats of *Tup1* interact with the homeo domain protein *Alpha* 2. *Genes and Development*, **8**, 2857–67.

Koonin, E.V., Zhou, S.B., and Lucchesi, J.C. (1995). The chromo superfamily: new members, duplication of the chromo domain and possible role in delivering transcription regulators to chromatin. *Nucleic Acids Research*, **23**, 4229–33.

Landecker, H.L., Sinclair, D.A., and Brock, H.W. (1994). Screen for enhancers of *Polycomb* and *Polycomblike* in *Drosophila melanogaster*. *Developmental Genetics*, **15**, 425–34.

Lonie, A., D'Andrea, R., Paro, R., and Saint, R. (1994). Molecular characterisation of the *Polycomblike* gene of *Drosophila melanogaster*, a *trans*-acting negative regulator of homeotic gene expression. *Development*, **120**, 2629–36.

McCall, K. and Bender, W. (1996). Probes for chromatin accessibility in the *Drosophila bithorax* complex respond differently to *Polycomb* mediated repression. *EMBO Journal*, **15**, 569–80.

McKeon, J. and Brock, H.W. (1991). Interactions of the *Polycomb* group genes with homeotic loci of *Drosophila*. *Roux's Archives of Developmental Biology*, **199**, 387–96.

Messmer, S., Franke, A., and Paro, R. (1992). Analysis of the functional role of the *Polycomb* chromo domain in *Drosophila melanogaster*. *Genes and Development*, **6**, 1241–54.

Müller, J., Gaunt, S., and Lawrence, P.A. (1995). Function of the Polycomb protein is conserved in mice and flies. *Development*, **121**, 2847–52.

Nomura, M., Takihara, Y., and Shimada, K. (1994). Isolation and characterization of retinoic acid-inducible cDNA clones in F9 cells: one of the early inducible clones encodes a novel protein sharing several highly homologous regions with a *Drosophila Polyhomeotic* protein. *Differentiation*, **57**, 39–50.

Orlando, V., and Paro, R. (1993). Mapping *Polycomb*-repressed domains in the *bithorax* complex using in vivo formaldehyde cross-linked chromatin. *Cell*, **75**, 1187–98.

Paro, R. and Harte, P.J. (1996). The role of *Polycomb* group and *trithorax* group chromatin complexes in the maintenance of determined cell states. In *Epigenetic mechanisms of gene regulation* (ed. E. Russo, R. Martienssen, and A. Riggs). Cold Spring Harbor Press, Cold Spring Harbor, 507–28.

Paro, R. and Hogness, D. (1991). The *Polycomb* protein shares a homologous domain with a heterochromatin-associated protein of *Drosophila*. *Proceedings of the National Academy of Sciences of the USA*, **88**, 263–7.

Paro, R. and Zink, B. (1992). The *Polycomb* gene is differentially regulated during oogenesis and embryogenesis of *Drosophila melanogaster*. *Mechanisms of Development*, **40**, 37–46.

Pearce, J.J., Singh, P.B., and Gaunt, S. (1992). The mouse has a *Polycomb*-like chromobox gene. *Development*, **114**, 921–9.

Platero, J.S., Hartnett, T., and Eissenberg, J.C. (1995). Functional analysis of the chromo domain of HP1. *EMBO Journal*, **14**, 3977–86.

Rastelli, L., Chan, C.S., and Pirrotta, V. (1993). Related chromosome binding sites for *Zeste*, *Suppressors of zeste* and *Polycomb* group proteins in *Drosophila* and their dependence on *Enhancer of zeste* function. *EMBO Journal*, **12**, 1513–22.

Reijnen, M.J., Hamer, K.M., den Blaauwen, J.L., Lambrechts, C., Schoneveld, I., van Driel, R., et al. (1995). *Polycomb* and *bmi*-1 homologs are expressed in overlapping patterns in *Xenopus* embryos and are able to interact with each other. *Mechanisms of Development*, **53**, 35–46.

Sathe, S.S. and Harte, P.J. (1995). The *Drosophila Extra sex combs* protein contains WD motifs essential for its function as a repressor of homeotic genes. *Mechanisms of Development*, **52**, 77–87.

Saunders, W.S., Chue, C., Goebl, M., Craig, C., Clark, R.F., Powers, J.A. *et al.* (1993). Molecular cloning of a human homologue of *Drosophila* heterochromatin protein HP1 using anti-centromere autoantibodies with anti-chromo specificity. *Journal of Cell Science*, **104**, 573–82.

Schlossherr, J., Eggert, H., Paro, R., Cremer, S., and Jack, R.S. (1994). Gene inactivation in *Drosophila* mediated by the *Polycomb* gene product or by position-effect variegation does not involve major changes in the accessibility of the chromatin fiber. *Molecular and General Genetics*, **243**, 453–62.

Simon, J., Chiang, A., Bender, W., Shimell, M.J., and O'Connor, M. (1993). Elements of the *Drosophila bithorax* complex that mediate repression by *Polycomb* group products. *Developmental Biology*, **158**, 131–44.

Simon, J., Bornemann, D., Lunde, K., and Schwartz, C. (1995). The *extra sex combs* product, contains WD40 repeats and its time of action implies a role distinct from other *Polycomb* group products. *Mechanisms of Development*, **53**, 197–208.

Singh, P.B., Miller, J.R., Pearce, J., Kothary, R., Burton, R.D., Paro, R., *et al.* (1991). A sequence motif found in a *Drosophila* heterochromatin protein is conserved in animals and plants. *Nucleic Acids Research*, **19**, 789–94.

Tschiersch, B., Hofmann, A., Krauss, V., Dorn, R., Korge, G., and Reuter, G. (1994). The protein encoded by the *Drosophila* position-effect variegation suppressor gene *Su(var)3–9* combines domains of antagonistic regulators of homeotic gene complexes. *EMBO Journal*, **13**, 3822–31.

Van der Lugt, N.M.T., Domen, J., Linders, K., Van Roon, M., Robanus-Maandag, E., Te Riel, H., *et al.* (1994). Posterior transformation, neurological abnormalities and severe hemopoietic defects in mice with a targeted deletion of the *bmi-1* proto-oncogene. *Genes and Development*, **8**, 757–69.

Van Lohuizen, M., Frasch, M., Wientjens, E., and Berns, A. (1991). Sequence similarity between the mammalian *bmi*-1 protooncogene and the *Drosophila* regulatory genes *Psc* and *Su(z)2*. *Nature*, **353**, 353–5.

Wismar, J., Lüffler, T., Habtermichael, N., Vef, O., Geissen, M., Zirwes, R. *et al.* (1995). The *Drosophila melanogaster* tumor suppressor gene *lethal* (3) malignant brain tumor encodes a proline-rich protein with novel zinc finger. *Mechanisms of Development*, **53**, 141–54.

Wreggett, K.A., Hill, F., James, P.S., Hutchings, A., Butcher, G.W., and Singh, P.B. (1994). A mammalian homologue of *Drosophila* heterochromatin protein 1 (HP1) is a component of constitutive heterochromatin. *Cytogenetics and Cell Genetics*, **66**, 99–103.

Zink, B. and Paro, R. (1989). *In vivo* binding pattern of a trans-regulator of homoeotic genes in *Drosophila melanogaster*. *Nature*, **337**, 468–71.

Zink, D. and Paro, R. (1995). *Drosophila Polycomb*-group regulated chromatin inhibits the accessibility of a *trans*-activator to its target DNA. *EMBO Journal*, **11**, 5660–71.

15

Heterochromatin and regulation of gene expression in *Saccharomyces cerevisiae*

Miriam Braunstein, Scott G. Holmes, and James R. Broach

Introduction

Heterochromatin was defined by Heitz in 1928 as those chromosomal regions or entire chromosomes that remain in a condensed state throughout the cell cycle (Heitz 1928; John 1988). This definition is cytological and therefore only applicable to organisms in which individual chromosomes can be visualized. However, heterochromatin also exerts a characteristic transcriptional effect, which permits its detection and analysis even in the absence of cytological cues. Specifically, heterochromatin can exert transcriptional repression in a position-specific but gene-nonspecific manner, a process that contrasts with examples of transcriptional repression involving sequence-specific DNA binding repressor proteins and gene-specific promoter elements. The molecular basis of this heterochromatin-promoted repression remains to be determined, but likely results from exclusion of the transcriptional apparatus from the compacted chromatin structure of heterochromatin. Two examples of position effects on transcription exist in the yeast *Saccharomyces cerevisiae*: the silent mating-type cassettes and telomeres. Substantial evidence exists indicating that these position effect loci are enshrouded in the yeast equivalent of metazoan heterochromatin. In this chapter we will examine how studies of these position effect loci in yeast have enhanced our knowledge of what produces and comprises heterochromatin and how heterochromatin elicits transcriptional repression.

Heterochromatin is associated with position effect repression in metazoans

Two well-studied heterochromatic regions in metazoans are the inactive X chromosome in female mammalian cells and centromeric heterochromatin in *Drosophila*. Many features of the transcriptional repression observed at these heterochromatic locations are shared with each other and with cases of yeast position effect repression and are briefly described below (see also Chapters 12, 13 and 14).

The inactive X chromosome is produced early in female mammalian development, when each cell randomly and independently inactivates one of its two X chromosomes and generates the heterochromatic Barr body (Cattanach 1975; Riggs and Pfeiffer 1992; Rastan 1994). Most, but not all, genes resident on the inactive X become transcriptionally repressed and this repressed state exhibits epigenetic inheritance. That is, the repression state is passed from one cell to its progeny without any alteration in the genetic sequence of DNA. The random nature of X chromosome inactivation, combined with the epigenetic inheritance of the repressed state, leads to a variegated (mosaic) expression of X-linked alleles throughout the developed organism. X inactivation-induced repression is not limited only to those genes normally located on the X chromosome, but can extend to essentially any autosomal gene that becomes positioned next to the heterochromatic X chromosome by translocation. This repression occurs by spreading of the heterochromatic domain into the translocated autosomal domain in a distance-dependent fashion. Those genes positioned closer to the heterochromatic breakpoint are more likely to be repressed than those located further away (see Chapter 12).

Similar position effects on transcription occur at heterochromatin in *Drosophila* (Eissenberg 1989; Henikoff 1990; Wilson *et al.* 1990; see Chapter 13). Translocation of virtually any active gene to a site adjacent to heterochromatin can yield transcriptional repression of that gene. Repression may not be absolute. Rather, a variegated expression phenotype is often observed as a result of repression in some clusters of cells but expression in others. The variegated phenotype arises from stable maintenance of the repressed state through multiple cell divisions by an epigenetic mechanism, but with switching between the repressed and expressed state occurring in a small but detectable number of cell divisions. As with mammalian X chromosome translocations, position effect variegation in *Drosophila* exhibits a distance effect: genes closer to heterochromatin are more frequently repressed than genes positioned greater distances from heterochromatin. Supporting the hypothesis that position effect variegation results from heterochromatin spreading, heterochromatin can be observed cytologically at genes exhibiting variegated expression, and restoration of a variegating gene to a non-heterochromatin location restores expression.

Yeast contain position effect loci

Both the silent mating-type cassettes of yeast (*HML* and *HMR*) and the expressed *MAT* locus contain copies of mating-type genes (see Fig. 15.1). At *MAT* these genes are expressed to determine cell type (Herskowitz 1989; Herskowitz *et al.* 1992). The *MAT* locus harbours either of two co-dominant alleles: **a** or α. Expression of either allele alone dictates the corresponding haploid mating type, **a** or α, while co-expression of both alleles yields the diploid **a**/α cell type. Despite the fact that the promoters, coding sequences, and even flanking sequences resident at the silent cassettes are identical to those at *MAT*, the genes present at *HML* (normally the α allele) and *HMR* (normally the **a** allele) are fully

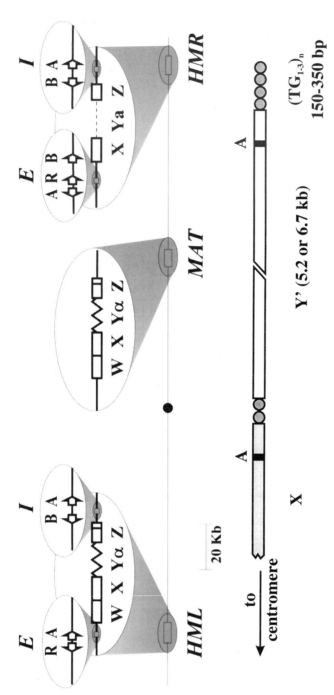

Fig. 15.1 Organization of transcriptionally silenced domains in yeast. (a) Diagram of the mating-type loci on chromosome III of *Saccharomyces*. The active (*MAT*) and two silenced (*HML* and *HMR*) mating-type loci share extensive sequence identity (boxes), as well as allele-specific domains (jagged and dashed lines). The silenced loci are bracketed by E and I silencers that are required in *cis* for transcriptional repression. These domains are comprised of DNA binding sites for Rap1p (R), Abf1p (B), and the origin replication complex (ORC) (A). Filled circle represents the centromere. (b) Diagram of a yeast telomere (adapted from Palladino and Gasser (1994). TG$_{1-3}$ repeats (shaded circles) are present at the ends of all telomeres, with shorter internal repeats sometimes present at the junction between X and Y' elements. Multiple Rap1p binding sites are present within the TG$_{1-3}$ repeats. X and Y elements are present at various copy numbers on different telomeres and each contains a site for binding the ORC (A).

repressed (Klar *et al.* 1981; Nasmyth *et al.* 1981). This repression is position specific but gene non-specific. Insertion of different RNA polymerase II and RNA polymerase III transcribed genes at the silent cassettes results in their repression, and transposition of the mating-type genes out of the silent cassettes results in their activation (Brand *et al.* 1985; Schnell and Rine 1986; Mahoney and Broach 1989; Herskowitz *et al.* 1992). Transcriptional repression of mating-type genes at the *HM* loci, often referred to as transcriptional silencing, is complete and stable. This is critical to maintenance of the cellular identity, as derepression of the silent cassettes would result in simultaneous expression of both **a** and α genes and a non-mating phenotype.

A similar position effect on transcription is observed at telomeres in *Saccharomyces* (Gottschling *et al.* 1990; Sandell and Zakian 1992). *Saccharomyces* chromosomes terminate with approximately 300 bp of the telomeric repeat $C_{1-3}A$ (Fig. 15.1). Placement of a transcriptionally active gene adjacent to the telomeric repeat results in transcriptional repression of the gene. As observed at the silent loci, repression at telomeres is gene non-specific. Unlike the silent cassettes, the degree of repression observed with telomere position effects is generally less than 100 per cent (see below).

A large number of *trans*- and *cis*-acting elements are required for silencing. The majority of these elements are required for silencing at both the silent cassettes and telomeres, which implicates a common transcriptional repression mechanism. *Trans*-acting factors that are required for transcriptional repression of both the position effect loci are listed in Table 15.1(a), and those required for repression of the silent cassettes, but not for telomere position effect, are listed in Table 15.1(b). *TLC1*, which encodes the RNA component of yeast telomerase, is the only gene so far identified that affects telomeric repression, but does not affect repression of the silent cassettes (Singer and Gottschling 1994). *Cis*-acting silencer elements (E and I) flank both *HML* and *HMR* and are also essential for the transcriptional repression of the silent cassettes (Abraham *et al.* 1984; Feldman *et al.* 1984). The silencers are composed of various combinations of at least two of three distinct elements: a Rap1p binding site, an Abf1p binding site, and an ARS (autonomously replicating sequence) element, which is bound by the multisubunit origin recognition complex (ORC) (Shore and Nasmyth 1987; Shore *et al.* 1987; Buchman *et al.* 1988a,b; Diffley and Stillman 1988). Telomeric repeats do not contain a classically defined silencer element, but do contain multiple Rap1p binding sites: approximately 20 per telomere (Wang and Zakian 1990; Gilson *et al.* 1993).

Position effect loci in yeast are packaged in an unusual chromatin structure

Heterochromatic regions of the yeast genome cannot be classically defined due to difficulties in visualizing, identifying, and determining the degree of compaction of individual chromosomes. Although newly developed fluorescence *in situ*

Table 15.1 Genes required for position effect repression in yeast

Gene	Function	References
(a) Required for silencing at both telomeres and silent cassettes		
SIR2		Aparicio *et al.* (1991), Klar *et al.* (1979), Rine and Herskowitz (1987)
SIR3		Haber and George (1979), Hartwell (1980), Klar *et al.* (1981), Rine and Herskowitz (1987)
SIR4		Hartwell (1980), Rine and Herskowitz (1987)
NAT1, ARD1	N-terminal acetyltransferase	Aparicio *et al.* (1991), Lee *et al*, (1989), Mullen *et al.* (1989), Whiteway *et al.* (1987)
HHF1,2	Histone H4	Aparicio *et al.* (1991), Johnson *et al.* (1990, 1992), Kayne *et al.* (1988), Megee *et al.* (1990), Park and Szostak (1990)
HHT1,2	Histone H3	Thompson *et al.* (1994)
RAP1	Repressor/activator protein	Kurtz and Shore (1991), Kyrion *et al.* (1993), Liu *et al.* (1994), Sussel and Shore (1991)
(b) Required for silencing at silent cassettes only		
SIR1		Ivy *et al.* (1986), Pillus and Rine (1989), Rine and Herskowitz (1987), Rine *et al.* (1979)
ORC2,5	Components of the replication origin recognition complex	Foss *et al.* (1993) Loo *et al.* (1995*a*), Micklem *et al.* (1993)
ABF1	ARS consensus binding factor	Buchman *et al.* (1988*a*), Diffley and Stillman (1988), Shore *et al.* (1987), Loo *et al.* (1995*b*)
RIF1	RAP1 interacting factor	Hady *et al.* (1992)

hybridization (FISH) techniques for yeast hold promise that a classical definition of heterochromatic regions in yeast will be achieved in the near future (Scherthan *et al.* 1992; Guacci *et al.* 1994; Weiner and Kleckner 1994), the identification of heterochromatin in yeast currently relies on evaluation of correlated characteristics. The following section summarizes experimental evidence in support of an altered chromatin structure existing at the silent cassettes and telomeric regions. These experiments either assayed the accessibility of the DNA sequence of these loci to different reagents, or characterized the nucleosomes associated with the loci. These experiments have all led to the same conclusion: position effect loci in yeast are packaged in an unusual chromatin structure that renders them less accessible, and this altered chromatin structure is strictly associated with position effect repression.

Accessibility to the *HO* endonuclease

In homothallic, or *HO*⁺, strains of *Saccharomyces* the mating type of cells can switch as often as every generation (Herskowitz *et al.* 1992). This occurs by means of a transposition event in which the mating-type genes at the *MAT* locus are replaced by a copy of the genes resident at either *HML* or *HMR*. The switching event is initiated by an *HO* endonuclease-induced double strand break at the *MAT* locus, which can be detected by Southern blot analysis on the DNA of exponentially growing *HO*⁺ yeast (Strathern *et al.* 1982). Complete *HO* cleavage sites are also present at *HML* and *HMR*, but are not cut by the endonuclease. This inaccessibility of the silent cassettes to cleavage by the *HO* endonuclease is dependent on transcriptional silencing (Klar *et al.* 1984). A *sir2* mutation that fully eliminates transcriptional repression at the silent cassettes renders the cassettes susceptible to *HO* endonuclease cleavage. The inaccessibility of the silent cassettes to cleavage by the *HO* endonuclease has also been demonstrated *in vitro*. Treatment of nuclei isolated from an *SIR*⁺ strain with partially purified *HO* endonuclease resulted in cleavage of the *MAT* locus, but not of the *HM* loci (Loo and Rine 1994). In contrast, the *HM* loci in nuclei isolated from an *sir⁻* strain were as accessible to cleavage as the *MAT* locus. This relative sensitivity was not simply a consequence of differential transcription of *HM* versus *MAT*, since identical results were obtained with an *HMR* locus deleted for its promoter. These results are consistent with the presence of a special chromatin structure associated with transcriptional silencing.

DNAase I and micrococcal nuclease sensitivity

Nuclease sensitivity, a classic technique for probing chromatin structure, has been assessed at the silent cassettes (Nasmyth 1982). The most striking result from these studies was the identification of a DNAase I-hypersensitive site at or near the *HO* cleavage site within the *MAT* locus, which was not evident at the same site within the *HM* loci. Elimination of transcriptional silencing elicited the hypersensitive site at *HM* loci. With the exception of this one site, the regions comprising the silent cassettes in an *SIR* strain did not show dramatic differences in nuclease sensitivity profiles when compared to the *MAT* locus or the *HM* loci in *sir* strains.

Differential repair of DNA damage

Pyrimidine dimers induced at the *MAT* locus are repaired *in vivo* preferentially in comparison to those induced at *HML* (Terleth *et al.* 1989). This preferential repair of UV damage at the *MAT* locus reflects a delay in the repair of the same damage at *HML*. The difference in repair rate is dependent on transcriptional silencing, as pyrimidine dimers at *HML* and *MAT* are repaired at the same rate in an *sir3* mutant. This work demonstrates that for most of the cell cycle the silent cassettes are less accessible to the repair process than are other regions of the genome. However, it raises the intriguing possibility, reinforced by work from Aparicio and Gottschling (1994), that the silenced domains may normally become accessible at a particular stage in the cell cycle.

Accessibility to DNA methylating enzymes

Another approach that has been used to probe the chromatin structure of genomic sequences in yeast is to determine the *in vivo* accessibility of regions of DNA to ectopically expressed bacterial DNA methylases. Since yeast contains no detectable DNA methylation, all the methylation in these experiments is catalysed by the ectopically expressed enzyme. Methylation is assayed by isolating genomic DNA, digesting it with restriction enzymes that are sensitive to the methylation status of their recognition sites, and then identifying cleavage by Southern analysis. Application of this technique to the silent cassettes suggested that in general the *HM* loci in the repressed state are less accessible to methylases than the loci in a derepressed state (Singh and Klar 1992). However, most of the methylation sites within the *HM* loci did not show dramatic differences in methylation in the repressed versus the derepressed state. Rather, the clearest examples of sequences less accessible to methylases in *SIR* strains were found at or adjacent to the silencer elements. In a similar study (Gottschling 1992; Kyrion *et al.* 1993) the single methylation site in the coding sequence of a *URA3* gene, placed adjacent to the telomere, was shown to be relatively inaccessible to an ectopically expressed methylase. Restricted access was not complete: 25 per cent of the cells were not methylated at the site, consistent with fact that only a partial repression phenotype is generally observed with telomeric silencing. In *sir* strains the *URA3* gene is methylated in all cells. These experiments support the idea that a compacted chromatin structure exists over these regions. However, the inaccessibility at the silent cassettes is only detectable at specific locations in the loci, and the existence of a single methylation site in the *URA3* gene precluded analysis of other regions in the telomere studies.

Accessibility to restriction enzymes

The ability of restriction enzymes to digest DNA of the *HMR* cassette in isolated nuclei provides yet another example of the inaccessibility of the silent cassettes to protein molecules in an *SIR*-dependent manner (Loo and Rine 1994). Unlike the previous studies this study revealed inaccessibility throughout the locus. *HMR* was protected from digestion with restriction enzymes in nuclei isolated from *SIR* strains, but was readily digested in nuclei isolated from *sir* strains. Restriction enzyme inaccessibility was not limited to sequences within the *HMR* locus, but was also detected in flanking DNA extending past the E and I silencers. Consistent with the interpretation that the altered chromatin structure present at the yeast position effect loci is similar to heterochromatin, Wallrath and Elgin (1995) reported that the *hsp26* gene, positioned in heterochromatin and subject to position effect variegation in *Drosophila*, is inaccessible to restriction enzyme cleavage in isolated nuclei.

Altered nucleosomal structure

The structure of nucleosomes over position effect loci is different from those elsewhere in the genome. In metazoan chromatin the accessibility to thiol-reactive reagents of a particular histone H3 cysteine residue located at the centre of the nucleosome core provides a means of monitoring nucleosome structure. Use of this approach in yeast first required construction of a yeast mutant in which all histone H3 protein in the cell carried a substitution of cysteine for alanine at position 110. Mercury affinity chromatography of chromatin from an *SIR* strain carrying the histone H3 substitution indicated that nucleosomes across the *MAT* locus and other actively transcribed genes, as well as those genes poised for transcription, could bind to the column. In contrast, those nucleosome across the *HM* loci do not bind (Chen *et al.* 1991). This result indicates that the nucleosomes positioned on active and poised genes have a conformation that allows the cysteinyl SH group to form a covalent linkage with the mercury column. In the nucleosomes at the silent cassettes this cysteine is inaccessible. This distinct nucleosomal structure associated with the silent cassettes is dependent on transcriptional silencing of the *HM* loci (Chen-Cleland *et al.* 1993). Nucleosomes across the *HM* loci in chromatin isolated from an *sir3* strain bind to a mercury column with the same affinity as those associated with the *MAT* locus. The distinct chromatin structure associated with the silent cassettes cannot be attributed solely to the lack of transcription, since nucleosomes on poised but transcriptionally repressed genes bind the mercury column. However, other regions that are subject to long-term repression, such as genes controlled by cell-type repression, are also packaged in nucleosomes that fail to bind to the mercury column. Thus, this measure of compact chromatin structure correlates with long-term repression of which transcriptional silencing is one example.

Altered acetylation status

Histones undergo a post-translational acetylation of several highly conserved internal lysine residues in the amino terminal tails of all of the four core histones (Turner 1991; see Chapter 3). The acetylation state of nucleosomes across a gene appears to correlate with the transcriptional activity of the gene. In general, transcriptionally active and potentially active genes are packaged in acetylated nucleosomes, while genes permanently repressed or subject to long-term repression are associated with hypoacetylated nucleosomes (Hebbes *et al.* 1988, 1992; Lin *et al.* 1989; Clayton *et al.* 1993; O'Neill and Turner 1995). A striking example of this relationship is provided by heterochromatic regions in metazoan genomes. The heterochromatic inactive X chromosome of mammals is associated with hypoacetylated histones H3 and H4 (Jeppeson and Turner 1993; Belyaev *et al.* 1996; B. Boggs and C.D. Allis, unpublished results). Similarly, histone H4 in *Drosophila* centromeric heterochromatin is hypoacetylated relative to that in active chromatin, with lysines 5, 8 and 16 exhibiting a greater reduction in acetylation than lysine 12 (Turner *et al.* 1992).

The acetylation state of nucleosomes spanning different genomic sequences in yeast, including the silent cassettes, has been probed by immunofractionation of yeast chromatin with antisera that recognize specific acetylated isoforms of either histones H3 or H4 (Braunstein *et al.* 1993, 1996). The abundance of a specific genomic sequence in the immunoprecipitated chromatin fraction reflects the acetylation status of histones spanning that particular sequence. For instance, the chromatin fraction immunoprecipitated by antibodies against acetylated histone H4 contained significant levels of transcribed genes, including *MAT*, as well as transcriptionally repressed but inducible genes. In contrast, the silent cassettes and the telomere-associated sequence (Y') were underrepresented in this acetylated chromatin fraction, indicating that the nucleosomes packaging the silent cassettes and sequences adjacent to telomeres have a reduced level of histone H4 acetylation. This hypoacetylation of histone H4 at the position effect loci is dependent on transcriptional silencing, since these loci are associated with normally acetylated histone H4 in *sir* strains. Similar experiments, using antisera directed at acetylated or unacetylated histone H3, indicate that nucleosome packaging position loci are also hypoacetylated at histone H3. Association of reduced nucleosome acetylation with the silent cassettes does not simply reflect reduced transcription: *SIR*-dependent changes in histone H4 acetylation over *HML* are equally dramatic in strains in which *HML* is incapable of transcription due to a deletion of promoter/enhancer elements. In addition, other genes subject to long-term repression are packaged in nucleosomes with normal levels of acetylation. Thus, histone hypoacetylation correlates strictly with transcriptional silencing in yeast.

Probing the acetylation state of the individual lysine residues in histone H4 molecules located at the silent cassettes has revealed a striking similarity between metazoan heterochromatin and chromatin packaging the silent cassettes (Braunstein *et al.* 1996). Using four different antisera, each specific for one of the acetylated lysines of histone H4, to immunoprecipitate the portion of yeast chromatin whose nucleosomes are acetylated at a specific histone H4 lysine, demonstrated that the *HM* loci are associated with histone H4 molecules that are significantly underacetylated at lysines 5, 8, and 16. In contrast, lysine 12 showed near-normal levels of acetylation at histone H4 at the silent cassettes. This pattern of histone H4 acetylation at the silent cassettes is identical to that at centromeric heterochromatin in *Drosophila*. This represents compelling evidence that the *HM* loci are associated with the yeast equivalent of heterochromatin. Although the causal relationship between acetylation and silencing has not been clarified, these results and additional genetic studies suggest that reduced histone acetylation may play a significant role in heterochromatin-induced repression.

Histones H3 and H4

Genetic analysis of histone H4 has demonstrated that this histone plays a critical role in transcriptional silencing in yeast. Two regions of histone H4, a basic domain (R1) composed of residues 16–19 and a non-basic domain (R2) composed of residues 21–29, are required specifically for transcriptional silencing (Kayne *et al.* 1988; Johnson *et al.* 1990, 1992; Megee *et al.* 1990; Park and Szostak 1990). A single mutation that results in a reduction in the positive charge of domain R1 results in transcriptional derepression, while a similar mutation that maintains the positive charge of R1 does not abolish transcriptional silencing. As acetylation results in the neutralization of the positive charge of a lysine, the mutational analysis of R1 suggests that lysine 16 must be in a deacetylated state in order for transcriptional silencing to exist. As noted in the preceding section, lysine 16 is in fact significantly underacetylated in histone H4 at the silent cassettes (Braunstein *et al.* 1996). In addition to the R1 and R2 domains, additional residues in histone H4 may be important in promoting transcriptional silencing. The first three acetylatable lysines in histone H4 (lysines 5, 8, and 12) were initially deemed unimportant for transcriptional silencing. However, in the context of significantly underacetylated histone H4 molecules, as is the case for heterochromatin, either lysine 5 or lysine 12 must be neutral in order to maintain efficient transcriptional silencing (Braunstein *et al.* 1996). At the silent cassettes this is achieved by acetylation of lysine 12 (Braunstein *et al.* 1996).

Mutations in the amino terminus of histone H3 can also affect transcriptional silencing. This is most clearly evident with telomeric silencing. Certain deletions in the amino terminus of histone H3 result in derepression of telomeric silencing of a *URA3* gene placed adjacent to a telomere (Thompson *et al.* 1994). Deletion of residues 4–10 yields a modest derepression, while deletion of residues 4–15 results in a significant level of derepression. The same histone H3 deletions do not have significant effect on silencing at the silent cassettes. However, these deletion alleles are synergistic with other mutations, such as *sir1*, that weaken silencing. This suggests that although the amino terminus of histone H3 is not essential for silencing, it may participate in a redundant way to help ensure that repression at the silent cassettes is efficient.

Position effects in yeast have characteristics in common with heterochromatin-induced position effects in metazoans

The numerous features in common between position effect repression in yeast and heterochromatin-associated position effects in metazoans argue that the regions in yeast subject to position effect are the yeast equivalent of metazoan heterochromatin. These features are elaborated in the following.

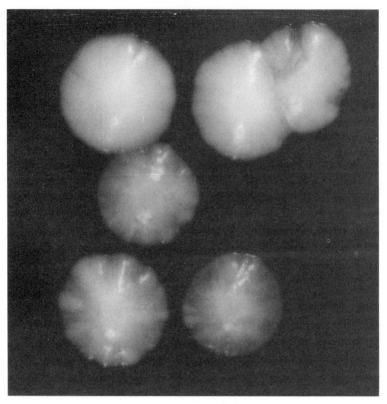

Fig. 15.2   Epigenetic inheritance of transcriptional silencing. Shown are colonies of an *ade2* strain carrying an *ADE2* gene inserted adjacent to the 5R telomere. Repression of the ectopic *ADE2* gene results in accumulation of a red pigment in cell, while expression of *ADE2* renders cells white. Note the white sectors within red sectors (shown in grey here), indicating sequential switches from derepressed to repressed and to derepressed in successive generations of cells. (Photograph kindly provided by J.J. Lin and V. Zakian.)

Epigenetic inheritance

Repression at position effect loci in yeast exhibits epigenetic inheritance. Genes placed adjacent to telomeres in yeast are not fully silenced. Rather, they exhibit a partial repression phenotype that reflects a population of genetically identical cells in which the gene is fully expressed in some cells and fully repressed in others. This can be visually appreciated in cells in which the *ADE2* gene is positioned adjacent to a telomere (Gottschling *et al.* 1990). Cells grown on plates containing low levels of adenine are white if wild-type levels of *ADE2* are produced and red if transcription of *ADE2* is repressed. The colonies which form from a strain containing *ADE2* at the telomere contain red and white sectors (see Fig. 15.2). These sectored colonies are composed of clusters of cells that have stably maintained the same expression state, either active or repressed, through multiple rounds of cell divisions. The uniform colour of the sectors, therefore, demonstrates epigenetic inheritance of the repressed state. These colonies additionally reveal that a cell can switch between the expressed and repressed state in that white sectors can appear within red sectors and vice versa.

The highly efficient transcriptional repression at the *HM* loci normally precludes observing its epigenetic inheritance. However, specific mutations in certain *trans*-acting factors, *sir1* and *rap1s* (Pillus and Rine 1989; Sussel *et al.* 1993) and in *cis*-acting elements, *HMLE*[P] and *hmrΔA* alleles (Mahoney and Broach 1989; Sussel *et al.* 1993), reveal an epigenetic component of the repression. All of these mutations lead to partial repression of the *HM* loci due to the presence of two populations of genetically identical cells, one of which exhibits full expression of the HM locus and the other exhibiting complete repression. For instance, *rap1s* or *hmrΔA* mutations in strains that contain the *ADE2* gene at the *HMR* cassette (*HMR::ADE2*) exhibit sectored colonies identical to that observed for telomeric silencing (Sussel *et al.* 1993). For *sir1* and *HMLE*[P] strains the expression phenotype of individual cells has been examined (Pillus and Rine 1989; Mahoney *et al.* 1991). By using resistance to mating pheromone as a monitor of expression of the *HML* locus the pedigree of individual *sir1* or *HMLE*[P] cells could be followed for a number of generations. These studies demonstrated that the particular state of expression at *HML* is stably maintained through multiple cell generations, but that switches between the two expression states occur at a low but detectable rate. Thus, the stable persistence of two distinct populations in a culture reflects the equilibrium established by the slow but finite rates of interconversion between the two expression states at the locus.

Distance effects

A gene positioned closer to the telomeric repeats or to the *HM* loci has a greater chance of being silenced then do genes placed further away (Renauld *et al.* 1993; M. Braunstein and J.R. Broach, unpublished results). The degree of transcriptional repression of *URA3* decreases exponentially as the distance between *URA3* and the telomere increases. The identical pattern is observed for the *URA3* gene placed at increasing distances away from the *HML* cassette. These observations suggest that a stochastic process underlies the spread of repression outward from position effect loci in yeast.

Peripheral nuclear localization

Metazoan heterochromatin is often located at the periphery of the nucleus. The inactive X chromosome in mammals, heterochromatic regions in *Drosophila*, and telomeres from a number of organisms, which are generally heterochromatic, are all located at the nuclear periphery (Comings 1968; Mathog *et al.* 1984; Hochstrasser *et al.* 1986; see Chapter 9). The localization of yeast telomeres in the nucleus has been inferred indirectly (Palladino *et al.* 1993). Yeast telomeres contain multiple Rap1p binding sites. Although Rap1p binds at additional sites in the genome the density of Rap1p sites at the telomeres suggests that Rap1p immunofluorescence reflects the localization of telomeric DNA. By immuno-fluorescence Rap1p exhibits punctate nuclear staining of roughly seven to eight patches per nucleus, all located at or near the nuclear periphery. This observation has been interpreted to indicate that telomeres are clustered, since diploid yeast contain 64 telomeres, and that these telomeric clusters are localized to the nuclear periphery. This localization is dependent on the presence of Sir3p and Sir4p. *sir3* or *sir4* cells exhibit a larger number of patches, which are no longer all sequestered to the periphery. The subnuclear localization of the *HM* loci has not been determined. The role of nuclear localization in position effect repression is not clear. The fact that mutation of a gene, *RLF6*, disrupts peripheral localization of Rap1p without perturbing transcriptional silencing (Konkel *et al.* 1995), and that derepression of silencing can occur even when Rap1p remains at the nuclear periphery (Cockell *et al.* 1995), suggests that subnuclear localization is not an essential attribute of position effect repression in yeast.

Late replication

An additional characteristic of heterochromatin is its late replication in the S phase of the cell cycle. This is true both for the inactive X chromosome and heterochromatin in *Drosophila* (Lima-de-Faria and Jaworska 1968). The *HM* loci and telomeres of chromosomes share this property of replicating late in the S phase of the cell cycle (McCarroll and Fangman 1988; Reynolds *et al.* 1989; Ferguson and Fangman 1992). However, the association between late replication and transcriptional silencing has not been investigated and, in fact, transcriptional silencing can be established at sites in the genome that are not normally late replicating (Shei and Broach 1995). However, since the replication timing of these ectopically silenced loci has not been determined, the contribution of replication timing to the mechanism of transcriptional silencing remains unre-solved.

Dosage effects

Genetic analysis of position effect variegation in *Drosophila* has identified a number of modifier loci that can influence the degree of transcriptional repression of a variegating gene. Some of these modifier loci demonstrate a dosage effect. For example, some modifiers suppress position effect variegation when present in single copy per diploid cell (i.e. one less than normal), and enhance position effect variegation when present in three copies (Locke *et al.* 1988; Grigliatti 1991). The dosage dependence exhibited by these modifiers has led to models in which these loci encode either structural components of a multimeric hetero-chromatin complex, heterochromatin assembly factors, or modifiers of hetero-chromatin proteins.

Dosage effects have been identified in silencing at the silent cassettes and telomeres. For instance, overexpression of *SIR4* abolishes silencing at both the silent cassettes and telomeres (Marshall *et al.* 1987; Cockell *et al.* 1995). However, in the presence of certain silencing defects different dosage effects of *SIR4* are observed. For instance, an extra copy of the *SIR4* enhances transcrip-tional repression of an *hmrΔA::ADE2* allele, while reducing the dosage of *SIR4* by one-half results in derepression of the locus (Sussel *et al.* 1993). Similarly, extra copies of *SIR1* can increase the degree of transcriptional repression at the silent cassettes in certain strains with silencing defects (Stone *et al.* 1991; Sussel and Shore 1991; Chien *et al.* 1993; Sussel *et al.* 1993). Finally, increased dosage of the *SIR3* gene has been shown to enhance telomeric silencing (Renauld *et al.* 1993). This is most notable in examining the distance effect associated with telomeric silencing; extra copies of *SIR3* extend the distance over which repression of an ectopically inserted *URA3* gene can be achieved. Observations of dosage effects are certainly consistent with the heterochromatic nature of position effect loci in yeast. More significantly, they help define the rate-limiting components involved in formation and maintenance of the repressed state, as described in the next section.

Models for silencing by yeast heterochromatin

Nucleation centres and the establishment of a silenced state

A nucleation site is required for formation of a silenced domain in yeast. Silencer elements, internal tracts of $C_{1-3}A$, and telomeres can all serve as nucleation sites that induce transcriptional silencing in the adjacent chromosomal domain (Lee and Gross 1993; Renauld *et al.* 1993; Stavenhagen and Zakian 1994; Shei and Broach 1995), a conclusion that emerges from induction of ectopic silencing by insertion of these nucleation sites adjacent to heterologous genes (see Chapter 13). All of these nucleation sites encompass well-defined functional elements. Silencers contain various combinations of Abf1p binding sites, Rap1p binding sites, and ARS elements, while internal tracts of $C_{1-3}A$ and telomeres contain multiple Rap1p binding sites. Despite this structural similarity not all nucleation sites behave identically. As revealed by ectopic silencing, different silencers have different strengths (Shei and Broach 1995). This difference appears primarily quantitative, rather than qualitative, since increasing the number of copies of a weak silencer enhances its potency. Similarly, the extent of repression of *URA3*, induced by inserting internal tracks of $C_{1-3}A$ next to the locus, increased with increasing copies of the internal tracks (Stavenhagen and Zakian 1994). Further, in strains expressing a fusion of Rap1p to the Gal4p DNA binding domain, multiple Gal4p DNA binding sites (UAS^G) can functionally substitute for the

HMR E silencer. In this case the level of repression of *HMR* increases with increasing copies of UAS^G (Buck and Shore 1995). Thus, we can conclude from these studies that nucleation sites, i.e. silencers, telomeres, and internal $C_{1-3}A$ tracks, dictate the location of silenced domains and do so as a consequence of their ability to bind specific DNA binding proteins.

Additional features of silencers and silencer binding protein have been noted. First, silencers are directional. That is, the transcriptional silencing is established to a significantly greater degree to one side of a silencer than to the other (Shei and Broach 1995). This suggests that the specific organization and orientation of elements within a nucleation site are critical for function, although the organizational rules for silencer function have not been determined. Second, the silencer binding protein Rap1p induces substantial DNA bending (Vignais and Sentenac 1989). The fact that Rap1p fused to the Gal4p DNA binding domain can function as a silencer suggests that DNA bending is not critical for silencer function. However, since the efficiency of this construct is substantially less than that of a normal silencer, specific DNA topology may contribute to full silencer activity.

Formation of the heterochromatin domain

A model for silencer-induced heterochromatin formation is presented in Fig. 15.3. Proteins that bind the nucleation sites, i.e. Rap1p, Abf1p, and ORC, recruit additional proteins, Sir2p, Sir3p, and Sir4p, that are themselves responsible for formation of the heterochromatic structure in the region adjacent to the nucleation site. We propose that these Sir proteins modify the adjacent nucleosomes in a way that increases their affinity for the Sir proteins, which then bind to the newly modified nucleosomes. In this manner the silencing domain can be propagated outward from the nucleation site. The Sir proteins remain associated with the nucleosomes and participate directly, either as a structural component or in a catalytic capacity, in the heterochromatin structure.

Support for this model emerges from definition of the interactions between and among components of the silencer apparatus. A number of silencer apparatus proteins have been shown to interact with Rap1p. Rap1p binds both Sir3p and Sir4p, as defined by the two-hybrid assay (Moretti *et al.* 1994), and displays a genetic interaction with Sir3p (Liu *et al.* 1994). Binding between Rap1p and Sir3p is direct as it is detected *in vitro* in the absence of additional proteins (Moretti *et al.* 1994). An interaction between Rap1p and Sir4p has been demonstrated by co-immunoprecipitation, which delimited the interaction to the carboxyl terminus of Sir4p (Cockell *et al.* 1995). This interaction could not be detected *in vitro* and, thus, may be indirect (Moretti *et al.* 1994). Sir3p and Sir4p additionally interact both with themselves and with each other based on the two-hybrid assay (Chien *et al.* 1991; Moretti *et al.* 1994), confirming an earlier report of genetic interaction between *SIR3* and *SIR4* (Marshall *et al.* 1987). Thus, Rap1p could function to recruit a complex containing at least Sir3p and Sir4p.

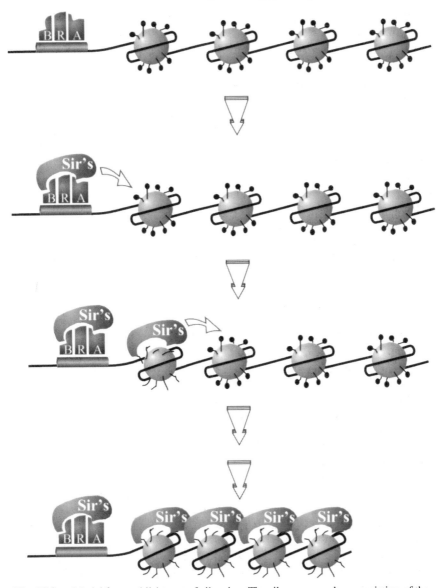

Fig. 15.3   Model for establishment of silencing. The silencer complex, consisting of the silencer DNA (shaded cylinder) plus associated Rap1p (R), Abf1p (B), and origin recognition complex (A) proteins, recruits a complex of proteins including Sir2p, Sir3p, and Sir4p. This Sir complex modifies the adjacent nucleosome, in this model represented by deacetylation of lysines in histone amino termini, to enhance the affinity of the nucleosome for the Sir complex. The Sir complex then binds to the modified nucleosome and catalyses modification of the next nucleosome. This promotes spreading of the Sir complex along the DNA. The Sir proteins remain associated with the chromatin as structural components of the newly formed heterochromatin.

The other silencer binding proteins, Abf1p and ORC, likely also participate in this recruitment, but their interactions with silencer apparatus proteins have not been explored as extensively. In this context, the homology between Orc1p, a subunit of the ORC, and Sir3p suggests the possibility of silencer apparatus recruitment through subunit substitution.

Sir3p and Sir4p bind directly to the tails of histones H3 and H4 *in vitro* (Hecht *et al.* 1995). The ability of Sir3p and Sir4p to interact both with a silencer binding protein and histones suggests that these proteins both recognize the nucleation site and promote the polymerization of heterochromatin in the adjacent domain. As noted earlier, the R1 and R2 domains of histone H4 are required for transcriptional silencing. Mutations of the R1 region of histone H4 eliminate the ability of Sir3p and Sir4p to bind the histone H4 tail. In contrast, mutations in the R2 domain did not eliminate Sir3p and Sir4p binding, suggesting an additional role for histone H4 in silencing. Sir3p and Sir4p bind to histone tails in which all lysines are in a positively charged, deacetylated state. The affinity of Sir3p and Sir4p for acetylated histones has not been assessed. As noted above, all the acetylatable lysines of histones H3 and H4, with the exception of lysine 12, are hypoacetylated at the silent cassettes (Braunstein *et al.* 1993, 1996), while those of histones elsewhere in the genome are predominantly acetylated. Thus, Sir3p and Sir4p may bind with greater affinity to those nucleosomes in the silenced domain than to other regions of the genome. If the Sir complex also promoted deacetylation, a possibility discussed below, then this differential affinity could provide a positive feedback loop that facilitates silent domain spreading, as shown in Fig. 15.3, as well as epigenetic inheritance.

The third protein essential for transcriptional silencing, Sir2p, promotes histone deacetylation in cells in which it is overexpressed (Braunstein *et al.* 1993). Although Sir2p does not appear to be a histone deacetylase, it may be a direct or indirect inhibitor of an acetyltransferase. For instance, Sir2p could indirectly promote deacetylation by inhibiting the majority of acetylation/deacetylation reactions that occur following histone deposition. *SIR2* and *SIR3* exhibit a genetic interaction. Overexpression of *SIR2* or *SIR3* alone mildly inhibits yeast cell growth, while simultaneous overexpression of both genes is toxic (Holmes *et al.* 1997). The interaction between the two proteins suggested by the synergistic effect of overexpression was confirmed by co-immunoprecipitation of Sir2p and Sir3p from yeast extract using either anti-Sir2p antibodies or anti-Sir3p antibodies (Holmes *et al.* 1997). Whether this physical interaction is direct or indirect is not known. The interaction between Sir2p and Sir3p and the effect of Sir2p on histone acetylation fulfil the remaining requirements for the model presented in Fig. 15.3. Sir3p and Sir4p, recruited to the nucleation site through interaction with one or more of the silencer binding proteins, in turn recruit Sir2p, which stimulates deacetylation of the adjacent nucleosomes. We propose that this deacetylation facilitates binding of Sir3p and Sir4p to the newly modified nucleosome, Sir3p and Sir4p again recruit Sir2p to modify the adjacent nucleosome, and the process continues. A multicomponent heterochromatin complex thus polymerizes along the silent locus. What stops this precessive process is discussed below.

We propose that formation and persistence of the compacted chromatin that comprises the heterochromatin structure of the silenced locus requires the continuous association or activity of Sir2p, Sir3p, and Sir4p. This requirement has been shown for Sir3p, in that shifting strains carrying an *sir3*[ts] allele to the non-permissive temperature results in prompt derepression of the silent cassettes both in an asynchronous population of cells and a population of cells that are arrested in the G_1 phase of the cell cycle (Miller and Nasymth 1984; Holmes and Broach 1996). Similar studies have not been done with the other *SIR* genes, but because of their genetic interconnectedness we would anticipate that Sir2p, Sir3p, and Sir4p all function in essentially the same process in silencing. Nonetheless, how these proteins induce heterochromatin and how the resulting heterochromatin structure prevents access of the transcriptional apparatus to the underlying DNA remains to be elucidated.

Maintenance of the silenced state

Once established, the silenced state must be maintained throughout the cell cycle and propagated efficiently to progeny cells. This function is apparently served by the silencer. Unlike the Sir2, 3, and 4 proteins, the silencers are not needed for maintaining the silenced state in absence of cell cycle progression. Deletion of the silencer by site-specific recombination *in vivo* in cells arrested in the G_1 phase of the cell cycle does not cause derepression of the adjacent silenced domain (Holmes and Broach 1996). However, silencers are required for the stable inheritance of the repressed state. Cells deleted for the silencer most often become derepressed for silencing following a single passage through the cell cycle. Thus, the silencer appears to be required both for establishment and for efficient inheritance of the silenced state. The role the silencer plays in propagation of the silenced state remains to be determined, but it could stabilize the silenced chromatin through replication and mitosis, or it could promote heterochromatin reassembly following disassembly during some stage of the cell cycle.

The organization of the silenced domain

The eukaryotic genome is likely divided into independently regulated domains (Eissenberg and Elgin 1991). How domains are delineated or prevented from influencing neighbouring regions is still unresolved, although small boundary elements, such as the scs (specialized chromatin structure) elements (Kellum and Schedl 1991) and suppressor of hairy wing (*su(Hw)*) binding sites (Corces and Geyer 1991; Roseman *et al.* 1993), both in *Drosophila*, or the insulator element of the chicken β-globin locus, may play a role in establishing the limits of such domains (Chung *et al.* 1993; see Chapter 5). The organization of the silenced domains at the telomeres and the silent cassettes provides a means of probing this issue. As described earlier, repression of the *URA3* decreases exponentially with increasing distances from telomeres and the *HML* locus (Renauld *et al.* 1993; M. Braunstein and J.R. Broach, unpublished results). In contrast to the domains adjacent to the telomeres and *HML*, repression within the silent cassette is not affected by distance (M. Braunstein and J.R. Broach, unpublished results). Insertion of *URA3* gene anywhere within *HML*, including within 150 bp from the I silencer, results in full repression of the gene. In addition, insertion of a single copy of the Ty1 transposon within the silent cassettes, i.e. increasing the distance between the E site and the $\alpha 1/\alpha 2$ promoters by five-fold, does not diminish silencing of these mating-type genes (Mastrangelo *et al.* 1992). Thus, over relatively large distances repression of genes lying between the silencers is complete, while repression of genes outside the silencers drops off exponentially with distance. In this respect silencers resemble boundary elements, acting to restrict a heterochromatic state to within a defined region. Transcriptional repression observed outside the domain may simply reflect limited spillage of the repression beyond the boundary, as is seen with insulator elements at the boundary of the chicken globin locus domain. These observations, in conjunction with the fact that the orientation of a silencer can affect its ability to repress genes (Shei and Broach 1995), may indicate that the silencers act as nucleation centres in one direction and as boundary elements in the opposite direction, in a manner similar to that proposed for the *su(Hw)* insulator (Gerasimova *et al.* 1995). Thus, silencers may serve both to delimit as well as establish a heterochromatin domain in yeast.

Future directions

Our understanding of how transcriptional silencing occurs at the silent cassettes and telomeres has increased rapidly over the last few years. The most significant questions that remain to be answered regarding transcriptional silencing in yeast are defining the precise nature of the heterochromatin structure and determining how heterochromatin causes transcriptional repression. Does the heterochromatin extend throughout the region and physically prevent interactions between transcription factors and the underlying DNA sequences? Is the repression a result of localization of heterochromatin complexes to the nuclear periphery? Does the heterochromatin complex lead to late replication of the region, which in turn leads to repression? *In vitro* reconstitution of silencing has begun (Loo and Rine 1994; Hecht *et al.* 1995) and the ability to address these questions at a molecular level is approaching.

Summary

Heterochromatin in metazoan chromosomes induces gene-nonspecific, position-specific transcriptional repression. The yeast *Saccharomyces cerevisiae* presents two examples of similar position-specific but gene-nonspecific repression: telomeric domains and the silent mating-type cassettes. By a variety of criteria, chromatin packaging of these regions renders the underlying DNA less accessible than at other regions of the genome. In addition, the nucleosomal structure in chromatin spanning these regions is distinct from that of active chromatin and exhibits a number of signature features of metazoan heterochromatin. Through extensive analysis of the genes and proteins that contribute to transcriptional silencing the mechanism underlying the establishment, maintenance, and inheritance of these heterochromatic domains has been extensively detailed. These conclusions provide insights into the mechanisms underlying heterochromatin-associated position effect repression in larger organisms.

References

Abraham, J., Nasmayth, K.A., Strathern, J.N., Kalr, A.J.S., and Hicks, J.B. (1984). Regulation of mating-type information in yeast: negative control requiring sequences both 5' and 3' to the regulated region. *Journal of Molecular Biology*, **176**, 307–31.

Aparicio, O.M. and Gottschling, D.E. (1994). Overcoming telomeric silencing: a trans-activator competes to establish gene expression in a cell cycle-dependent way. *Genes and Development*, **8**, 1133–46.

Aparicio, O.M., Billington, B.L., and Gottschling, D.E. (1991). Modifiers of position effect are shared between telomeric and silent mating-type loci in *S. cerevisiae*. *Cell*, **66**, 1279–87.

Belyaev, N.D., Keohane, A.M., and Turner, B.M. (1996). Differential underacetylation of histones H2A, H3 and H4 on the inactive X chromosome in human female cells. *Human Genetics*, **97**, 573–8.

Brand, A., Breeden, L., Abraham, J., Sternglanz, R., and Nasmyth, K.A. (1985). Characterization of a silencer in yeast: a DNA sequence with properties opposite to those of a transcriptional enhancer. *Cell*, **41**, 41–8.

Braunstein, M., Rose, A.B., Holmes, S.G., Allis, C.D., and Broach, J.R. (1993). Transcriptional silencing in yeast is associated with reduced nucleosome acetylation. *Genes and Development*, **7**, 592–604.

Braunstein, M., Allis, C.D., Turner, B.M., and Broach, J.R. (1996). Transcriptional silencing in yeast requires a heterochromatin histone acetylation pattern. *Molecular and Cellular Biology*, **16**, 4349–56.

Buchman, A.R., Kimmerly, W.J., Rine, J., and Kornberg, R.D. (1988*a*). Two DNA-binding factors recognize specific sequences at silencers, upstream activating sequences, automomously replicating sequences, and telomeres in *Saccharomyces cerevisiae*. *Molecular and Cellular Biology*, **8**, 210–25.

Buchman, A.R., Lue, N.F., and Kornberg, R.D. (1988*b*). Connections between transcriptional activators, silencers, and telomeres as revealed by functional analysis of a yeast DNA-binding protein. *Molecular and Cellular Biology*, **8**, 5086–99.

Buck, S.W. and Shore, D. (1995). Action of a RAP1 carboxy-terminal silencing domain reveals an underlying competition between HMR and telomeres in yeast. *Genes and Development*, **9**, 370–84.

Cattanach, B.M. (1975). Control of chromosome inactivation. *Annual Review of Genetics*, **9**, 1–18.

Chen, T.A., Smith, M.M., Le, S., Sternglanz, R., and Allfrey, V.G. (1991). Nucleosome fractionation by mercury affinity chromatography. *Journal of Biological Chemistry*, **266**, 6489–98.

Chen-Cleland, T.A., Smith, M.M., Le, S., Sternglanz, R., and Allfrey, V.G. (1993). Nucleosome structural changes during derepression of silent mating-type loci in yeast. *Journal of Biological Chemistry*, **268**, 1118–24.

Chien, C.T., Bartel, P.L., Sternglanz, R., and Fields, S. (1991). The two-hybrid system: a method to identify and clone genes for proteins that interact with a protein of interest. *Proceedings of the National Academy of Sciences of the USA*, **88**, 9578–82.

Chien, C.T., Buck, S., Sternglanz, R., and Shore, D. (1993). Targeting of SIR1 protein establishes transcriptional silencing at HM loci and telomeres in yeast. *Cell*, **75**, 531–41.

Chung, J.H., Whiteley, M., and Felsenfeld, G. (1993). A 5' element of the chicken β-globin domain serves as an insulator in human erythroid cells and protects against position effect in *Drosophila*. *Cell*, **74**, 505–14.

Clayton, A.L., Hebbes, T.R., Thorne, A.W., and Crane-Robinson, C. (1993). Histone acetylation and gene induction in human cells. *FEBS Letters*, **336**, 23–6.

Cockell, M., Palladino, F., Laroche, T., Kyrion, G., Liu, C., Lustig, A.J., *et al.* (1995). The carboxy termini of Sir4 and Rap1 affect Sir3 localization: evidence for a multi-component complex required for yeast telomeric silencing. *Journal of Cell Biology*, **129**, 909–24.

Comings, D.E. (1968). The rationale for an ordered arrangement of chromatin in the interphase nucleus. *American Journal of Human Genetics*, **20**, 440–60.

Corces, V.G. and Geyer, P.K. (1991). Interactions of retrotransposons with the host genome: the case of the *gypsy* element of *Drosophila*. *Trends in Genetics*, **7**, 69–73.

Diffley, J.F. and Stillman, B. (1988). Purification of a yeast protein that binds to origins of DNA replication and a transcriptional silencer. *Proceedings of the National Academy of Sciences of the USA*, **85**, 2120–4.

Eissenberg, J.C. (1989). Position effect variegation in *Drosophila*: towards a genetics of chromatin assembly. *BioEssays*, **11**, 14–17.

Eissenberg, J.C. and Elgin, S.C.R. (1991). Boundary functions in the control of gene expression. *Trends in Genetics*, **7**, 335–40.

Feldman, J.B., Hicks, J.B. and Broach, J.R. (1984). Identification of the sites required for repression of a silent mating type locus in yeast. *Journal of Molecular Biology*, **178**, 815–34.

Ferguson, B.M. and Fangman, W.L. (1992). A position effect on the time of replication origin activation in yeast. *Cell*, **68**, 333–9.

Foss, M., McNally, F.J., Laurenson, P., and Rine, J. (1993). Origin recognition complex (ORC) in transcriptional silencing and DNA replication in *S. cerevisiae*. *Science*, **262**, 1838–44.

Gerasimova, T.I., Gdula, D.A., Gerasimov, D.V., Simonova, O., and Corces, V.G. (1995). A *Drosophila* protein that imparts directionality on a chromatin insulator is an enhancer of position-effect variegation. *Cell*, **82**, 587–97.

Gilson, E., Roberge, M., Giraldo, R., Rhodes, D., and Gasser, S. (1993). Distortion of the DNA double helix by RAP1 at silencers and multiple telomeric binding sites. *Journal of Molecular Biology*, **231**, 293–310.

Gottschling, D.E. (1992). Telomere-proximal DNA in *Saccharomyces cerevisiae* is refractory to methyltransferase activity *in vivo*. *Proceedings of the National Academy of Sciences of the USA*, **89**, 4062–5.

Gottschling, D.E., Aparicio, O.M., Billington, B.L., and Zakian, V.A. (1990). Position effect at *S. cerevisiae* telomeres: reversible repression of pol II transcription. *Cell*, **63**, 751–62.

Grigliatti, T. (1991). Position-effect variegation—an assay for nonhistone chromosomal proteins and chromatin assembly and modifying factors. *Methods in Cell Biology*, **35**, 587–627.

Guacci, V., Hogan, E., and Koshland, D. (1994). Chromosome condensation and sister chromatid pairing in budding yeast. *Journal of Cell Biology*, **125**, 517–30.

Haber, J.E. and George, J.R. (1979). A mutation that permits the expression of normally silent copies of mating-type infromation in *Saccharomyces cerevisiae*. *Genetics*, **93**, 13–35.

Hardy, C.F.J., Sussel, L., and Shore, D. (1992). A Rap1-interacting protein involved in transcriptional silencing and telomere length regulation. *Genes and Development*, **6**, 801–14.

Hartwell, L.H. (1980). Mutants of *Saccharomyces cerevisiae* unresponsive to cell divison control by polypeptide mating hormone. *Journal of Cell Biology*, **85**, 811–22.

Hebbes, T.R., Thorne, A.W., Clayton, A.L., and Crane-Robinson, C. (1992). Histone acetylation and globin gene switching. *Nucleic Acids Research*, **20**, 1017–22.

Hebbes, T.R., Thorne, A.W., and Crane-Robinson, C. (1988). A direct link between core histone acetylation and transcriptionally active chromatin. *EMBO Journal*, **7**, 1395–402.

Hecht, A., Laroche, T., Strahl-Bolsinger, S., Gasser, S.M., and Grunstein, M. (1995). Histone H3 and H4 N-termini interact with SIR3 and SIR4 proteins: a molecular model for the formation of heterochromatin in yeast. *Cell*, **80**, 583–92.

Heitz, E. (1928). Das heterochromatin der moose. *Jahrbuecher fuer Wissenschaftliche Botanik*, **69**, 726–818.

Henikoff, S. (1990). Position-effect variegation after 60 years. *Trends in Genetics*, **6**, 422–6.

Herskowitz, I. (1989). A regulatory hierarchy for cell specialization in yeast. *Nature*, **342**, 749–57.

Herskowitz, I., Rine, J., and Strathern, J. (1992). Mating-type determination and mating type interconversion in *Saccharomyces cerevisiae*. In *The molecular and cellular biology of the yeast Saccharomyces* (ed. E.W. Jones, J.R. Pringle and J.R. Broach), pp. 583–656. Cold Spring Harbor Press, Cold Spring Harbor.

Hochstrasser, M., Mathog, D., Gruenbaum, Y., Saumweber, H., and Sedat, J. (1986). Spatial organization of chromosomes in the salivary gland nuclei of *Drosophila melanogaster*. *Journal of Cell Biology*, **102**, 112–23.

Holmes, S. and Broach, J.R. (1996). Silencers are required for inheritance of the repressed state in yeast. *Genes and Development*, **10**, 1021–32.

Holmes, S.G., Rose, A.B., Seuerle, K., Saez, E., Sayegh, S., Lee, Y.M., *et al.* (1997). Hyperactivation of the silencing proteins, Sir2p and Sir3p, cause genomic loss. *Genetics*, **145**, 605–14.

Ivy, J.M., Klar, A.J.S., and Hicks, J.B. (1986). Cloning and characterization of four *SIR* genes of *Saccharomyces cerevisiae*. *Molecular and Cellular Biology*, **6**, 688–702.

Jeppeson, P. and Turner, B.M. (1993). The inactive X chromosome in female mammals is distinguished by a lack of histone H4 acetylation, a cytogenetic marker for gene expression. *Cell*, **74**, 281–9.

John, B. (1988). The biology of heterochromatin. In *Heterochromatin: molecular and structural aspects* (ed. R.S. Verma), pp. 1–147. Cambridge University Press, Cambridge.

Johnson, L.M., Kayne, P.S., Kahn, E.S., and Grunstein, M. (1990). Genetic evidence for an interaction between *SIR3* and histone H4 in the repression of the silent mating loci in *Saccharomyces cerevisiae. Proceedings of the National Academy of Sciences of the USA*, **87**, 6286–90.

Johnson, L.M., Fisher-Adams, G., and Grunstein, M. (1992). Identification of a non-basic domain in the histone H4 N-terminus required for repression of the yeast silent mating loci. *EMBO Journal*, **11**, 2201–9.

Kayne, P.S., Kim, U.-J., Han, M., Mullen, J.R., Yoshizaki, F., and Grunstein, M. (1988). Extremely conserved histone H4 N terminus is dispensable for growth but essential for repressing the silent mating loci in yeast. *Cell*, **55**, 27–39.

Kellum, R. and Schedl, P. (1991). A position-effect assay for boundaries of higher order chromosomal domains. *Cell*, **64**, 1–20.

Klar, A.J.S., Fogel, S., and MacLeod, K. (1979). *MAR1*—a regulator of *HM*a and *HM*α loci in *Saccharomyces cerevisiae. Genetics*, **93**, 37–50.

Klar, A.J.S., Strathern, J.N., Broach, J.R., and Hicks, J.B. (1981). Regulation of transcription in expressed and unexpressed mating type cassettes of yeast. *Nature*, **289**, 239–44.

Klar, A.J.S., Strathern, J.N., and Abraham, J.A. (1984). Involvement of double-strand chromosomal breaks for mating-type switching in *Saccharomyces cerevisiae. Cold Spring Harbor Symposia on Quantitative Biology*, **49**, 77–88.

Konkel, L.M., Enomoto, S., Chamberlain, E.M., McCune-Zierath, P., Iyadurai, S.J., and Berman, J. (1995). A class of single-stranded telomeric DNA-binding proteins required for Rap1p localization in yeast nuclei. *Proceedings of the National Academy of Sciences of the USA*, **92**, 5558–62.

Kurtz, S. and Shore, D. (1991). RAP1 protein activates and silences transcription of mating-type genes in yeast. *Genes and Development*, **5**, 616–28.

Kyrion, G., Liu, K., Liu, C., and Lustig, A.J. (1993). RAP1 and telomere structure regulate telomere position effects in *Saccharomyces cerevisiae. Genes and Development*, **7**, 1146–59.

Lee, S. and Gross, D.S. (1993). Conditional silencing: the HMRE mating-type silencer exerts a rapidly reversible position effect on the yeast HSP82 heat shock gene. *Molecular and Cellular Biology*, **13**, 727–38.

Lee, F.-J. S., Lin, L.-W., and Smith, J.A. (1989). N^α Acetylation is required for normal growth and mating of *Saccharomyces cerevisiae. Journal of Bacteriology*, **171**, 5795–802.

Lima-de-Faria, A. and Jaworska, H. (1968). Late DNA synthesis in heterochromatin. *Nature*, **217**, 138–42.

Lin, R., Leone, J.W., Cook, R.G., and Allis, C.D. (1989). Antibodies specific to acetylated histones document the existence of deposition- and transcription-related histone acetylation in *Tetrahymena. Journal of Cell Biology*, **108**, 1577–88.

Liu, C., Mao, X., and Lustig, A.J. (1994). Mutational analysis defines a C-terminal tail domain of RAP1 essential for telomeric silencing in *Saccharomyces cerevisiae. Genetics*, **138**, 1025–40.

Locke, J., Kotarski, M.A., and Tartof, K.D. (1988). Dosage-dependent modifiers of position effect variegation in *Drosophila* and a mass action model that explains their effects. *Genetics*, **120**, 181–98.

Loo, S. and Rine, J. (1994). Silencers and domains of generalized repression. *Science*, **264**, 1768–71.

Loo, S., Fox, C.A., Rine, J., Kobayashi, R., Stillman, B., and Bell, S. (1995a). The origin recognition complex in silencing, cell cycle progression, and DNA replication. *Molecular Biology of the Cell*, **6**, 741–756.

Loo, S., Laurenson, P., Foss, M., Dillin, A., and Rine, J. (1995*b*). Roles of ABF1, NPL3 and YCL5 in silencing in *Saccharomyces cerevisiae*. *Genetics*, **141**, 889–902.

Mahoney, D.J. and Broach, J.R. (1989). The *HML* mating-type cassette of *Saccharomyces cerevisiae* is regulated by two separate but functionally equivalent silencers. *Molecular and Cellular Biology*, **9**, 4621–30.

Mahoney, D.J., Marquardt, R., Shei, G.J., Rose, A.B., and Broach, J.R. (1991). Mutations in the *HML* E silencer of *Saccharomyces cerevisiae* yield metastable inheritance of transcriptional repression. *Genes and Development*, **5**, 605–15.

Marshall, M., Mahoney, S., Rose, A., Hicks, J.B., and Broach, J.R. (1987). Functional domains of *SIR4*, a gene required for position effect regulation in *Saccharomyces cerevisiae*. *Molecular and Cellular Biology*, **7**, 4441–52.

Mastrangelo, M.F., Weinstock, K.G., Shafer, B.K., Hedge, A.-M., Garfinkel, D.J., and Strathern, J.N. (1992). Disruption of a silencer domain by a retrotransposon. *Genetics*, **131**, 529.

Mathog, D., Hochstrasser, M., Gruenbaum, Y., Saumweber, H., and Sedat, J. (1984). Characteristic folding pattern of polytene chromosomes in *Drosophila* salivary gland nuclei. *Nature*, **308**, 414–21.

McCarroll, R.M. and Fangman, W.L. (1988). Time of replication of yeast centromeres and telomeres. *Cell*, **54**, 505–13.

Megee, P.C., Morgan, B.A., Mittman, B.A., and Smith, M.M. (1990). Genetic analysis of histone H4: essential role of lysines subject to reversible acetylation. *Science*, **247**, 841–5.

Micklem, G., Rowley, A., Harwood, J., Nasmyth, K., and Diffley, J.F. (1993). Yeast origin recognition complex is involved in DNA replication and transcriptional silencing. *Nature*, **366**, 87–9.

Miller, A. and Nasymth, K. (1984). Role of DNA replication in the repression of silent mating type loci in yeast. *Nature*, **312**, 247–51.

Moretti, P., Freeman, K., Coodly, L., and Shore, D. (1994). Evidence that a complex of SIR proteins interacts with the silencer and telomere-binding protein RAP1. *Genes and Development*, **8**, 2257–69.

Mullen, J.R., Kayne, P.S., Moerschell, R.P., Tsunasawa, S., Gribskov, M., Colavito-Shepanski, M., *et al.* (1989). Identification and characterization of genes and mutants for an N-terminal acetyltransferase from yeast. *EMBO Journal*, **8**, 2067–75.

Nasmyth, K.A. (1982). The regulation of yeast mating-type chromatin structure by *SIR*: an action at a distance affecting both transcription and transposition. *Cell*, **30**, 567–78.

Nasmyth, K.A., Tatchell, K., Hall, B.D., Astell, C.R., and Smith, M. (1981). A position effect in the control of transcription at yeast mating type loci. *Nature*, **289**, 244–50.

O'Neill, L.P. and Turner, B.M. (1995). Histone H4 acetylation distinguishes coding regions of the human genome from heterochromatin in a differentiation-dependent but transcription-independent manner. *EMBO Journal*, **14**, 3946–57.

Palladino, F. and Gasser, S.M. (1994). Telomere maintenance and gene repression: a common end? *Current Opinion in Cell Biology*, **6**, 373–9.

Palladino, F., Laroche, T., Gilson, E., Axelrod, A., Pillus, L., and Gasser, S. M. (1993). SIR3 and SIR4 proteins are required for the positioning and integrity of yeast telomeres. *Cell*, **75**, 543–55.

Park, E.-C. and Szostak, J.W. (1990). Point mutations in the yeast histone H4 gene prevent silencing of the silent mating type locus *HML*. *Molecular and Cellular Biology*, **10**, 4932–4.

Pillus, L. and Rine, J. (1989). Epigenetic inheritance of transcription states in *S. cerevisiae*. *Cell*, **59**, 637–47.

Rastan, S. (1994). X chromosome inactivation and the Xist gene. *Current Opinion in Genetics and Development*, **4**, 292–7.

Renauld, H., Aparicio, O.M., Zierath, P.D., Billington, B.L., Chhablani, S.K., and Gottschling, D.E. (1993). Silent domains are assembled continuously from the telomere and are defined by promoter distance and strength, and by SIR3 dosage. *Genes and Development*, **7**, 1133–45.

Reynolds, A.E., McCarroll, R.M., Newlon, C.S., and Fangman, W.L. (1989). Time of replication of *ARS* elements along yeast chromosome III. *Molecular and Cellular Biology*, **9**, 4488–94.

Riggs, A.D. and Pfeiffer, G.P. (1992). X-chromosome inactivation and cell memory. *Trends in Genetics*, **8**, 169–74.

Rine, J. and Herskowitz, I. (1987). Four genes responsible for a position effect on expression from *HML* and *HMR* in *Saccharomyces cerevisiae*. *Genetics*, **116**, 9–22.

Rine, J.D., Strathern, J.N., Hicks, J.B., and Herskowitz, I. (1979). A suppressor of mating-type locus mutation in *Saccharomyces cerevisiae*: evidence for and identification of cryptic matingtype loci. *Genetics*, **93**, 877–901.

Roseman, R.R., Pirrotta, V., and P.K.Geyer (1993). The su(Hw) protein insulates expression of the *Drosophila melanogaster white* gene from chromosomal position-effects. *EMBO Journal*, **12**, 435–42.

Sandell, L.L. and Zakian, V.A. (1992). Telomeric position effect in yeast. *Trends in Cell Biology*, **2**, 10–14.

Scherthan, H., Loidl, J., Schuster, T., and Schweizer, D. (1992). Meiotic chromosome condensation and pairing in *Saccharomyces cerevisiae* studied by chromosome painting. *Chromosoma*, **101**, 590–5.

Schnell, R. and Rine, J. (1986). A position effect on the expression of a tRNA gene mediated by the *SIR* gene of *Saccharomyces cerevisiae*. *Molecular and Cellular Biology*, **6**, 494–501.

Shei, G.J. and Broach, J.R. (1995). Yeast silencers can act as orientation-dependent gene inactivation centers that respond to environmental signals. *Molecular and Cellular Biology*, **15**, 3496–506.

Shore, D. and Nasmyth, K. (1987). Purification and cloning of a DNA binding protein from yeast that binds to both silencer and activator elements. *Cell*, **51**, 721–32.

Shore, D., Stillman, D.J., Brand, A.H., and Nasmyth, K.A. (1987). Identification of silencer binding proteins from yeast: possible roles in *SIR* control and DNA replication. *EMBO Journal*, **6**, 461–7.

Singer, M.S. and Gottschling, D.E. (1994). TLC1: template RNA component of *Saccharomyces cerevisiae* telomerase. *Science*, **266**, 404–9.

Singh, J. and Klar, A.J.S. (1992). Active genes in budding yeast display enhanced *in vivo* accessibility to foreign DNA methylases: a novel *in vivo* probe for chromatin structure of yeast. *Genes and Development*, **6**, 186–96.

Stavenhagen, J.B. and Zakian, V.A. (1994). Internal tracts of telomeric DNA act as silencers in *Saccharomyces cerevisiae*. *Genes and Development*, **8**, 1411–22.

Stone, E. M., Swanson, M. J., Romeo, A. M., Hicks, J. B., and Sternglanz, R. (1991). The SIR1 gene of *Saccharomyces cerevisiae* and its role as an extragenic suppressor of several mating-defective mutants. *Molecular and Cellular Biology*, **11**, 2253–62.

Strathern, J. N., Klar, A.J.S., Hicks, J.B., Abraham, J.A., Ivy, J.M., Nasmyth, K.A., and McGill, C. (1982). Homothallic switching of yeast mating type cassettes is initiated by a double-stranded cut in the *MAT* locus. *Cell*, **31**, 183–92.

Sussel, L. and Shore, D. (1991). Separation of transcriptional activation and silencing functions of the RAP1-encoded repressor/activator protein 1: isolation of viable mutants affecting both silencing and telomere length. *Proceedings of the National Academy of Sciences of the USA*, **88**, 7749–53.

Sussel, L., Vannier, D., and Shore, D. (1993). Epigenetic switching of transcriptional states: cis- and trans-acting factors affecting establishment of silencing at the *HMR* locus in *Saccharomyces cerevisiae*. *Molecular and Cellular Biology*, **13**, 3919–28.

Terleth, C., Sluis, C.A. v., and Van de Putte, P.(1989). Differential repair of UV damage in *Saccharomyces cerevisiae*. *Nucleic Acids Research*, **17**, 4433–9.

Thompson, J.S., Ling, X., and Grunstein, M. (1994). Histone H3 amino terminus is required for telomeric and silent mating locus repression in yeast. *Nature*, **369**, 245–7.

Turner, B.M. (1991). Histone acetylation and control of gene expression. *Journal of Cell Science*, **99**, 13–20.

Turner, B.M., Birley, A.J., and Lavender, J. (1992). Histone H4 isoforms acetylated at specific lysine residues define individual chromosomes and chromatin domains in *Drosophila* polytene nuclei. *Cell*, **69**, 375–84.

Vignais, M.L. and Sentenac, A. (1989). Asymmetric DNA bending induced by the yeast multifunctional factor TUF. *Journal of Biological Chemistry*, **264**, 8463–6.

Wallrath, L.L. and Elgin, S.C. (1995). Position effect variegation in *Drosophila* is associated with an altered chromatin structure. *Genes and Development*, **9**, 1263–77.

Wang, S.-S. and Zakian, V.A. (1990). Sequencing of *Saccharomyces* telomeres cloned using T4 DNA polymerase reveals two domains. *Molecular and Cellular Biology*, **10**, 4415–19.

Weiner, B.M. and Kleckner, N. (1994). Chromosome painting via multiple interstitial interactions before and during meiosis in yeast. *Cell*, **77**, 977–91.

Whiteway, M., Freedman, R., Arsdell, S.V., Szostak, J.W., and Thorner, J. (1987). The yeast *ARD1* gene product is required for repression of cryptic mating-type information at the *HML* locus. *Molecular and Cellular Biology*, **7**, 3713–22.

Wilson, C., Bellen, H.J., and Gehring, W.J. (1990). Position effects on eukaryotic gene expression. *Annual Review of Cell Biology*, **6**, 679–714.

16

Homology-dependent gene silencing in transgenic plants and related phenomena in other eukaryotes

Marjori A. Matzke and Antonius J. M. Matzke

Introduction

Higher eukaryotic genomes contain substantial amounts of repetitive DNA. What limitations, if any, are placed on the quantity and distribution of repetitive sequences and do these parameters in turn influence gene expression? Some contend that non-genic repeated DNA sequences, which constitute ca. 30 per cent of mammalian genomes and 15–90 per cent of flowering plant genomes, are mostly non-functional 'junk' (John and Miklos 1988), while others suggest that they have essential structural and/or regulatory roles (Davidson and Britten 1979; Smyth 1991; see Chapter 13). In plants large disparities in DNA content arise primarily from differences in the repetitive fraction, as reflected in recent findings showing that even though DNA amounts can vary greatly in closely related species, gene composition and gene order remain relatively constant (synteny). This is particularly striking in the cereals (Moore *et al.* 1993). For example, the genetic linkage maps of rice and wheat are remarkably conserved, even though they differ 30-fold in DNA content (Kurata *et al.* 1994). These findings have prompted the remark that 'rice is wheat without the repeats' (G. Moore, quoted in Cherfas 1994). Noteworthy synteny has also been observed in related mammals (O'Brien and Graves 1991). Despite arguments to the contrary (John and Miklos 1988), it is difficult to abandon the notion that at least some of this repetitive DNA, particularly those sequences that are species specific, contributes to the differential regulation of a 'generic' protein-coding genome.

In this chapter we will consider a possible role for repeated sequences in gene silencing. Although differential gene activation usually receives the most attention, the problem of gene inactivation is not trivial: in differentiated cell types of multicellular organisms a formidable number of genes must be silenced. For a mammalian genome encoding 80 000 proteins, it has been estimated that a given differentiated cell type produces 30 000 different ubiquitous mRNAs and 10 000 tissue-specific mRNAs. This leaves 40 000 genes that must be repressed in a given cell type (Bird 1995). How is this massive degree of silencing achieved? While repressor proteins surely exist, it is unlikely that they are numerous or diverse enough to account for the extensive silencing that needs to take place (Johnson 1995; see Chapters 13 and 14).

Requirements for gene silencing also have an evolutionary dimension. Bird (1995) has suggested that major evolutionary transitions involving an increase in gene number must have been accompanied by novel silencing mechanisms that minimized the inappropriate expression of the additional genetic information in specific cells types. He has proposed that cytosine methylation fulfils this role in vertebrates. Similar arguments can be made for higher plants, which like vertebrates generally have large, highly methylated genomes (Bestor 1990; see also Chapters 11 an 12).

Silencing is thus an essential aspect of differential gene regulation in higher organisms as well as a prerequisite for the proliferation of gene number during the course of evolution. Cytosine methylation has consistently been associated with gene silencing in eukaryotes, and it appears to be essential for the normal development of mice (Li *et al.* 1992) and plants (Finnegan *et al.* 1996). Despite the importance of cytosine methylation in higher organisms, the trigger(s) for the extensive *de novo* methylation required to silence large portions of genomes have not been identified. Recall that the problem of *de novo* methylation is separate from that of preserving methylation once it has been acquired, because maintenance methylase will continue to methylate symmetrical cytosine residues (CpG in animals and both CpG and CpNpG in plants) during subsequent rounds of DNA replication.

As we will discuss, accumulating evidence from diverse eukaryotes suggests that homologous DNA sequences can pair to produce silenced genetic states *de novo*. Because pairing can conceivably occur between both linked and unlinked repeats, the inactive state can spread to other members of the repeat family dispersed throughout a genome. It is thus plausible that gene silencing in higher eukaryotes can be accomplished by interactions among repetitive sequences as long as they maintain sufficient sequence homology over a certain minimum length to allow pairing. This proposal derives from recent work on the behaviour of repeated or homologous sequences in transgenic plants as well as filamentous fungi and *Drosophila* (also see Chapter 13).

Pairing of DNA repeats can lead to *de novo* methylation and gene silencing: experimental evidence

One approach to the problem of repetitive DNA function is to study the influence of specific repeats on the expression of defined genes. There has until recently been surprisingly little experimental work done in this area, although such questions are admittedly difficult to approach experimentally because of the sheer bulk of repetitive DNA in many genetically tractable higher eukaryotes, such as maize and mice. Even studies with naturally occurring repeated sequences in *Drosophila*, which has a comparatively small genome and modest amounts of repetitive DNA, have led to conflicting results. One analysis has suggested that loss of repetitive DNA leads to reduced fitness (Wu *et al.* 1989), while others indicate that large regions of heterochromatin, which are enriched in repeated sequences, can be deleted with no ill effects (John and Miklos 1988). As often happens in biological science, conceptual advances are driven by techno-logical innovations. The ability to introduce foreign DNA sequences into organisms has led to new insights about how diverse taxa deal with newly duplicated or repeated DNA sequences.

Filamentous fungi

Two classic examples from filamentous fungi serve as models for how higher eukaryotes might treat repetitive sequences. These process are repeat-induced point mutation (RIP) in *Neurospora crassa* (Selker 1990) and methylation induced premeiotically (MIP) in *Ascobolus immersus* (Rossignol and Faugeron 1994). The central principle illuminated by these systems is that pairing of duplicated DNA regions can serve as the primary signal for *de novo* methylation of cytosine residues. Both examples owe their discovery to observations on the fate of transforming sequences during the sexual cycle. Filamentous fungi maintain a streamlined genome with little repetitive DNA, suggesting that they possess a means for actively avoiding the accumulation of excess DNA. If a gene is duplicated by a natural mechanism, or added by transformation, an organism can eliminate it by recombination. Although recombination of a tandem duplication will lead to deletion of one copy, recombination between unlinked (ectopic) repeats would lead to a translocation. Because *Neurospora* and *Ascobolus* exist as haploids, such rearrangements could be lethal. Therefore, even though both organisms often actively delete one copy of a tandem duplication, another strategy that has less physically disruptive consequences has evolved. During the sexual cycle both tandem and ectopic duplicated sequences are inactivated by heavy methylation at most or all cytosines (Selker 1990; Rossignol and Faugeron 1994). Although methylation in the duplicated region can infiltrate somewhat into adjacent unique sequences in *Neurospora*, it is exactly co-extensive with the duplication in *Ascobolus*, suggesting that paired DNA is the substrate recognized by the methylation machinery. In both fungi either both of the copies become methylated or none, again implying a pair-wise interaction.

With RIP in *Neurospora* there is an additional outcome to cytosine methyla-tion. Duplicated sequences become riddled with $C : G$ to $T : A$ transition mutations. Although the exact relationship between methylation and mutation in *Neurospora* is not completely understood, Selker (1990) has suggested that $C : G$ to $T : A$ mutations possibly represent a methylation attempt that is aborted by a deficiency of the methyl donor, S-adenosyl methionine (SAM). This prolongs the existence of an intermediate compound, 5,6-dihydrocytosine, which has an extremely high rate of spontaneous deamination, leading to the production of thymine. Mechanistically, therefore, RIP and MIP are probably the result of the same process, which under SAM-limiting conditions can be shunted toward C : G to T : A transitions and away from C-methylation in *Neurospora*. While methylation itself is probably sufficient to prevent recombination between duplicated sequences in *Ascobolus* (Rossignol and Faugeron 1994), RIP further reduces recombination efficiency in *Neurospora* by creating enough sequence divergence so that repeats can no longer pair.

Although RIP and MIP were originally discovered by studying duplications generated by transformation, subsequent work in *Neurospora* has demonstrated that endogenous duplications are also subject to RIP. Therefore, the process is not restricted to foreign DNA sequences. RIP does not act on all repeated sequences, however; the gene family encoding 18S and 28S rRNA in *Neurospora* remains unmutated and largely unmethylated, possibly because of a protected chromosomal location (Selker 1990). It has been argued that RIP and MIP are processes unique to a small number of haploid, dikaryotic cells at a specific (premeiotic) stage of the sexual cycle of *Neurospora* and *Ascobolus* (Selker 1990; Rossignol and Faugeron 1994). Recent results, however, suggest that this definition might be too restricted. First, methylation of multiple copies of transforming sequences has been shown to occur in vegetative (non-sexual) cells of *Neurospora* (Pandit and Russo 1992). Second, *Coprinus cinereus*, a basidiomycete fungus that is evolutionarily more advanced than the ascomycetes *Neurospora* and *Ascobolus*, has been found to possess a MIP-like activity (Freedman and Pukkila 1993). These data remove the rigid constraints from the basic MIP/RIP process and lend credence to the proposition that DNA–DNA pairing is a universal signal for *de novo* methylation that could also operate in higher eukaryotes. Indeed, Kricker and co-workers have inferred the action of RIP in vertebrate genomes by comparing CpG depletion in repetitive versus unique sequences. Consistent with a RIP-like process, fewer CpG dinucleotides were found in repeated sequences than in single copy sequences (Kricker *et al.* 1992). Recent work on the behaviour of artificially duplicated sequences in transgenic plants also suggests that repeated DNA sequences are preferential substrates for inactivation by methylation.

Transgenic plants

Ascobolus and *Neurospora* either delete one copy of a duplicated DNA sequence or methylate/diversify both copies by MIP/RIP, thus maintaining relatively uncluttered genomes and minimizing the problem of detrimental DNA rearrangements. In contrast, the genomes of higher plants contain substantial amounts of repetitive DNA, indicating that deletion strategies are not employed to prevent the accumulation of repeated sequences. Consequently, in the absence of active elimination mechanisms proliferation of DNA repeats via natural processes, such as duplicative transposition and unequal crossing over, will occur over time in plant lineages. Nevertheless, some protective defence against repeats is suggested, because repetitive sequences in plants are usually highly methylated. Taking the fungal systems as precedents, one can ask whether pairing between homologous DNA sequences has served as the trigger for *de novo* methylation of these regions. Transformation of plants with genes homologous to resident genes (this can be either an endogenous plant gene or a previously introduced transgene) has yielded information on how plants deal with newly duplicated DNA sequences.

Homology-dependent or repeat-induced gene silencing

Homology-dependent or repeat-induced gene silencing (RIGS) (Assaad *et al.* 1993) refers to the inactivation of either multiple copies of transgenes or a transgene and homologous endogenous gene (endogene). This phenomenon, which has been reviewed extensively (Finnegan and McElroy 1994; Flavell 1994; Matzke and Matzke 1995; Meyer 1995, 1996), has generated considerable interest, in large part because it was so unexpected. One type of experiment that led to the discovery of homology-dependent silencing was the introduction of sense copies of genes in attempts to enhance the synthesis of flower pigments. Instead of the expected overexpression, however, a complete collapse of expression was often observed, leading to pure white or patterned flowers instead of deep purple ones (reviewed by Jorgensen 1995). Likewise, sense constructs introduced as controls for antisense experiments often were more effective at silencing endogenes than the antisense versions of the transgene. Repeated copies of transgenes were also found to affect one another's expression. Particularly intriguing were cases in which one transgene locus could influence the state of methylation and expression of a second, partially homologous transgene locus on a different chromosome. These results suggested a previously unsuspected ability for unlinked homologous sequences to cross-talk in somatic cells (Matzke *et al.* 1989; Vaucheret 1993).

Two general classes of homology-dependent gene silencing have been identified in transgenic plants (Matzke and Matzke 1995). One of these is a post-transcriptional process involving RNA turnover (Van Blokland *et al.* 1994; Niebel *et al.* 1995) and will not be considered further here (for a review see Jorgensen 1995). The second class includes silencing phenomena that are due to transcriptional inactivation and are correlated with cytosine methylation in promoter regions (Meyer *et al.* 1993; Neuhuber *et al.* 1994; Park *et al.* 1996). Our discussion will focus on these latter effects because they possibly involve direct interactions between homologous DNA sequences.

Methylation of repeated transgenes in plants

Work from a number of labs has demonstrated a marked propensity for repeated transgene sequences, either linked or unlinked, to become methylated and silenced in different plant species (Matzke *et al.* 1989, 1994b; Linn *et al.* 1990; Assaad *et al.* 1993; Vaucheret 1993; Park *et al.* 1996). This methylation is not simply a response to 'foreign' or invading DNA (Bestor 1990) because it can be reduced when the transgene copy number at a locus is decreased by recombination (Assaad *et al.* 1993) or reversed when unlinked repeats are segregated in progeny (Matzke *et al.* 1989; Vaucheret 1993; Park *et al.* 1996). Although there is no direct proof that DNA–DNA pairing is involved, the fact that duplicated or repeated sequences are particularly inclined to methylate suggests that DNA interactions are occurring. Moreover, the cases in which methylation of a target locus is provoked by a methylated 'silencing' locus require a *trans*-acting methylation signal, which could possibly involve physical contact of unlinked DNA repeats (see below).

It is important to note that in transgenic plants there are cases where repeated sequences can remain unmethylated. The chromosomal or nuclear location is possibly an important factor, as it apparently is in the immunity of *Neurospora* 18S and 28S rRNA genes to RIP. The arrangement of multiple copies might also play a role. For example, inverted repeats are generally more prone to silencing than are direct repeats (Van Blokland *et al.* 1994).

Trans-silencing

Homology-dependent gene silencing in transgenic plants can involve not only linked repeats (*cis*-inactivation), but also repeated sequences on non-homologous chromosomes. Here one transgene locus, which autonomously acquires methylation, can direct the silencing and methylation of an unlinked homologous sequence, which normally would maintain a hypomethylated state (*trans*-inactivation).

Two *trans*-acting 'silencing' loci have been characterized. Both contain multiple, methylated copies of the respective transgene construct (Vaucheret 1993; Matzke *et al.* 1994b). It has been proposed that the autonomous methylation of a silencing locus is due to pairing of the multiple transgene copies present at the locus. When a homologous unlinked 'target' locus is introduced by sexual crossing into the silencer line, the target becomes inactivated and acquires a degree of methylation comparable to the silencing locus. Thus, a *trans*-acting signal originating at the silencing locus can lead to *de novo* methylation of the target. How might this occur? Of the possible signals for *de novo* methylation in plants (Matzke and Matzke 1995) only two could potentially act in *trans*: a diffusible RNA molecule or DNA–DNA pairing. Because of the structure of the transgene construct used in the original experiments, the silencing and the target locus share DNA sequence homology primarily in the promoter regions. The promoters are not transcribed, therefore the diffusible RNA originating at the silencing locus cannot be the *trans*-acting signal for *de novo* methylation of the target locus in these cases (as it could if homology were also present in transcribed regions). Pairing between homologous promoters at the silencing locus and the target locus is currently the most plausible explanation for the observed *trans*-silencing and the associated methylation. DNA pairing can thus account for both the spontaneous methylation of the multicopy silencing locus and for imposition of methylation on the target in the presence of the silencer.

A model for the pairing-dependent imposition of methylation from a methylated sequence to a homologous partner has been proposed (Matzke *et al.* 1994b). It takes advantage of the fact that maintenance methylase acts preferentially on a hemi-methylated substrate, which could conceivably form when methylated and unmethylated homologous partners pair. The term 'epigene conversion', originally introduced by Sabl and Laird (1992), has been suggested as a general term for phenomena in which the epigenetic state at one allele or locus can be imposed on a homologous partner (Matzke *et al.* 1994b).

Non-transgenic plant systems: paramutation and transposable elements

Transgenes reveal the behaviour of newly introduced repeated genes in plant nuclei. Is there any evidence that endogenous plant genes are subject to similar epigenetic silencing and methylation that may be due to pairing of homologous DNA sequences? The phenomenon of paramutation has many parallels to the *trans*-silencing of transgenes. Paramutation was first identified decades ago for several endogenous genes in maize and tomato. It refers to an interaction between a sensitive (paramutable) allele and an inducing (paramutagenic) allele in the heterozygote. The interaction weakens the activity of the paramutable allele in a way that persists even after the two alleles have segregated in progeny (i.e. the decreased expression is meiotically heritable). Although paramutation has long been considered an esoteric process affecting only a handful of plant genes, the remarkable similarities to *trans*-silencing of transgenes, which also involves a directed reduction in the activity of one gene by another, has reinvigorated interest in the phenomenon. Models for paramutation also invoke pairing or direct physical contact of alleles (Meyer *et al.* 1993; Patterson and Chandler 1995). Paramutation at the *r* locus in maize shows particularly striking similarities to transgene interactions. First, sensitive and inducing *R* alleles are complex and contain multiple copies of the *r* gene (Walker *et al.* 1995; Kermicle *et al.* 1995; Eggleston *et al.* 1995). Second, a clear correlation between copy number of the *r* gene and paramutagenicity (i.e. silencing ability) has been established (Kermicle *et al.* 1995). Finally, changes in cytosine methylation in promoter regions are associated with *r* genes involved in paramutation (Eggleston *et al.* 1995; M. Alleman, personal communication).

The behaviour of certain plant transposable elements resembles *trans*-inactivation processes in transgenic plants. For example, the numerous copies of *Mutator* elements that are dispersed throughout the maize genome can become coordinately inactivated and methylated. Bennetzen and co-workers have suggested that multiple rounds of ectopic pairing between modified and unmodified copies of the homologous *Mu* elements leads to serial transmission of methylation from one copy to the next (Brown *et al.* 1994). This proposal is similar to the epigene conversion process suggested for *trans*-silencing effects involving unlinked transgenes (Matzke *et al.* 1994b).

Drosophila

Drosophila lacks significant cytosine methylation and inactive genetic states are associated with the formation of heterochromatin. However, in a clear parallel with the observations made in fungi and plants, recent work with transgenes in *Drosophila* has suggested that heterochromatinization might be driven by pairing between homologous sequences. Multiple copies of a *white* transgene, in contrast to single copies at the same locus, were found to spontaneously form heterochromatin and become silenced (Dorer and Henikoff 1994). The authors discussed the possibility that pairing of the transgene repeats served as the trigger for heterochromatinization (see Chapter 13).

Position effect variegation (PEV) is a natural phenomenon in *Drosophila* that has relevance for homology-dependent gene silencing. PEV refers to the variegated or mosaic expression of a euchromatic gene when it is juxtaposed next to heterochromatin following a chromosome rearrangement (Henikoff 1995). A role for pairing of repeats in conferring susceptibility to PEV was suggested by experiments showing that multiple copies of a *brown* transgene were more likely to exhibit variegated expression in the presence of adjacent heterochromatin than were single copies of *brown* (Sabl and Henikoff 1995). Note that PEV is not restricted to *Drosophila*. Plant transgenes that are deliberately placed downstream of tandemly repeated sequences can be expressed in a variegated manner (Ten Lohuis *et al.* 1995; M. Matzke and A. Matzke, unpublished results).

Finally, *trans*-silencing has also been observed in *Drosophila*. The classic case concerns dominant PEV at the *brown* locus. Here, the inactive, heterochromatinized state of the rearranged allele can be imposed in *trans* on the unrearranged allelic partner in a pairing-dependent process (Dreesen *et al.* 1991).

Mammals

Work on parental imprinting in mice is beginning to suggest a role for DNA repeats in provoking epigenetic modifications in these organisms. A recent survey revealed that short repeats are invariably associated with imprinted genes, and the possibility that these repeats could pair with each other and induce heterochromatin formation has been discussed (Neumann *et al.* 1995).

Do homologous DNA sequences pair in somatic cells?

We have reviewed evidence suggesting that pairing of repeated sequences can lead to the *de novo* generation of inactive genetic states (methylation in fungi and plants; heterochromatinization in *Drosophila*). The phenomenon of PEV indicates that this heterochromatinization/methylation can lead to *cis*-inactivation of adjacent structural genes. *Trans*-inactivation phenomena, such as dominant PEV, paramutation, and silencing of unlinked transgene loci, demonstrate that an inactive state can be imposed from one gene to another on a different DNA molecule, possibly through a pairing-dependent process.

If, as the studies described above suggest, DNA pairing indeed plays a major role in the acquisition and spreading of the silenced state, an important consideration is whether the requisite pairing actually occurs in somatic cells. Moreover, to be implicated in differential gene regulation the pairing of specific repeat families would have to be developmentally controlled. These points bear on the issue of the arrangement of chromatin in interphase nuclei, which by all indications so far does not appear to be random (Manuelidis 1990; Heslop-Harrison and Bennett 1990; see Chapter 9). Although many questions remain unanswered, there are some relevant observations on the occurrence and control of pairing of homologous DNA sequences in higher eukaryotes. In *Drosophila* stable somatic pairing of homologous chromosomes occurs early in development (Hiraoka *et al.* 1993). Homologous sequences at ectopic locations can also interact at an appreciable frequency in somatic cells, even in relatively large genomes that contain significant amounts of heterochromatin and many dispersed families of moderately repeated sequences. This is best illustrated by work in *Drosophila* on the repair of double-stranded DNA breaks that are incurred when P transposable elements excise from the genome. The repair of these breaks apparently involves a search for homology by the free DNA ends at the breakpoint, followed by strand invasion and repair synthesis across the intact template DNA. Remarkably, even short regions of homology, perhaps as little as 100 bp, can be detected anywhere in the genome (Nassif and Engels 1993; Nassif *et al.* 1994). In plants there is no indication for stable pairing of homologous chromosomes, but there might be some preferential association of non-homologous chromosomes (Hilliker and Appels 1989). Some evidence indicates that double strand breaks can be repaired from ectopic templates in plants, pointing towards associations of homologous DNA sequences on non-homologous chromosomes (Rommens *et al.* 1993). Collectively, these data suggest that eukaryotic genomes possess a homology searching/pairing mechanism that operates efficiently in diploid somatic cells.

With respect to developmental control of homologous associations, certain repetitive sequences have been shown to interact and occupy distinct domains in interphase cells (Toledo *et al.* 1992; Abbo *et al.* 1995). The phenomenon of 'nuclear repatterning', in which different heterochromatic regions interact variably in specific tissues of certain plant species, suggests that differential association of repeated sequences might underlie development (Ceccarelli and Cionini 1993). In animal cells distinct repeat families have been shown to acquire methylation at specific times during development (Matzke *et al.* 1994a and references therein), suggesting that interactions between repeats follow a pattern of temporal control.

Escape from homology-dependent silencing

As we have discussed, the pairing of homologous DNA regions can apparently provide a signal for the formation of inactive genetic states, defined as increased cytosine methylation and/or heterochromatinization. Transgenes have been particularly useful in illuminating this behaviour, because after integration they are initially 100 per cent homologous and subsequently fall readily into the 'homology trap', meaning that multiple transgene copies would be able to undergo homologous pairing and become silenced. Which natural DNA sequences might be involved in these interactions? It is important to note that multigene families encoding essential proteins can avoid homology-dependent silencing if they are kept physically separate in the nucleus (for example by attachment to the nuclear matrix or envelope), or they are divided into exons that are too short to pair efficiently (Kricker *et al.* 1992). Although many endogenous non-genic repeats have diverged to varying degrees, a number of them share sufficient sequence homology to interact by DNA pairing. Therefore, these sequences are probably the most affected by homology-dependent processes, and their associations could serve to establish silenced chromosomal domains that influence the expression of neighbouring structural genes.

Repeated sequences: homogenization vs diversification

One can distinguish two opposing forces, both dependent on DNA sequence homology, that operate in eukaryotic genomes to deal with repeats that are not eliminated by recombination. First, there are mechanisms, such as unequal crossing over and gene conversion, that can homogenize repeats ('concerted evolution') (Zimmer *et al.* 1980). These processes and others, such as transposition and slippage, are embodied in the concept of 'molecular drive' (Dover 1986), which attempts to account for the spread of repeated DNA sequences throughout genomes and the preservation of sequence homology. On the other hand, homology-dependent silencing, provoking hypermethylation of repeats, will eventually serve to diversify them either through a greatly accelerated mutational process such as RIP, or because 5-methylcytosine is inherently mutagenic (Jones *et al.* 1992). When sequences become too dissimilar, they can no longer pair and are free to evolve into something new, or truly become nuclear refuse.

Such processes would lead to an accumulation of repeats and an increase in genome size, while having the additional consequence of gene silencing when specific, highly homologous repeats are able to interact by pairing. The size, organization, and associations of different repeat families would thus serve to modify the expression of a generic genome that encodes proteins. As has been noted previously (Dorer and Henikoff 1994), the exact DNA sequence of the repeats would not matter. The only crucial feature would be whether the degree and length of sequence homology were insufficient to allow pairing. In this view, non-conservation of a DNA sequence between species does not automatically imply lack of function: if the function in question is gene silencing, then the critical factor would not be a specific sequence but the intraspecific sequence homogeneity of repeats. The existence of similar but divergent repeats that have become fixed in related species (Matzke *et al.* 1992) accurately fulfils this expectation.

Future directions

The phenomenon of homology-dependent gene silencing is revealing novel interactions between both linked and unlinked DNA sequences and is suggesting a new role for repetitive sequences in eukaryotic genomes. Future work will establish further the incidence of this type of silencing and the exact nature and genomic distribution of the DNA sequences involved. A long-term goal will be to understand the dynamic interactions among members of repetitive sequence families. This problem clearly focuses attention on the three-dimensional disposition of genes in the interphase nucleus of somatic cells and on the ways in which this arrangement can change in a regulated manner to allow new contacts between homologous sequences as development unfolds.

Summary

Transgenic organisms allow systematic studies of the fate of newly duplicated sequences in both higher and lower eukaryotes. In this way they enrich our view of the behaviour and functions of repetitive sequences. The ready inactivation and epigenetic modification of repeated transgenes in plants, fungi, and *Drosophila* suggest that the pairing of homologous DNA sequences can serve as a signal for the *de novo* generation of inactive genetic states, defined as cytosine methylation and/or heterochromatinization. The resultant methylated or heterochromatic regions could then spread in *cis* and in *trans* to silence structural genes in the vicinity of a repeat family member. The ability of repeats to interact by pairing would be determined by the non-random arrangement of interphase chromatin, which could differ in different cell types. Homology-dependent interactions could thus play a powerful role in provoking the extensive gene silencing that occurs in differentiated cells of higher eukaryotes.

Acknowledgements

Our research has been supported by the Austrian Fonds zur Förderung der wissenschaftlichen Forschung and the Bundesministerium für Wissenschaft und Forschung.

References

Abbo, S., Dunford, R.P., Foote, T.N., Reader, S.M., Flavell, R.B., and Moore, G. (1995). Organization of retro-element and stem-loop repeat families in the genomes and nuclei of cereals. *Chromosome Research*, **3**, 5–15.

Assaad, F., Tucker, K.L, and Signer, E.R. (1993). Epigenetic repeat-induced gene silencing (RIGS) in *Arabidopsis*. *Plant Molecular Biology*, **22**, 1067–85.

Bestor, T.H. (1990). DNA methylation: evolution of a bacterial immune function into a regulator of gene expression and genome structure in higher eukaryotes. *Philosophical Transactions of the Royal Society of London*, **B326**, 179–87.

Bird, A.P. (1995). Gene number, noise reduction and biological complexity. *Trends in Genetics*, **11**, 94–100.

Brown, W.E., Springer, P.S., and Bennetzen, J.L. (1994). Progressive modification of *Mu* transposable elements during development. *Maydica*, **39**, 119–26.

Ceccarelli, M. and Cionini, P.G. (1993).Tissue-specific nuclear repatterning in plant cells. *Genome*, **36**, 1092–8.

Cherfas, J. (1994). Give us the dough to crack wheat genome. *New Scientist*, **141**, 10.

Davidson, E.H. and Britten, R.J. (1979). Regulation of gene expression: possible role of repetitive sequences. *Science*, **204**, 1052–9.

Dorer, D.R. and Henikoff, S. (1994). Expansions of transgene repeats cause heterochromatin formation and gene silencing in *Drosophila*. *Cell*, **77**, 993–1002.

Dover, G.A. (1986). Molecular drive in multigene families: how biological novelties arise, spread and are assimilated. *Trends in Genetics*, **2**, 159–65.

Dreesen, T.D., Henikoff, S., and Loughney, K. (1991). A pairing-sensitive element that mediates trans-inactivation is associated with the *Drosophila brown* gene. *Genes and Development*, **5**, 331–40.

Eggleston, W., Alleman, M., and Kermicle, J. (1995). Molecular organization and germinal instability of *R-stippled* maize. *Genetics*, **141**, 347–60.

Finnegan, J. and McElroy, D. (1994). Transgene inactivation: plants fight back! *Biotechnology*, **12**, 883–8.

Finnegan, J., Genger, R.K., Peacock, W.J., and Dennis, E.S. (1996). Reduced DNA methylation in *Arabidopsis thaliana* results in abnormal plant development. *Proc. Natl. Acad. Sci. USA*, **93**, 8449–54.

Flavell, R.B. (1994). Inacitvation of gene expression in plants as a consequence of novel sequence duplications. *Proceedings of the National Academy of Sciences of the USA*, **91**, 3490–6.

Freedman, T. and Pukkila, P.J. (1993). *De novo* methylation of repeated sequences in *Coprinus cinereus*. *Genetics*, **135**, 357–66.

Henikoff, S. (1995). A pairing-looping model for position-effect variegation in *Drosophila*. In *Genomes (Proceedings of the 22nd Stadler Genetics Symposium)* (ed. J.P. Gustafson and R.B. Flavell), in press. Plenum Press, New York.

Heslop-Harrison, J.S. and Bennett, M.D. (1990). Nuclear architecture in plants. *Trends in Genetics* **6**, 401–5.

Hilliker, A.J. and Appels, R. (1989). The arrangement of interphase chromosomes: structural and functional aspects. *Experimental Cell Research*, **185**, 297–318.

Hiraoka, Y., Dernburg, A.F., Parmelee, S.J., Rykowski, M.C., Agard, D.A., and Sedat, J.W. (1993). The onset of homologous chromosome pairing during *Drosophila melanogaster* embryogenesis. *Journal of Cell Biology*, **120**, 591–600.

John, B. and Miklos, G. (1988). *The Eukaryote Genome in Development and Evolution*. Allen and Unwin, London.

Johnson, A.D. (1995). The price of repression. *Cell*, **81**, 655–8.

Jones, P.A., Rideout, W.M., Shen, J.C., Spruck, C.H., and Tsai, Y.C. (1992). Methylation, mutation and cancer. *BioEssays*, **14**, 33–6.

Jorgensen, R. (1995). Cosuppression, flower color patterns, and metastable gene expression states. *Science*, **268**, 686–91.

Kermicle, J., Eggleston, W.B., and Alleman, M. (1995). Organization of paramutagenicity in *R-stippled* maize. *Genetics*, **141**, 361–72.

Kricker, M.C., Drake, J.W., and Radman, M. (1992). Duplication-targeted DNA methylation and mutagenesis in the evolution of eukaryotic chromosomes. *Proceedings of the National Academy of Sciences of the USA*, **89**, 1075–9.

Kurata, N., Moore, G., Nagamura, Y., Foote, T., Yano, M., Minobe, Y., *et al.* (1994). Conservation of genome structure between rice and wheat. *Biotechnology*, **12**, 276–8.

Li, E., Bestor, T.H., and Jaenisch, R. (1992). Targeted mutation of the DNA methyltransferase gene results in embryonic lethality. *Cell*, **69**, 915–26.

Linn, F., Heidmann, I., Saedler, H., and Meyer, P. (1990). Epigenetic changes in the expression of the maize A1 gene in *Petunia hybrida*: role of numbers of integrated gene copies and state of methylation. *Molecular and General Genetics*, **222**, 329–36.

Manuelidis, L. (1990). A view of interphase chromosomes. *Science*, **250**, 1533–40.

Matzke, M.A. and Matzke, A.J.M. (1995). How and why do plants inactivate homologous (trans)genes? *Plant Physiology*, **107**, 679–85.

Matzke, M.A., Primig, M., Trnovsky, J., and Matzke, A.J.M. (1989). Reversible methylation and inactivation of marker genes in sequentially transformed tobacco plants. *EMBO Journal*, **8**, 643–9.

Matzke, A.J.M., Varga, F., Gruendler, P., Unfried, I., Berger, H., Mayr, B., *et al.* (1992). Characterization of a new repetitive sequence that is enriched on microchromosomes of turkey. *Chromosoma*, **102**, 9–14.

Matzke, M.A., Matzke, A.J.M., and Mittelsten Scheid, O. (1994a). Inactivation of repeated genes—DNA–DNA interaction? In *Homologous Recombination and Gene Silencing in Plants* (ed. J. Paszkowski), pp. 271–307. Kluwer, Dordrecht.

Matzke, A.J.M., Neuhuber, F., Park, Y.-D., Ambros, P.F., and Matzke, M.A. (1994b). Homology-dependent gene silencing in transgenic plants: epistatic silencing loci contain multiple copies of methylated transgenes. *Molecular and General Genetics*, **244**, 219–29.

Meyer, P. (1995). Understanding and controlling transgene expression. *Trends in Biotechnology*, **13**, 332–7.

Meyer, P. (1996). Repeat-induced gene silencing: common mechanisms in plants and fungi. *Biological Chemistry Hoppe-Seyler*, **377**, 87–95.

Meyer, P., Heidmann, I., and Niedenhof, I. (1993). Differences in DNA-methylation are associated with a paramutation phenomenon in transgenic petunia. *Plant Journal*, **4**, 89–100.

Moore, G., Gale, M.D., Kurata, N., and Flavell, R.B. (1993). Molecular analysis of small grain cereal genomes: current status and prospects. *Biotechnology*, **11**, 584–9.

Nassif, N. and Engels, W. (1993). DNA homology requirements for mitotic gap repair in *Drosophila*. *Proceedings of the National Academy of Sciences of the USA*, **90**, 1262–6.

Nassif, N., Penney, J., Pal, S., Engels, W.R., and Gloor, G.B. (1994). Efficient copying of nonhomologous sequences from ectopic sites via P-element-induced gap repair. *Molecular and Cellular Biology*, **14**, 1613–25.

Neuhuber, F., Park, Y.-D., Matzke, A.J.M., and Matzke, M.A. (1994). Susceptibility of transgene loci to homology-dependent gene silencing. *Molecular and General Genetics*, **244**, 230–41.

Neumann, B., Kubicka, P., and Barlow, D.P. (1995). Characteristics of imprinted genes. *Nature Genetics*, **9**, 12–3.

Niebel, F.C., Frendo, P., Van Montagu, M., and Cornelissen, M. (1995). Post-transcriptional cosuppression of β-1,3-glucanase genes does not affect accumulation of transgene nuclear mRNA. *Plant Cell*, **7**, 347–58.

O'Brien, S.J. and Graves, J.A.M. (1991). Report on the committee on comparative gene mapping. *Cytogenetics and Cell Genetics*, **58**, 1124–51.

Pandit, N. and Russo, V.E.A. (1992). Reversible inactivation of a foreign gene, hph, during the asexual cycle in *Neurospora crassa* transformants. *Molecular and General Genetics*, **234**, 412–22.

Park, Y.-D., Papp, I., Moscone, E.A., Iglesias, V.A., Vaucheret, H., Matzke, M.A., *et al.* (1996). Gene silencing mediated by promoter homology occurs at the level of transcription and results in meiotically heritable alterations in methylation and gene activity. *Plant Journal*, **9**, 183–94.

Patterson, G.I. and Chandler, V.L. (1995). Paramutation in maize and related allelic interactions. In *Gene Silencing in Higher Plants and Related Phenomena in Other Eukaryotes* (ed. P. Meyer), pp. 121–41. Springer Verlag, Berlin.

Rommens, C.M.T., van Haaren, M.J.J., Nijkamp, H.J.J., and Hille, J. (1993). Differential repair of excision gaps generated by transposable elements of the '*Ac* family'. *BioEssays*, **15**, 507–12.

Rossignol, J.-L. and Faugeron, G. (1994). Gene inactivation triggered by recognition between DNA repeats. *Experientia*, **50**, 307–17.

Sabl, J.F. and Laird, C.L. (1992). Epigene conversion: a proposal with implications for gene mapping in humans. *American Journal of Human Genetics*, **50**, 1171–7.

Sabl, J.F. and Henikoff, S. (1995). Copy number and orientation determine the susceptibility of a gene to silencing by nearby heterochromatin in *Drosophila*. *Genetics*, **142**, 447–58.

Selker, E.U. (1990). Premeiotic instability of repeated sequences in *Neurospora crassa*. *Annual Review of Genetics*, **24**, 579–613.

Smyth, D.R. (1991). Dispersed repeats in plant genomes. *Chromosoma*, **100**, 355–9.

Ten Lohuis, M., Müller, A., Heidmann, I., Niedenhof, I., and Meyer, P. (1995). A repetitive DNA-fragment carrying a hot spot for *de novo* DNA methylation enhances expression variegation in tobacco and petunia. *Plant Journal*, **8**, 919–32.

Toledo, F., Le Roscouet, D., Buttin, G., and Debatisse, M. (1992). Co-amplified markers alternate in megabase long chromosomal inverted repeats and cluster independently in interphase nuclei at early steps of mammalian gene amplification. *EMBO Journal*, **11**, 2665–73.

Van Blokland, R., Van der Geest, N., Mol, J.N.M., and Kooter, J.M. (1994). Transgene-mediated suppression of chalcone synthase expression in *Petunia hybrida* results from an increase in RNA turnover. *Plant Journal*, **6**, 861–77.

Vaucheret, H. (1993). Identification of a general silencer for 19S and 35S promoters in a transgenic tobacco plant: 90 bp of homology in the promoter sequence are sufficient for trans-inactivation. *Comptes Rendus de l'Academie de Science de Paris*, **316**, 1471–83.

Walker, E.L., Robbins, T.P., Bureau, T.E., Kermicle, J., and Dellaporta, S.L. (1995). Transposon-mediated chromosomal rearrangements and gene duplications in the formation of the maize R-r complex. *EMBO Journal*, **14**, 2350–63.

Wu, C.-I., True, J.R., and Johnson, N. (1989). Fitness reduction associated with the deletion of a satellite DNA array. *Nature*, **341**, 248–51.

Zimmer, E.A., Martin, S.L., Beverley, S.M.., Kan, Y.W., and Wilson, A.C. (1980). Rapid duplication and loss of genes coding for the alpha chains of hemoglobin. *Proceedings of the National Academy of Sciences of the USA*, **77**, 2158–62.

Index